Studies in Computational Intelligence 410

For further volumes:
http://www.springer.com/series/7092

Petia Georgieva, Lyudmila Mihaylova,
and Lakhmi C. Jain (Eds.)

Advances in Intelligent Signal Processing and Data Mining

Theory and Applications

Springer

Editors
Dr. Petia Georgieva
Department of Electronics
Telecommunications and Informatics
Signal Processing Lab, IEETA
University of Aveiro
Aveiro
Portugal

Dr. Lakhmi C. Jain
School of Electrical and Information
Engineering
University of South Australia
Adelaide
South Australia
Australia

Dr. Lyudmila Mihaylova
School of Computing and Communications
InfoLab21, South Drive
Lancaster University
Lancaster
UK

ISSN 1860-949X e-ISSN 1860-9503
ISBN 978-3-642-43980-3 ISBN 978-3-642-28696-4 (eBook)
DOI 10.1007/978-3-642-28696-4
Springer Heidelberg New York Dordrecht London

Printed on acid-free paper

Springer is part of Springer Science+Business Media (www.springer.com)

Preface

"In the depth of winter, I finally learned that within me there lay an invincible summer."

Albert Camus
French existentialist author & philosopher (1913–1960)

Dealing with large amounts of data and reasoning in real time are some of the challenges that our every day life poses to us. The answer to these questions can be given by advanced methods in signal processing and data mining which is the scope of this book. The book presents theoretical and application achievements on some of the most efficient statistical and deterministic methods for information processing (filtering, clustering, decomposition, modelling) in order to extract targeted information and find hidden patterns. The techniques presented range from Bayesian approaches such as sequential Monte Carlo methods, Markov Chain Monte Carlo filters, Rao Blackwellization, to the biologically inspired paradigm of Neural Networks and decomposition techniques such as Empirical Mode Decomposition, Independent Component Analysis (ICA) and Singular Spectrum Analysis. Advances and new theoretical interpretations related with these techniques are detailed and illustrated on a variety of real life problems as multiple object tracking, group object tracking, localization in wireless sensor networks, brain source localization, behavior reasoning, classification, clustering, video sequence processing, and others.

This research book is directed to the research students, professors, researchers and practitioners interested in exploring the advanced techniques in intelligent signal processing and data mining paradigms.

We are grateful to the contributions and reviewers for their time and vision. We appreciate the support provided by the University of Aveiro, Lancaster University and the University of South Australia. The editorial assistance provided by Springer-Verlag is acknowledged.

Petia Georgieva
Portugal

Lyudmila Mihaylova
UK

Lakhmi C. Jain
Australia

Contents

Chapter 1
Introduction to Intelligent Signal Processing and Data Mining

Lyudmila Mihaylova, Petia Georgieva, and Lakhmi C. Jain

Abstract. Intelligent signal processing and data mining are the key components of present advances in many disciplines including science and engineering. This chapter presents a brief outline of contributions from some of the leading researchers in the field of intelligent signal processing and data mining.

1.1 Inroduction

Recent years have been witnessed by an increased interest in processing large amounts of data from heterogeneous sensors and for high-dimensional systems, in real time. Such tasks face various challenges in processing noisy data under

Lyudmila Mihaylova
School of Computing and Communications
InfoLab21, South Drive
Lancaster University
Lancaster LA1 4WA,
UK

Petia Georgieva
Department of Electronics Telecommunications and Informatics
Signal Processing Lab, IEETA
University of Aveiro
3800 - 193 Aveiro
Portugal

Lakhmi C. Jain
School of Electrical and Information Engineering
University of South Australia
Adelaide
Mawson Lakes Campus
South Australia SA 5095
Australia

P. Georgieva et al. (Eds.): Adv. in Intelligent Signal Processing and Data Mining, SCI 410, pp. 1–5.
springerlink.com

uncertainties, or even more coping with incomplete or conflicting data. Sequential Monte Carlo methods have gained a significant popularity in the recent decades and hence a purpose of this book is to show their potential for solving problems such as reasoning in real time, group and extended object tracking and localization in sensor networks. The approaches are general. Their application areas are numerous and include but are not limited to surveillance, navigation and positioning systems, brain computer interfaces and intelligent transportation systems. The first part of the book presents novel methods for intelligent signal processing, mainly based on sequential Monte Carlo methods, whilst the second part of the book gives approaches for data mining and machine learning, such as clustering of data, anomaly detection, neural networks, classification and time series analysis. Research is focused on developing approaches able to fuse (aggregate) information from multiple sources. From methodological point of view the Bayesian approaches have received a lot of attention. A number of theoretical milestones have been achieved, including the design of proposal distributions that ultimately improve the mixing properties of Markov Chain Monte Carlo (MCMC) approaches, especially for smoothing, filtering and interpolation in high dimensional systems. More details can be found in the surveys of the chapters in the light of localization and navigation, sensor data fusion, tracking and other applications.

Data mining, also popularly referred to as *knowledge discovery from data (KDD)*, is the automated extraction of patterns representing knowledge implicitly stored in massive information repositories or data streams. Data mining is a subfield of artificial intelligence, drawing work from areas including database technology, machine learning, statistics, pattern recognition, information retrieval, neural networks, knowledge-based systems, artificial intelligence, high-performance computing, and data visualization. This multidisciplinary field emerged during the late 1980s, made great advances during the 1990s, and continues to attract increasing attention over the last decade.

In this book, we present recent advances in data mining techniques focused on empirical mode decomposition, hierarchical clustering, recurrent neural networks for forecasting, spatial filtering, and neural networks training by reinforcement learning. A variety of typical applications are covered ranging from biomedical data processing, through financial time series prediction to board games. As a result, the selected chapters present a state of the art in the field, reporting achievements both in data mining techniques and challenging applications. Future research directions are also outlined. We hope that this book will encourage people with different backgrounds and experiences to exchange their views regarding intelligent signal processing and data mining so as to contribute toward the further promotion and shaping of this exciting and dynamic field.

1.2 Chapters Included in the Book

The book includes 12 chapters. A range of topics in the field of intelligent signal processing and data mining is presented. Since each chapter is designed to be a stand-alone work, the reader can focus on the topics that most interest him/her. The remaining chapters cover the following material.

Chapter 2 is focused on Sequential Monte Carlo methods and Markov Chain Monte Carlo methods for tracking a number of objects moving in a coordinated and interacting fashion, i.e. forming a group or a pattern. Two types of methods are presented: 1) an evolving network-based multiple object tracking algorithm that is capable of categorizing objects into groups, and 2) a multiple cluster tracking algorithm for dealing with prohibitively large number of objects. A causality inference framework is proposed for identifying dominant agents based on their observed trajectories.

Chapter 3 is focused on a related problem, on multi-target tracking. The approach relies on Poisson point processes, leading to the intensity filter (iFilter). The iFilter is a powerful approach, especially suitable for solving problems with various types of measurement uncertainty and unknown number of targets. The iFilter is implemented as a Sequential Monte Carlo filter and its performance is studied in clutter environments with various parameters.

Chapter 4 presents another group of Sequential Monte Carlo methods, for localization and positioning in wireless sensor networks. Multiple model particle filters are developed and their performance is evaluated based on received signal strength signals by accounting for and without considering the measurement noize time correlation. The unknown parameters of the measurement noize are estimated with Gibbs sampling.

Chapter 5 proposes a solution to the Electroencephalography (EEG) source localization problem based on a Sequential Monte Carlo algorithm and spatial filtering by beamforming. The localization and the waveform at the dynamically changing most active zones are performed based on uncorrelated dipoles.

Chapter 6 is focused on the application of Sequential Monte Carlo methods in the automotive industry: lane detection and tracking. Model-based lane detection algorithms are separated into lane modelling, feature extraction and model parameter estimation. Each of these steps is discussed in detail with examples and results. Three different types of particle filters are compared and the influence of feature maps on the tracking performance is shown.

Chapter 7 presents methods for detecting anomalies in sensor data. Such methods are used in situation awareness systems based on patterns characterizing situations of interest. The patterns can vary over time. It is shown that continuous queries allow for an ongoing automatic evaluation of search patterns.

Chapter 8 provides a tutorial overview of hierarchical clustering. Data visualization methods, based on hierarchical clustering, are demonstrated and the scaling of hierarchical clustering in time and memory is analyzed. A new method for speeding up hierarchical clustering with cluster seeding is introduced. Its performance is compared with a traditional agglomerative hierarchical, average link clustering algorithm using several internal and external cluster indices.

Chapter 9 focuses on the problem of automatic identification of real world objects in complex scenes. Shape context representation and many-to-one matching is employed to extract the point correspondence between a complex scene and a target object. Subtractive clustering and data mining allow for extracting significant information for the object identification.

Chapter 10 presents a new type of recurrent neural networks, known as historical consistent neural networks. Historical consistent neural networks are convenient models for high-dimensional systems made up of a number of subdynamics interacting. The models are symmetrical, that is there is no distinction between the input and output variables. Historical consistent neural networks are illustrated in financial markets forecasting and risk analysis.

Chapter 11 introduces the concept of reinforcement learning as a neural network mechanism. Through repeated interactions with the environment and rewards the agents learn which actions are associated with the greatest cumulative re-ward. Typical board game examples illustrate the performance of the neural networks trained by reinforcement learning.

Chapter 12 considers data analysis and modeling of biomedical non-linear and non-stationary time series. On-line variant of Empirical Mode Decomposition (EMD) is proposed, termed weighted Sliding EMD (SEMD). SEMD in conjunction with Hilbert-Huang Transform provides a fully adaptive and data-driven technique to extract Intrinsic Mode Functions. Applications to biomedical time series recorded during neuromonitoring are presented.

1.3 Conclusion

Being areas with intensive developments, intelligent signal processing and data mining are influenced by practical challenges and unresolved theoretical problems. Although significant results have been achieved during the last decades, there is still a long way to go in order to develop systems which are able to cope with various uncertainties and provide autonomous decisions. This book provides answers to some of the important questions and guides the readers how solutions can be achieved in:

- Monte Carlo-Based Bayesian Group Object Tracking and Causal Reasoning
- A Sequential Monte Carlo Method for Multi-Target Tracking with the Intensity Filter
- Sequential Monte Carlo Methods for Localization in Wireless Sensor Networks
- A Sequential Monte Carlo Approach for Brain Source Localization
- Computational Intelligence in Automotive Applications
- Detecting Anomalies in Sensor Signals Using Database Technology
- Hierarchical Clustering of Large Data Sets
- A Novel Framework for Object Recognition Under Severe Occlusions
- Historical Consistent Neural Networks for Forecasting and Risk Analysis in Economics
- Reinforcement Learning with Neural Networks: Tricks of the Trade
- Sliding Empirical Mode Decomposition – Brain Status Data Analysis and Modeling.

1.4 Resources

The following books may prove useful for exploring the field of intelligent signal processing and data mining further. This list is neither complete, nor exclusive. It can be seen as a starting point to explore the field further.

References

[1] Bandyopadhyay, S., et al. (eds.): Advanced Methods for Knowledge Discovery from Complex Data. Springer, London (2005)
[2] Bianchini, M., Maggini, M., Scarselli, F., Jain, L.C.: Innovations in Neural Information Processing Paradigms. Springer, Germany (2009)
[3] Echizen, I., et al. (eds.): Intelligent Information Hiding and Multi-media Signal Processing. IEEE Press, USA (2010)
[4] Ghosh, A., Jain, L.C. (eds.): Evolutionary Computation in Data Mining. Springer, Germany (2005)
[5] Holmes, D., Jain, L.C.: Data Mining: Foundations and Intelligent Paradigms, vol. 1. Springer, Heidelberg (2012)
[6] Holmes, D., Jain, L.C.: Data Mining: Foundations and Intelligent Paradigms, vol. 2. Springer, Heidelberg (2012)
[7] Holmes, D., Jain, L.C.: Data Mining: Foundations and Intelligent Paradigms, vol. 3. Springer, Heidelberg (2012)
[8] Haykin, S., Kosko, B. (eds.): Intelligent Signal Processing. Wiley-IEEE Press (2001)
[9] Maloof, M.A. (ed.): Machine Learning and Data Mining for Computer Security. Springer, London (2006)
[10] Motoda, H.: Active Mining, vol. 79. IOS Press, The Netherlands (2002)
[11] Pal, N., Jain, L.C. (eds.): Advanced Techniques in Knowledge Discovery and Data Mining. Springer, London (2005)
[12] Pan, J.S., Chen, Y.W., Jain, L.C. (eds.): Intelligent Information Hiding and Multimedia Signal Processing. IEEE Computer Society Press, USA (2009)
[13] Pan, J.S., Niu, X.M., Huang, H.C., Jain, L.C. (eds.): Intelligent Information Hiding and Multimedia Signal Processing. IEEE Computer Society Press, USA (2008)
[14] Liao, B.-H., Pan, J.-S., Jain, L.C., Liao, M., Noda, H., Ho, A.T.S.: Intelligent Information Hiding and Multimedia Signal Processing, vol. 1. IEEE Computer Society Press, USA (2007)
[15] Liao, B.-H., Pan, J.-S., Jain, L.C., Liao, M., Noda, H., Ho, A.T.S.: Intelligent Information Hiding and Multimedia Signal Processing, vol. 2. IEEE Computer Society Press, USA (2007)
[16] Ruano, A., Várkonyi-Kóczy, A.R. (eds.): New Advances in Intelligent Signal Processing. Springer (2011)
[17] Wang, J.T.L., et al. (eds.): Data Mining in Bioinformatics. Springer, London (2005)
[18] Wang, L., Fu, X.: Data Mining with Computational Intelligence. Springer, London (2005)
[19] Zaknich, A.: Neural Networks for Intelligent Signal Processing. World Scientific (2003)
[20] Zhang, S., et al.: Knowledge Discovery in Multiple Databases. Springer, London (2004)

Chapter 2
Monte Carlo-Based Bayesian Group Object Tracking and Causal Reasoning

Avishy Y. Carmi, Lyudmila Mihaylova, Amadou Gning,
Pini Gurfil, and Simon J. Godsill

Abstract. We present algorithms for tracking and reasoning of local traits in the subsystem level based on the observed emergent behavior of multiple coordinated groups in potentially cluttered environments. Our proposed Bayesian inference schemes, which are primarily based on (Markov chain) Monte Carlo sequential methods, include: 1) an evolving network-based multiple object tracking algorithm that is capable of categorizing objects into groups, 2) a multiple cluster tracking algorithm for dealing with prohibitively large number of objects, and 3) a causality inference framework for identifying dominant agents based exclusively on their observed trajectories. We use these as building blocks for developing a unified tracking and behavioral reasoning paradigm. Both synthetic and realistic examples are provided for demonstrating the derived concepts.

2.1 Overview

In recent years there has been an increasing interest in tracking a number of interacting objects moving in a coordinated fashion. There are many fields in which such

Avishy Y. Carmi
Department of Mechanical and Aerospace Engineering,
Nanyang Technological University, Singapore
e-mail: acarmi@ntu.edu.sg

Lyudmila Mihaylova · Amadou Gning
School of Computing and Communications, Lancaster University, United Kingdom
e-mail: {mila.mihaylova,e.gning}@lancaster.ac.uk

Pini Gurfil
Department of Aerospace Engineering, Technion - Israel Institute of Technology,
Haifa 32000, Israel
e-mail: pgurfil@technion.ac.il

Simon J. Godsill
Department of Engineering, University of Cambridge, United Kingdom
e-mail: sjg@eng.cam.ac.uk

P. Georgieva et al. (Eds.): Adv. in Intelligent Signal Processing and Data Mining, SCI 410, pp. 7–53.
springerlink.com

situations are frequently encountered: video surveillance, feature tracking in video sequences, biomedicine, neuroscience and meteorology, to mention only a few. Although individual targets in the group can exhibit independent movement at a certain level, overall the group will move as one whole, by synchronizing the movement of the individual entities and avoiding collisions. In most of the multi-target tracking methods, as opposed to groups tracking methods, tracking of individual objects is the common approach.

It is obvious that any inference in such environments would have to cope with an ever growing complexity proclaimed by the spatio-temporal interrelations among constituents in the scene. In some cases where the number of objects becomes excessively large, it might be impractical to individually track them all (e.g., tracking features in a video sequence). An efficient approach would thereby consist of tracking the grouping structure formed by object concentrations rather than individual entities.

Groups are often referred to as structured objects, a term which reflects the ingrained interplay between their components. These endogenous forces give rise to group hierarchies and are eminent in producing emergent phenomena. Fortunately, these are exactly the factors essential for maintaining coordination within and between groups, a premise which to some extent allows us to treat them as unite entities at a high level tracking paradigm. Any knowledge of existence of such interrelations facilitates sophisticated agent-based behavioral modeling which, in practice, comprises of a set of local interaction rules or mutually interacting processes (e.g., Boids system [1], causality models [2, 3]) - an approach which by itself provides insightful justifications of characteristic behaviors in the fundamental subsystem level and likewise of group hierarchies and emergent social patterns (see [2, 4, 5]).

2.1.1 Reasoning About Behavioral Traits

Being the underlying driving mechanism evoking emergent phenomena, hierarchies and principle behavior patterns, the ingrained interactions between agents are possibly the most pivotal factors that should be scrutinized in high level scene understanding. Such interrelations can take the form of a causal chain in which agent's decisions and behavior are affected by its neighbors and likewise have either direct or indirect influence on other agents. The ability to fully represent these interrelations based exclusively on passive observations such as velocity and position, lays the ground for the development of sophisticated reasoning schemes that can potentially be used in applications such as activity detection, intentionality prediction, and artificial awareness, to name only a few.

In this work we demonstrate this concept by developing a causality reasoning framework for ranking agents with respect to their cumulative contribution in shaping the collective behavior of the system. In particular, our framework is able to distinguish leaders and followers based exclusively on their observed trajectories.

2.1.2 Novelties and Contributions

The objective of this work is three fold. In the first few sections (2.2 – 2.3), efficient Bayesian methods are proposed for tracking multiple groups of possibly large number of coordinated objects (e.g., clusters) (Section 2.4). In this respect, the derived tracking schemes either involve a combination of sequential Monte Carlo (SMC) and Markov chain Monte Carlo (MCMC) methods, or exclusively rely on MCMC techniques for approximating the underlying state statistics in potentially high dimensional settings. In the second part of this work (Section 2.5) a novel Bayesian causality detection method is suggested for behavioral reasoning. In particular, the newly derived scheme is aimed at ranking agents with respect to their decision-making capabilities (dominance) as substantiated by the observed emergent behavior. Dominant agents in that sense are considered to have a prominent influence on the collective behavior and are experimentally shown to coincide with actual leaders in groups. Both these methodologies, namely, tracking and behavioral reasoning, are consolidated to form a unified tracking and reasoning paradigm (Section 2.5.6). The underlying methodologies are demonstrated throughout this work and in Section 2.6. Concluding remarks and some open issues are discussed in Section 2.7. Finally, the algorithms for the update of evolving graphs are given in the Appendix 2.8.

2.1.3 Multiple Group Tracking

Over the past decade various methods have been developed for group tracking. These can be divided into two broad classes, depending on the underlying complexities:

1. Methods for a relatively small number of groups, with a small number of group components [6, 7, 8, 9, 10, 11, 12, 13].
2. Methods for groups comprised of hundreds or thousands of objects (normally referred to as cluster/crowd tracking techniques) [14, 15].

In the second case the whole group is usually considered as an extended object (an ellipse or a circle) whose center position is estimated, together with the parameters of the shape.

Different models of groups of objects have been proposed in the literature, such as particle models for flocks of birds [16, 17, 18, 19], and leader-follower models [20]. However, estimating the dynamic evolution of the group structure has not been widely studied in the literature, although there are similarities with methods used in evolving network models [21, 22].

Methods for group object tracking also vary widely: from Kalman filtering approaches, Joint Probability Data Association (JPDA) [23, 24] to Probability Hypothesis Density (PHD) filtering [25,26,27], and others [12,28,29,30,11]. The influence of the 'negative' information on group object tracking is considered in [31] and ground moving target indicator tracking based on particle filtering in [32]. In [10] a coordinated group tracking model is presented, comprising a continuous-time motion of the group and a group structure transition model. A Markov chain Monte Carlo (MCMC) particle filter algorithm is proposed to approximate the posterior probability density function (PDF) of the high dimensional state.

Typically tracking many targets (hundreds or thousands) can be solved by clustering techniques or other methods where the aggregated motion is estimated, as it is in the case of vehicular traffic flow prediction/ estimation, with fluid dynamics type of models combined with particle filtering techniques [33, 34]. For thousands of targets forming a group, the only practical solution currently is to consider them as an extended target. The extended target tracking problem reduces then to joint state and parameter estimation.

Estimation of parameters in general nonlinear non-Gaussian state-space models is a long-standing problem. Since particle filters (PFs) are known with the challenges they face for parameter estimation and for joint state and parameter estimation [35], most solutions in the literature split the problems into two parts: *i*) state estimation, followed by *ii*) parameter estimation (see e.g., [11, 36]). In [36] an extended object tracking problem is solved when the static parameters are estimated using Monte Carlo methods (data augmentation and particle filtering), whereas the states are estimated with a Mixture Kalman filter or with an interacting multiple model filter. In [11] the group object tracking problem again is split into two parts: the extent of the group is estimated with the approach of random matrices whereas the state estimation is performed with Kalman filtering techniques.

A different approach for extended object tracking relies on combined set-theoretic and stochastic fusion. Group and extended objects are characterized with multiple measurements originated from different locations of the object surface. In [37] the set of measurements are considered as belonging to a bounded region and a SMC approach with a novel likelihood function is derived for general nonlinear measurements. In [38], the true shape of the target object is modeled as the smallest circular disc including the extended object which actually is the problem of tracking a circular disc with unknown radius and center in the two-dimensional space. In [39,38] the approach of random hypersurface model is compared with the approach of random matrices [13].

Multiple Object Tracking via Random Finite Sets

The optimal filtering scheme involves the propagation of the joint probability density of target states conditioned on the data. Following the conventional approach, in which all states are concatenated to form an augmented vector, leads to a problematic statistical representation owing to the fact that the objects themselves are unlabeled and thus can switch positions within the resulting joint state vector [40]. Furthermore, objects may appear and disappear, thereby yielding inconsistencies in the joint state dimension. These problems can be circumvented by adopting one of the following approaches: 1) introducing some sort of labeling mechanism which identifies existing targets within the augmented vector [41], or 2) considering the joint state as a random finite set. The latter approach provides an elegant and natural way to make statistical inference in multi object scenarios. Nevertheless, its practical implementation as well as its mathematical subtleties need to be carefully considered [42].

Random sets can be thought of as a generalization of random vectors. The elements of a set may have arbitrary dimensions, and as opposed to vectors, the ordering of their elements is insignificant. These properties impose difficulties in constructing probability measures over the space of sets. This has led some researchers to develop new concepts based on belief mass functions such as the set derivative and set integral for embedding notions from measure theoretic probability within random set theory (e.g., Bayes rule). As part of this, point process statistics are commonly used for deriving probabilistic quantities [43]. The PHD filter presented in [44] is the first attempt to implement finite set statistics concepts for multi-object tracking. This algorithm uses a Poisson point process formulation to derive a semi closed-form recursion for propagating the first moment of the random set's intensity (i.e., the set's cardinality). A brief summary of the PHD algorithm can be found in [42]. An extension of the PHD for group object tracking is provided in [45].

SMC Algorithms for Tracking in Variable State Dimensions

An extension of the PF technique to a varying number of objects is introduced in [42], [10] and [8]. In [42, 46] a PF implementation of the PHD filter is derived. This algorithm maintains a representation of the filtering belief mass function using random set realizations (i.e., particles of varying dimensions). The samples are propagated and updated based on a Bayesian recursion consisting of set integrals. Both works of [10] and [8] develop a MCMC PF scheme for tracking varying numbers of interacting objects. The MCMC approach does possess a reported advantage over conventional PF due to its efficient sampling mechanism. Nevertheless, in its traditional non-sequential form it is inadequate for sequential estimation. The techniques used by [10] and [8] amend the MCMC for sequential filtering (see also [47]). The work in [8] copes with inconsistencies in state dimension by utilizing the reversible jump MCMC method introduced in [48]. The approach proposed in [10], on the other hand avoids the computation of the marginal filtering distribution as in [47] and operates on a fixed dimension state space through use of indicator variables for labeling of active object states (the two approaches being essentially equivalent, see [49]).

2.2 Group Tracking by SMC Methods and Evolving Networks

The approach described in this section models the group of objects as a random undirected graph in which the nodes of the graph correspond to the targets. An edge between two nodes reflects the fact that there is a relation between these two objects.

2.2.1 Problem Formulation

Consider the problem of tracking the motion of groups of targets. Each target i is characterized by its state vector $\mathbf{x}_{t,i}$ Targets which are close to one another tend to form a group. The Mahalanobis distance $d_{i,k}$ ir another metric can be chosen as a criterion of closeness between the targets within a group. At each time instant k, the set of objects tracked in a group $\mathbf{g}$ can be modeled by a *Random Finite Set* (RFS, see [20]) that incorporates the state vectors of the group members,

$$\mathbf{X}_k^{\mathbf{g}} = \{\mathbf{x}^{\mathbf{g}}{}_{k,1}, \mathbf{x}^{\mathbf{g}}{}_{k,2}, \ldots, \mathbf{x}^{\mathbf{g}}{}_{k,n_g}\} \tag{2.1}$$

where n_g is the random size of group $\mathbf{g}$. Knowing the group structure

$$\mathbf{G}_k = \{\mathbf{g}_1, \ldots, \mathbf{g}_{n_G}\} \tag{2.2}$$

where n_G is the number of groups, the joint state vector for all the targets in the n_G groups can be written in the form

$$\mathbf{X}_k = \{\mathbf{X}_k^{\mathbf{g}_1}, \ldots, \mathbf{X}_k^{\mathbf{g}_{n_G}}\}. \tag{2.3}$$

At time k a measurement vector $\mathbf{z}_k$ is received which can be described as a function of the state $\mathbf{X}_k = \{\mathbf{X}_k^{\mathbf{g}_1}, \ldots, \mathbf{X}_k^{\mathbf{g}_{n_G}}\}$. Assuming that the measurement likelihood function $p(\mathbf{z}_k|\mathbf{X}_k)$ can be calculated, the purpose is to compute sequentially the state PDF for each group of objects. The changes of the groups such as merging and splitting are taken into account during the graph update process. Additionally, the groups' movements are assumed independent.

Under the Markovian assumption for the state transition, the Bayesian *prediction* and *filtering steps* can be written as follows:

$$\begin{aligned} p(\mathbf{X}_k, \mathbf{G}_k|\mathbf{Z}_{1:k-1}) &= p(\mathbf{G}_k|\mathbf{X}_k, \mathbf{Z}_{1:k-1})p(\mathbf{X}_k|\mathbf{Z}_{1:k-1}) \\ &= \int p(\mathbf{G}_k|\mathbf{X}_k,\mathbf{G}_{k-1})p(\mathbf{X}_k|\mathbf{X}_{k-1},\mathbf{G}_{k-1})p(\mathbf{X}_{k-1},\mathbf{G}_{k-1}|\mathbf{Z}_{1:k-1})d\mathbf{X}_{k-1}d\mathbf{G}_{k-1}, \end{aligned} \tag{2.4}$$

$$p(\mathbf{X}_k, \mathbf{G}_k|\mathbf{Z}_{1:k}) = \frac{p(\mathbf{z}_k|\mathbf{X}_k, \mathbf{G}_k) \times p(\mathbf{X}_k, \mathbf{G}_k|\mathbf{Z}_{1:k-1})}{p(\mathbf{z}_k|\mathbf{Z}_{1:k-1})}, \tag{2.5}$$

where $\mathbf{Z}_{1:k}$ is the set of measurements up to time k and $\mathbf{z}_k$ is the current vector of measurements.

The transition PDF $p(\mathbf{G}_k|\mathbf{X}_k,\mathbf{G}_{k-1})$ of the group structure can be calculated knowing the prediction of the target state and group structure in the previous time instant, and using the graph evolution model from [9], which is also detailed in the *Appendix*, Section 2.8.1. The transition PDF $p(\mathbf{X}_k|\mathbf{X}_{k-1},\mathbf{G}_{k-1})$ of the state of all targets is calculated knowing the previous time target states and group structure

PDF $p(\mathbf{X}_{k-1},\mathbf{G}_{k-1}|\mathbf{Z}_{1:k-1})$. With the assumption of independence between group motions, the PDF $p(\mathbf{X}_k|\mathbf{X}_{k-1},\mathbf{G}_{k-1})$ can be decomposed in the form

$$p(\mathbf{X}_k|\mathbf{X}_{k-1},\mathbf{G}_{k-1}) = \prod_{\mathbf{g}_i \in \mathbf{G}_{k-1}} p(\mathbf{X}_k^{\mathbf{g}_i}|\mathbf{X}_{k-1}^{\mathbf{g}_i}), \tag{2.6}$$

where $p(\mathbf{X}_k^{\mathbf{g}_i}|\mathbf{X}_{k-1}^{\mathbf{g}_i})$ is the transition density of the set of targets from the group $\mathbf{g}_i$.

In order to perform the correction step, the likelihood function $p(\mathbf{z}_k|\mathbf{X}_k,\mathbf{G}_k)$ of the whole state vector has to be evaluated by means of a data association approach. In this paper, the JPDA algorithm [24] is used to resolve the measurement origin uncertainty.

The gating process in the JPDA algorithm is enhanced by using other information about the graph structure, such as the distance between groups. Note that the graph $\mathbf{G}'_k$ estimated at each time instant is applied in the edge updating process and in the node incorporation steps which leads to a reduction of the computation time. At each time step, the graph $\mathbf{G}'_k$ can also be used in the gating process. Indeed, groups of the same graph $\mathbf{G}'_k$'s connected component can be gathered in a separate data association process: the graph $\mathbf{G}'_k$ offers a straightforward method of clustering the targets for the data association process.

Denote by $\{\mathbf{g}'_1,\dots,\mathbf{g}'_{n_{G'}}\}$, the set of $n_{G'}$ connected components in graph $\mathbf{G}'$. Any connected component $\mathbf{g}'_i$ can model a set of groups that are close enough to be treated in the same data association algorithm. Under the independence assumption between the $\mathbf{g}'_i$, the following equation can be written

$$p(\mathbf{z}_k|\mathbf{X}_k,\mathbf{G}_k) = p(\mathbf{z}_k|\mathbf{X}_k,\mathbf{G}_k,\mathbf{G}'_k) = \prod_{i=1,\dots,n_{G'}} p(\mathbf{z}_k^{\mathbf{g}'_i}|\mathbf{X}_k^{\mathbf{g}'_i}), \tag{2.7}$$

where $\mathbf{X}_k^{\mathbf{g}'_i}$ is the set of target states belonging to the groups in $\mathbf{g}'_i$. The vector $\mathbf{z}_k^{\mathbf{g}_i}$ comprises the subset of measurements related with the group in $\mathbf{g}'_i$. For example, $\mathbf{z}_k^{\mathbf{g}_i}$ can be chosen by gating measurements using the set $\mathbf{X}_k^{\mathbf{g}'_i}$ of targets' states.

In practice, two gating procedures are performed. First, using the center of the groups and the graph $\mathbf{G}'_k$, the measurements are clustered. Then the classical JPDA is used for each cluster.

2.2.2 A Nearly Constant Velocity Model for Individual Targets

The nearly constant velocity model [50, 51] is used for the update of each node of the graph, i.e., for modelling the motion of each target within a group. In two dimensions, the state of the ith target is given by:

$$\mathbf{x}_{k,i} = \mathbf{A}\mathbf{x}_{k-1,i} + \boldsymbol{\Gamma}\eta_{k-1}, \tag{2.8}$$

where $\mathbf{x}_{k,i} = (x_{k,i},\dot{x}_{k,i},y_{k,i},\dot{y}_{k,i})'$ is the state vector comprising the positions and speeds in x and y directions, respectively, $\mathbf{A} = \mathrm{diag}(\mathbf{A}_1,\mathbf{A}_1)$, $\mathbf{A}_1 = \begin{pmatrix} 1 & T \\ 0 & 1 \end{pmatrix}$,

$\boldsymbol{\Gamma} = \begin{pmatrix} T/2 & T & 0 & 0 \\ 0 & 0 & T/2 & T \end{pmatrix}'$, $'$ denotes the transpose operation, T is the sampling interval and η_{t-1} is the system dynamics noise. In order to cover a wide range of motions, the velocity should be approximately constant over straight line trajectories and the velocity change should be abrupt at each turn (especially for the direction of the velocity). The system dynamics noise η_{t-1} is represented as a sum of two Gaussian components

$$p(\eta_{k-1}) = \alpha \mathcal{N}(0, \mathbf{Q}_1) + (1-\alpha)\mathcal{N}(0, \mathbf{Q}_2), \tag{2.9}$$

where $\mathbf{Q}_1 = \text{diag}(\sigma^2, \sigma_1^2)$, $\mathbf{Q}_2 = \text{diag}(\sigma^2, \sigma_2^2)$; σ is a standard deviation assumed common and constant for x and y; $\sigma_1 \ll \sigma_2$ are standard deviations allowing to model respectively smooth and abrupt changes in the velocity. The fixed coefficient α has values in the interval $[0,1]$.

In addition, to model the interactions between objects in each group, the average velocity of group objects is used in (2.8) instead of the velocity of each group component. For each group $\mathbf{g}$, in the group structure $\mathbf{G}$ and for each $\mathbf{x^g}_{k,i} \in \mathbf{X}_k^{\mathbf{g}} = \{\mathbf{x^g}_{k,1}, \mathbf{x^g}_{k,2}, \ldots, \mathbf{x^g}_{k,n_g}\}$ we have the following equation

$$\mathbf{x}_{k,i}^{\mathbf{g}} = \mathbf{x}_{k-1,i}^{\mathbf{g}} + \sum_{j=1}^{n_g} (\mathbf{B}\mathbf{x}_{k-1,j}^{\mathbf{g}}) + \boldsymbol{\Gamma}\eta_{k-1}, \tag{2.10}$$

where $\mathbf{B} = \text{diag}(\mathbf{B}_1, \mathbf{B}_1)$ with $\mathbf{B}_1 = \begin{pmatrix} 0 & \frac{T}{n_g} \\ 0 & 0 \end{pmatrix}$. There is a resemblance between formula (2.10) and swarm optimization [52]. More sophisticated models can be adopted to model targets' interactions in each group such as the developed in [26].

2.2.3 Observation Model

Range and bearing observations from a network of low cost sensors are considered as measurements. The measurement vector $\mathbf{z}_{k,i}$ for the ith target contains the range $r_{k,i}$ to each target and the bearing $\beta_{k,i}$. The measurement equation is of the form:

$$\mathbf{z}_{k,i} = h(\mathbf{x}_{k,i}) + \mathbf{w}_{k,i}, \quad h(\mathbf{x}_{k,i}) = \left(\sqrt{x_{k,i}^2 + y_{k,i}^2}, \tan^{-1}\frac{y_{k,i}}{x_{k,i}}\right) \tag{2.11}$$

where the measurement noise $\mathbf{w}_{k,i}$ is supposed to be Gaussian with a known covariance matrix $\mathbf{R}$.

2.2.4 Particle Filtering Algorithms for Group Motion Estimation

In [9] two approaches for group tracking are proposed: with a deterministic graph and with a random evolving graph.

One way of considering the group structure is to propagate, at each time step, a deterministic group structure using the previous group structure $\mathbf{G}_{k-1}$ and the current estimate of all the target states denoted by $\widehat{\mathbf{X}}_k$, i.e., $\mathbf{G}_k = f(\mathbf{G}_{k-1}, \widehat{\mathbf{X}}_k)$. This approach has a relatively low complexity, but does not provide full information about the group structure uncertainty. This approach has been introduced in [53].

In this chapter we present an algorithm with an augmented state vector $(\mathbf{X}_k^g, \mathbf{G}_k)$, where the respective matrices are defined in (2.1) and (2.2). Hence, particles are defined with the augmented state $(\mathbf{X}_k^{\mathbf{g}}, \mathbf{G}_k)$.

We denote by N_p the number of particles and L the current index of a particle. A Metropolis-Hastings (MH) step is added to move particles in more likely regions. Due to the augmented state vector with the graph structure, in this approach, each particle contains the targets' state and the group structure. In general, the MH move steps, such as the move step proposed in [54] are known to provide accurate filter performance using a smaller number of particles than the number of particles in the classical particle filter. We are, then introducing these MH move steps in order to reduce the size of the particle cloud.

Having in mind (2.4)-(2.7), the implemented evolving group model is described as Algorithm 1, where the samples $\mathbf{X}_k^{\mathbf{g}_i,(L)}$ are drawn from the proposal PDF $q_{g_i}^{(L)}(\mathbf{X}_k^{\mathbf{g}_i,(L)} | \mathbf{X}_{0:k-1}^{\mathbf{g}_i,(L)}, \mathbf{z}_{0:k-1}) = p(\mathbf{X}_k^{\mathbf{g}_i,(L)} | \mathbf{X}_{k-1}^{\mathbf{g}_i,(L)})$. The samples $\mathbf{G}_k^{(L)}$ for the graph structure are drawn from the PDF $Q(\mathbf{G}_k | \mathbf{X}_{0:k}, \mathbf{G}_{k-1}) = p(\mathbf{G}_k | \mathbf{X}_k, \mathbf{G}_{k-1})$.

Algorithm 1. Particle filtering with a state augmented by the group structure

1. Prediction step
FOR $L = 1, \ldots, N_p$
  FOR all $\mathbf{g}_i^{(L)} \in \mathbf{G}_{k-1}^{(L)}$
    DRAW a sample $\mathbf{X}_k^{\mathbf{g}_i,(L)}$ from the proposal PDF $q_{g_i}^{(L)}$:
    $\mathbf{X}_k^{\mathbf{g}_i,(L)} \sim q_{g_i}^{(L)}(\mathbf{X}_k^{\mathbf{g}_i,(L)} | \mathbf{X}_{0:k-1}^{\mathbf{g}_i,(L)}, \mathbf{z}_{0:k-1})$
  END
  DRAW a sample $\mathbf{G}_k^{(L)}$ from a proposal PDF Q
  $\mathbf{G}_k^{(L)} \sim Q(\mathbf{G}_k | \mathbf{X}_{0:k}^{(L)}, \mathbf{G}_{k-1}^{(L)})$
END

2. Measurement Update Step
FOR $L = 1, \ldots, N_p$
  CALCULATE the likelihood function according to (2.7) and using a JPDA algorithm [24]
END
RUN the Metropolis-Hastings algorithm with m steps (see Algorithm 2)
UPDATE and NORMALIZE the weights
CALCULATE the estimate $\widehat{\mathbf{X}}_k$ of the current state vector $\mathbf{X}_k$

To sample from the proposal PDF q_{g_i}, a nearly constant velocity model (2.8)-(2.9) is used for each component $\mathbf{X}_{k-1}^{\mathbf{g}_i,(L)}$ of a particle $\mathbf{X}_{k-1}^{(L)}$ to obtain $\mathbf{X}_k^{\mathbf{g}_i,(L)}$. To sample a group structure $\mathbf{G}_k^{(L)}$ from $p(\mathbf{G}_k | \mathbf{X}_k, \mathbf{G}_{k-1})$, the previous time group structure $\mathbf{G}_{k-1}^{(L)}$, for each particle, is propagated by using the new evolving graph model,

$\mathbf{G}_k = f(\mathbf{G}_{k-1}, \mathbf{X}_k)$; for each particle graph structure, the edge updating step, the new node incorporation step and the old node suppression step are processed. The interactions within each group are modeled based on the mean velocity of group components (from the constant velocity model instead of the velocity of each group component).

To sample from the proposal PDF Q, the group structure evolution model $\mathbf{G}_k = f(\mathbf{G}_{k-1}, \mathbf{X}_k)$ introduced in the *Appendix*, Section 2.8.1, is used. In step 2 of Algorithm 1, the likelihood is calculated by assuming independence between clusters of measurements corresponding to each group. The MH move step is described in Algorithm 2 and is iterated m time steps (m being chosen beforehand). The MH algorithm is introduced to sample from the joint PDF $p(\mathbf{X}_k, \mathbf{G}_k | \mathbf{X}_{k-1}, \mathbf{Z}_{1:k})$. In step 2 the likelihood and the weight update is performed, similarly to Algorithm 1, using the JPDA algorithm. Finally, in step 3, for each target we estimate the corresponding efficient components in the particles $\mathbf{X}_k^{(L)}$.

Algorithm 2. Metropolis-Hastings step with the group structure

FOR $L = 1 \ldots N_p$
  FOR all $\mathbf{g}_i^{(L)} \in \mathbf{G}_{k-1}^{(L)}$
    DRAW a new sample $\mathbf{X}_k^{\mathbf{g}_i(prop)}$ using the proposal PDF $q_{\mathbf{g}_i}^{(L)}$ (see Algorithm 1):
      $\mathbf{X}_k^{\mathbf{g}_i(prop)} \sim q_{\mathbf{g}_i}^{(L)}(\mathbf{X}_k^{\mathbf{g}_i,(L)} | \mathbf{X}_{0:k-1}^{\mathbf{g}_i,(L)}, \mathbf{z}_{0:k-1})$
  END
  DRAW a new sample $\mathbf{G}_k^{prop}$ using a the proposal PDF Q (see Algorithm 1):
    $\mathbf{G}_k^{prop} \sim Q(\mathbf{G}_k | \mathbf{X}_k^{(prop)}, \mathbf{G}_{k-1}^{(L)})$
  CALCULATE the likelihood for $\mathbf{X}_k^{(prop)}$
  CALCULATE the acceptance ratio
  $\rho = min(1, \frac{p(\mathbf{z}_k | \mathbf{X}_k^{(prop)})}{p(\mathbf{z}_k | \mathbf{X}_k^{(L)})})$
  UPDATE $(\mathbf{X}_k^{(L)}, \mathbf{G}_k^{(L)})$ and its likelihood
END

JPDA Combined with the Estimated Group Structure

In step 2 of Algorithm 1, the data association problem is resolved by the JPDA algorithm [24]. The graph structure is used in the first step of the JPDA algorithm. Information contained in the graph structure is used to cluster the data association problem into distinct subproblems (Eq. (2.7)). This clustering stage helps reducing the computational time during the gating process. This gating process, in turn, is important for reducing the number of data associations hypotheses. Once, the hypotheses are listed, the classical JPDA likelihood calculation is performed (see e.g., [30], p. 263-268). The weight update is then performed by multiplying the likelihood with the previous time weights (Eq. (2.5)).

This approach has been extensively validated and simulation results comparing it with a deterministic graph are presented in [9]. The validation over real GMTI radar data shown in Figure 2.1 provided to us by QinetiQ, UK. Two groups of targets are moving on the ground by crossing their paths which constitutes an additional ambiguity for the group tracking algorithm. The GMTI measurements are obtained by an embedded radar on a moving airborn platform. There is a measurement origin uncertainty which requires the solution of the data association problem. As seen from Figure 2.1, there is clutter noise in the measured bearing angles and measured distances to the targets.

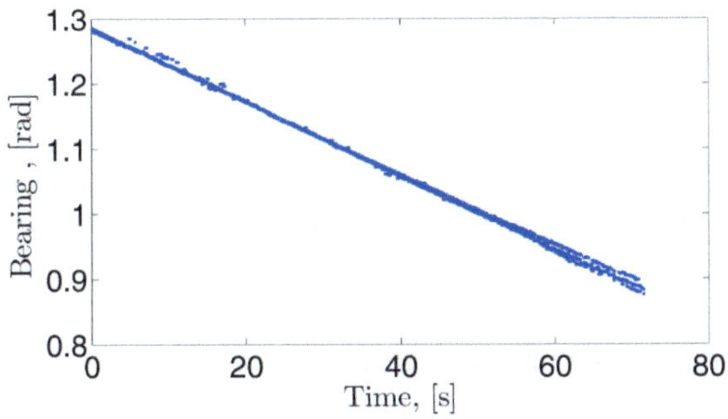

(a) Bearing

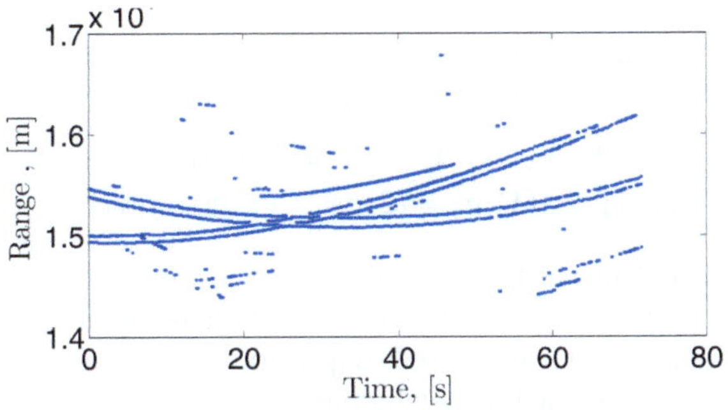

(b) Range

Fig. 2.1. Measured bearing and measured range for two groups.

The developed approach provides good estimation accuracy of each vehicle trajectory positions. Figure 2.2 shows additionally that the estimated x coordinates of the groups are close to x coordinates calculated from the measurements. The proposed algorithm is able to cope with the crossed trajectories of the groups.

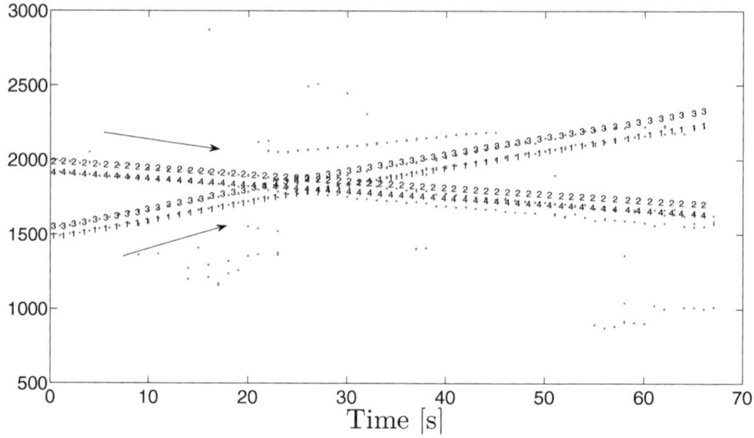

Fig. 2.2. The estimated X coordinates for the 2 groups are shown along with the X coordinates calculated from the measurements (converted from range and bearing).

Figure 2.3 presents the group structures estimated by the particle filter. In the real scenario vehicles 1 and 2 are forming group 1 and vehicles 3 and 4 are forming the second group. To plot Figure 2.3, only four relevant group structures appearing during the estimation process are labeled from 1 to 4, respectively

$$\mathbf{G}_1 : \{\mathbf{g}_1 = (1,2), \mathbf{g}_2 = (3,4)\}, \tag{2.12}$$

$$\mathbf{G}_2 : \{\mathbf{g}_1 = (1), \mathbf{g}_2 = (2), \mathbf{g}_3 = (3,4)\}, \tag{2.13}$$

$$\mathbf{G}_3 : \{\mathbf{g}_1 = (1,2), \mathbf{g}_2 = (3), \mathbf{g}_3 = (4)\}, \tag{2.14}$$

$$\mathbf{G}_4 : \{\mathbf{g}_1 = (1), \mathbf{g}_2 = (2), \mathbf{g}_3 = (3), \mathbf{g}_4 = (4)\} \tag{2.15}$$

and a probability is calculated for each group at each time step. From Figure 2.3 one can conclude that the group structure is well estimated by the introduced graph evolution model. In addition, we can deduce precious information about the group structure uncertainty during the time evolution.

2.3 The Cluster Tracking Problem

Assume that at time k there are l_k clusters, or objects at unknown locations. Each cluster may produce more than one observation yielding the measurement set realization $\mathbf{z}_k = \{\mathbf{y}_k(i)\}_{i=1}^{m_k}$, where typically $m_k >> l_k$. At this point we assume that the observation concentrations (clusters) can be adequately represented by a parametric statistical model.

Letting $\mathbf{Z}_{1:k} = \{\mathbf{Z}_1, \ldots, \mathbf{Z}_k\}$ and $\mathbf{z}_{1:k} = \{\mathbf{z}_1, \ldots, \mathbf{z}_k\}$ be the measurement history up to time k and its realization, respectively, the cluster tracking problem may be defined as follows. We are concerned with estimating the posterior distribution of

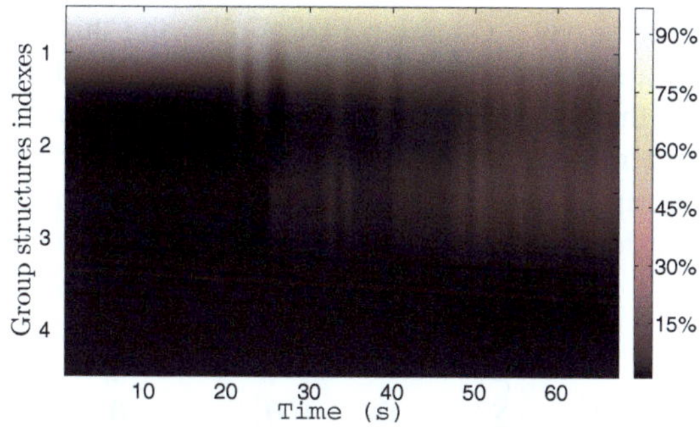

Fig. 2.3. Group structure evolution estimated by the PF. The 4 more relevant group structures are labeled from 1 to 4 (see (2.12)-(2.15)) with a probability calculated for each group at each time step.

the random set of unknown parameters, i.e. $p(\theta_k \mid \mathbf{z}_{1:k})$, from which point estimates for θ_k and posterior confidence intervals can be extracted.

For reasons of convenience we consider an equivalent formulation of the posterior that is based on existence variables. Thus, following the approach adopted in [10] the random set θ_k is replaced by a fixed dimension vector, $\boldsymbol{\theta}_k$, coupled to a vector of indicator variables $\boldsymbol{e}_k = [e_k^1, \ldots, e_k^n]^T$ showing the activity status of elements (i.e., $e_k^j = 1,\ j \in [1, n]$ indicates the existence of the jth element where n stands for the total number of elements).

2.4 Bayesian Formulation

Following the Bayesian filtering approach while assuming that the observations are conditionally independent given $(\boldsymbol{\theta}_k, \boldsymbol{e}_k)$ the density $p(\boldsymbol{\theta}_k, \boldsymbol{e}_k \mid \mathbf{z}_{1:k})$ is obtained recursively using the conventional Bayesian recursion [55]. Thus, the filtering PDF is completely specified given some prior $p(\boldsymbol{\theta}_0, \boldsymbol{e}_0)$, a transition kernel $p(\boldsymbol{\theta}_k, \boldsymbol{e}_k \mid \boldsymbol{\theta}_{k-1}, \boldsymbol{e}_{k-1})$ and a likelihood PDF $p(\mathbf{z}_k \mid \boldsymbol{\theta}_k, \boldsymbol{e}_k, m_k)$. These are derived next for the cluster tracking problem.

2.4.1 Likelihood Derivation

Recalling that a single observation $\mathbf{y}_k(i)$ is conditionally independent given $(\boldsymbol{\theta}_k, \boldsymbol{e}_k)$ yields

$$p(\mathbf{z}_k \mid \boldsymbol{\theta}_k, \boldsymbol{e}_k, m_k) \propto \prod_{i=1}^{m_k} p(\mathbf{y}_k(i) \mid \boldsymbol{\theta}_k, \boldsymbol{e}_k) \tag{2.16}$$

In the above equation the PDF $p(\mathbf{y}_k(i) \mid \boldsymbol{\theta}_k, \boldsymbol{e}_k)$ describes the statistical relation between a single observation and the cluster parameters. An explicit expression of (2.16) is derived in [55] assuming a spatial Poisson distribution for the expected number of observations. In this work we restrict ourselves to clusters in which the shape can be modeled via a Gaussian PDF. Following this only the first two moments, namely the mean and covariance, need to be specified for each cluster (under these restrictions, the cluster tracking problem is equivalent to that of tracking an evolving Gaussian mixture model with a variable number of components). Note however, that our approach does not rely on the Gaussian assumption and other parameterized density functions could equally be adopted in our framework. Thus,

$$\boldsymbol{\theta}_k^j = \begin{bmatrix} \boldsymbol{\mu}_k^j \\ \dot{\boldsymbol{\mu}}_k^j \\ \boldsymbol{\Sigma}_k^j \\ w_k^j \\ \boldsymbol{B}_k^j \end{bmatrix}, \quad \boldsymbol{\theta}_k = \left[(\boldsymbol{\theta}_k^1)^T, \dots, (\boldsymbol{\theta}_k^n)^T \right]^T \tag{2.17}$$

where $\boldsymbol{\mu}_k^j$, $\dot{\boldsymbol{\mu}}_k^j$, $\boldsymbol{\Sigma}_k^j$ and w_k^j denote the jth cluster's mean, velocity, covariance matrix entries and associated unnormalised mixture weight at time k, respectively. The additional vector $\boldsymbol{B}_k^j$ consists of any other motion parameters that might affect the cluster's behavior. In this work we have set $\boldsymbol{B}_k^j = \rho_k^j$ the local turning radius of the jth cluster's mean at time k.

Following the argument in [55] while recalling that the clusters are modeled via Gaussian PDFs, the likelihood (2.16) further assumes the following form

$$p(\mathbf{z}_k \mid \boldsymbol{\theta}_k, \boldsymbol{e}_k, m_k) \propto \prod_{i=1}^{m_k} \left[\sum_{j=0}^{n} e_k^j \tilde{w}_k^j \mathcal{N}\left(\mathbf{y}_k(i) \mid \boldsymbol{\mu}_k^j, \boldsymbol{\Sigma}_k^j\right) \right] \tag{2.18}$$

where $\tilde{w}_k^j = w_k^j / (\sum_{l=0}^{n} e_k^l w_k^l)$ is the jth normalised weight. The formulation (2.18) explicitly accounts for the clutter group (otherwise known as the bulk or null group [55]) for which the shape parameters are $w_k^0, \boldsymbol{\mu}_k^0, \boldsymbol{\Sigma}_k^0$ and $e_k^0 = 1$. An equivalent formulation of (2.18) designates the clutter probability as $v_0 := \tilde{w}_k^0 \mathcal{N}\left(\mathbf{y}_k(i) \mid \boldsymbol{\mu}_k^0, \boldsymbol{\Sigma}_k^0\right)$, yielding

$$p(\mathbf{z}_k \mid \boldsymbol{\theta}_k, \boldsymbol{e}_k, m_k) \propto \prod_{i=1}^{m_k} \left[v_0 + \sum_{j=1}^{n} e_k^j \tilde{w}_k^j \mathcal{N}\left(\mathbf{y}_k(i) \mid \boldsymbol{\mu}_k^j, \boldsymbol{\Sigma}_k^j\right) \right] \tag{2.19}$$

2.4.2 Modeling Cluster Evolution

The cluster formation may exhibit a highly complex behavior resulting, amongst other things, from group interactions. This in turn may bring about shape deformations and may also affect the number of clusters involved in the formation (i.e., splitting and merging of clusters). At this stage, in order to maintain a generic modelling approach, the filtering algorithm assumes no such interactions which thereby yields the following independent cluster evolution model

$$p(\boldsymbol{\theta}_k, \boldsymbol{e}_k \mid \boldsymbol{\theta}_{k-1}, \boldsymbol{e}_{k-1}) = \prod_{j=1}^{n} p(\boldsymbol{\mu}_k^j \mid \boldsymbol{\mu}_{k-1}^j, \dot{\boldsymbol{\mu}}_k^j) p(\dot{\boldsymbol{\mu}}_k^j \mid \dot{\boldsymbol{\mu}}_{k-1}^j, \rho_k^j) p(\boldsymbol{\Sigma}_k^j \mid \boldsymbol{\Sigma}_{k-1}^j) \times \prod_{j=1}^{n} p(w_k^j \mid w_{k-1}^j) p(e_k^j \mid e_{k-1}^j) p(\rho_k^j \mid \rho_{k-1}^j) \quad (2.20)$$

The time propagation models underlying the unique decomposition (2.20) are described next.

Time Propagation Models

Both the mean position and velocity $\boldsymbol{\mu}_k^j$ and $\dot{\boldsymbol{\mu}}_k^j$ are modeled assuming a 2-dimensional nonuniform circular motion. The kinematics of this type of motion which is well established in classical mechanics can be described in the random setting by means of a stochastic differential equation (SDE).

Let $\boldsymbol{u}_t(s)$ and $\boldsymbol{u}_n(s)$ be the tangential and normal unit vectors along the cluster's mean trajectory s, that is, $\boldsymbol{u}_t$ is in the direction of $\dot{\mu}$ whereas $\boldsymbol{u}_n$ is in the perpendicular direction along s. Because s is a parametric curve in t, similarly, one can express both $\boldsymbol{u}_t$ and $\boldsymbol{u}_n$ as functions of t. For the sake of notational simplicity we shall designate $\boldsymbol{u}_t(s(t))$ and $\boldsymbol{u}_n(s(t))$ by $\boldsymbol{u}_t(t)$ and $\boldsymbol{u}_n(t)$, respectively. Following this it can be easily shown that the jth cluster's mean velocity and position along s can be described via the following set of SDE's

$$\begin{bmatrix} d\dot{\boldsymbol{\mu}}^j(t) \\ d\boldsymbol{\mu}^j(t) \end{bmatrix} = \begin{bmatrix} a^j(t)\boldsymbol{u}_t(t) + \frac{\|\dot{\boldsymbol{\mu}}^j(t)\|_2^2}{\rho_k^j} \boldsymbol{u}_n(t) \\ \dot{\boldsymbol{\mu}}^j(t) \end{bmatrix} dt + \begin{bmatrix} \sigma_1 \\ \sigma_2 \end{bmatrix} dW(t) \quad (2.21)$$

where $a^j(t)$ and ρ_k^j denote the magnitude of the tangential acceleration and the turning radius, respectively. The SDE's (2.21) are driven by a Wiener process $W(t)$ with diffusion coefficients σ_1, σ_2. A rather simplified and explicit variant of (2.21) can be obtained by assuming that the unspecified quantity $a^j(t)\boldsymbol{u}_t(t)$ can be compensated by a proper tuning of the diffusion coefficient σ_1. Further recognizing that $\boldsymbol{u}_n(t) = [-\dot{\mu}_2^j(t), \dot{\mu}_1^j(t)]^T / \parallel \dot{\boldsymbol{\mu}}^j(t) \parallel_2$ where $\dot{\mu}_l^j(t)$ denotes the lth element of $\dot{\boldsymbol{\mu}}^j(t)$ (i.e., $\dot{\boldsymbol{\mu}}^j(t) = [\dot{\mu}_1^j(t), \dot{\mu}_2^j(t)]^T$), the first equation in (2.21) can be rewritten as

$$d\dot{\boldsymbol{\mu}}^j(t) = \frac{\parallel \dot{\boldsymbol{\mu}}^j(t) \parallel_2}{\rho_k^j} \begin{bmatrix} -\dot{\mu}_2^j(t) \\ \dot{\mu}_1^j(t) \end{bmatrix} + \sigma_1 dW(t) \quad (2.22)$$

In practice, the models (2.21) are numerically integrated over the time interval $[t_{k-1}, t_k)$ with the initial conditions

$$\dot{\boldsymbol{\mu}}^j(t_{k-1}) = \dot{\boldsymbol{\mu}}_{k-1}^j, \quad \boldsymbol{\mu}^j(t_{k-1}) = \boldsymbol{\mu}_{k-1}^j \quad (2.23)$$

for simulating random samples from the transition kernels $p(\dot{\boldsymbol{\mu}}_k^j \mid \dot{\boldsymbol{\mu}}_{k-1}^j, \rho_k^j)$ and $p(\boldsymbol{\mu}_k^j \mid \boldsymbol{\mu}_{k-1}^j, \dot{\boldsymbol{\mu}}_k^j)$ in (2.20).

Covariance and Mixture Weights Propagation

Both the covariance matrices $\boldsymbol{\Sigma}_k^j \in \mathbb{R}^{2\times 2}$ and the mixture weights $w_k^j \in \mathbb{R}$ satisfy certain constraints that render their sampling procedure somewhat distinct compared to the previous motion parameters (see also [56]). Thus, any sampling scheme must maintain these parameters either positive or positive definite in the matrix case.

A natural approach for producing valid covariance samples assumes $p(\boldsymbol{\Sigma}_k^j \mid \boldsymbol{\Sigma}_{k-1}^j)$ obeys a Wishart distribution. New conditionally independent samples of $\boldsymbol{\Sigma}_k^j$ can then be drawn using

$$\boldsymbol{\Sigma}_k^j \sim \mathscr{W}\left(\cdot \mid \frac{1}{n_1}\boldsymbol{\Sigma}_{k-1}^j, n_1, n_2\right) \tag{2.24}$$

where $\mathscr{W}(\cdot \mid V, n_1, n_2)$ denotes the Wishart distribution with a scaling matrix V and parameters n_1 and n_2. Notice that following this approach the mean of the newly sampled matrices is exactly $\boldsymbol{\Sigma}_{k-1}^j$.

The mixture weights are sampled in an analogous manner in which the Wishart distribution is replaced by a Gamma distribution. Thus, similarly to (2.24) we have

$$w_k^j \sim \Gamma\left(\cdot \mid \frac{1}{2}w_{k-1}^j, 2\right) \tag{2.25}$$

where $\Gamma(\cdot \mid a, n_1)$ denotes the Gamma distribution with shaping and scaling parameters a and n_1. Note that (2.25) implies that the normalised weights $\tilde{w}_k^j$, $j \in [1, n]$ obey a Dirichlet distribution (see also [56]).

Turning Radius Propagation

The cluster turning radius is assumed to change its direction according to a Markov chain. Let ν be the probability of keeping the current direction, then

$$p(\rho_k^j \mid \rho_{k-1}^j) = \begin{cases} \nu, & \text{if } \rho_k^j = \rho_{k-1}^j \\ 1-\nu, & \text{if } \rho_k^j = -\rho_{k-1}^j \end{cases} \tag{2.26}$$

Note that (2.26) does not necessarily imply that ρ_k^j assumes only two distinct values. In fact, the magnitude of ρ_k^j remains fixed over the entire time interval, i.e., $|\rho_k^j| = |\rho_0^j|$, $\forall k > 0$, whereas its initial value is drawn from $p(\rho_0^j)$. As a side remark, we note that other propagation models may be adopted for ρ_k^j depending on the scenario.

Birth and Death Moves

The existence indicators, denoted here by e_k^j, $j = 1, \ldots, n$, are assumed to evolve according to a Markov chain. Denote γ_l the probability of staying in state $l \in \{0, 1\}$, then

$$p(e_k^j \mid e_{k-1}^j = l) = \begin{cases} \gamma_l, & \text{if } e_k^j = l \\ 1-\gamma_l, & \text{otherwise} \end{cases} \tag{2.27}$$

2.4.3 MCMC Particle Algorithm for Cluster Tracking

In practice the filtering PDF $p(\boldsymbol{\theta}_k, \mathbf{e}_k \mid \mathbf{z}_{1:k})$ cannot be obtained analytically and approximations should be made instead. In this section we introduce a sequential MCMC particle filtering algorithm for approximating $p(\boldsymbol{\theta}_k, \mathbf{e}_k \mid \mathbf{z}_{1:k})$. The MCMC particle filter is derived consisting of a secondary Gibbs refinement stage that is aimed at increasing the efficiency of the overall sampling procedure.

The following sequential scheme is partially based on the inference algorithm presented in [10] (see also [47]). Suppose that at time $k-1$ there are N samples $\{\boldsymbol{\theta}_{k-1}^{(i)}, \boldsymbol{e}_{k-1}^{(i)}\}_{i=1}^{N}$ drawn approximately from the filtering density $p(\boldsymbol{\theta}_{k-1}, \boldsymbol{e}_{k-1} \mid \mathbf{z}_{1:k-1})$ (i.e., the previous time target distribution). A new set of samples $\{\boldsymbol{\theta}_k^{(i)}, \boldsymbol{e}_k^{(i)}\}_{i=1}^{N}$ representing $p(\boldsymbol{\theta}_k, \boldsymbol{e}_k \mid \mathbf{z}_{1:k})$ can be then simulated using the following Metropolis Hastings scheme.

Metropolis Hastings Step

The MH algorithm generates samples from an aperiodic and irreducible Markov chain with a predetermined (possibly unnormalised) stationary distribution. This is a constructive method which specifies the Markov transition kernel by means of acceptance probabilities based on the preceding time outcome. As part of this, a proposal density is used for drawing new samples. In our case, setting the stationary density as the joint filtering PDF $p(\boldsymbol{\theta}_k, \boldsymbol{e}_k, \boldsymbol{\theta}_{k-1}, \boldsymbol{e}_{k-1} \mid \mathbf{z}_{1:k})$ (of which the marginal is the desired filtering PDF), a new set of samples from this distribution can be obtained after the MH burn-in period. This procedure is described next.

First, we simulate a sample from the joint propagated PDF $p(\boldsymbol{\theta}_k, \boldsymbol{e}_k, \boldsymbol{\theta}_{k-1}, \boldsymbol{e}_{k-1} \mid \mathbf{z}_{1:k-1})$ by drawing

$$(\boldsymbol{\theta}'_k, \boldsymbol{e}'_k) \sim p(\boldsymbol{\theta}_k, \boldsymbol{e}_k \mid \boldsymbol{\theta}'_{k-1}, \boldsymbol{e}'_{k-1}) \tag{2.28}$$

where $(\boldsymbol{\theta}'_{k-1}, \boldsymbol{e}'_{k-1})$ is uniformly drawn from the empirical approximation of the preceding time step filtering PDF, given by

$$\hat{p}(\boldsymbol{\theta}_{k-1}, \boldsymbol{e}_{k-1} \mid \mathbf{z}_{1:k-1}) = N^{-1} \sum_{i=1}^{N} \delta(\boldsymbol{\theta}_{k-1}^{(i)} - \boldsymbol{\theta}_{k-1}) \delta(\boldsymbol{e}_{k-1}^{(i)} - \boldsymbol{e}_{k-1}) \tag{2.29}$$

where $\delta(\cdot)$ denotes the Dirac delta measure. This sample is then accepted or rejected using the following Metropolis rule.

Let $(\boldsymbol{\theta}_k^{(i)}, \boldsymbol{e}_k^{(i)}, \boldsymbol{\theta}_{k-1}^{(i)}, \boldsymbol{e}_{k-1}^{(i)})$ be a sample from the realized chain of which the stationary distribution is the joint filtering PDF. Then the MH algorithm accepts the new candidate $(\boldsymbol{\theta}'_k, \boldsymbol{e}'_k, \boldsymbol{\theta}'_{k-1}, \boldsymbol{e}'_{k-1})$ as the next realization from the chain with probability

$$\alpha = \min\left\{1, \frac{p(\mathbf{z}_k \mid \boldsymbol{\theta}'_k, \boldsymbol{e}'_k, m_k)}{p(\mathbf{z}_k \mid \boldsymbol{\theta}_k^{(i)}, \boldsymbol{e}_k^{(i)}, m_k)}\right\} \tag{2.30}$$

That is,

$$(\boldsymbol{\theta}_k^{(i+1)}, \boldsymbol{e}_k^{(i+1)}, \boldsymbol{\theta}_{k-1}^{(i+1)}, \boldsymbol{e}_{k-1}^{(i+1)}) = \begin{cases} (\boldsymbol{\theta}'_k, \boldsymbol{e}'_k, \boldsymbol{\theta}'_{k-1}, \boldsymbol{e}'_{k-1}), & \text{If } u \leq \alpha \\ (\boldsymbol{\theta}_k^{(i)}, \boldsymbol{e}_k^{(i)}, \boldsymbol{\theta}_{k-1}^{(i)}, \boldsymbol{e}_{k-1}^{(i)}), & \text{otherwise} \end{cases} \tag{2.31}$$

where the uniform random variable $u \sim U[0,1]$. The converged output of this scheme simulates the joint density $p(\boldsymbol{\theta}_k, \boldsymbol{e}_k, \boldsymbol{\theta}_{k-1}, \boldsymbol{e}_{k-1} \mid \mathbf{z}_{1:k})$ of which the marginal is the desired filtering PDF.

It has already been noted that the above sampling scheme may be inefficient in exploring the sample space as the underlying proposal density of a well behaved system (i.e., of which the process noise is of low intensity) introduces relatively small moves. This drawback is alleviated here by using a secondary Gibbs refinement stage.

Gibbs Refinement

In this work the accepted cluster means undergo a refinement procedure for improving the algorithm's sampling efficiency. For each sample $\boldsymbol{\theta}_k^{(i)}$ we perform a series of successive MH steps with target distributions

$$\hat{p}(\boldsymbol{\mu}_k^{j,(i)} \mid \boldsymbol{\theta}_k^{/j,(i)}, \boldsymbol{e}_k^{(i)}, \mathbf{z}_{1:k}) \tag{2.32}$$

for all $\{j \mid e_k^{j,(i)} = 1, \quad j = 1, \dots, n\}$ where the superscript $j,(i)$ denotes the jth component of the ith particle, and $\boldsymbol{\theta}_k^{/j,(i)} := \{\boldsymbol{\theta}_k^{l,(i)}\}_{l=1}^n / \{\boldsymbol{\mu}_k^{j,(i)}\}$. Here $\hat{p}(\cdot)$ denotes the empirical approximation of the underlying PDF (i.e., based on its particle representation). The refined joint sample $\boldsymbol{\theta}_k^{(i)}$ is then taken as the output of a Gibbs routine.

In practice sampling from the conditionals in (2.32) is carried out based on the following factorization of $\hat{p}(\boldsymbol{\mu}_k^j \mid \boldsymbol{\theta}_k^{/j}, \boldsymbol{e}_k, \mathbf{z}_{1:k})$

$$\begin{aligned} \hat{p}(\boldsymbol{\mu}_k^j \mid \boldsymbol{\theta}_k^{/j}, \boldsymbol{e}_k, \mathbf{z}_{1:k}) &\propto p(\mathbf{z}_k \mid \boldsymbol{\mu}_k^j, \boldsymbol{\theta}_k^{/j}, \boldsymbol{e}_k, m_k) \hat{p}(\boldsymbol{\mu}_k^j \mid \boldsymbol{\theta}_k^{/j}, \boldsymbol{e}_k, \mathbf{z}_{1:k-1}) \\ &= p(\mathbf{z}_k \mid \boldsymbol{\mu}_k^j, \boldsymbol{\theta}_k^{/j}, \boldsymbol{e}_k, m_k) \frac{\hat{p}(\boldsymbol{\mu}_k^j, \boldsymbol{\theta}_k^{/j}, \boldsymbol{e}_k \mid \mathbf{z}_{1:k-1})}{\int \hat{p}(\boldsymbol{\mu}_k^j, \boldsymbol{\theta}_k^{/j}, \boldsymbol{e}_k \mid \mathbf{z}_{1:k-1}) d\boldsymbol{\mu}_k^j} \\ &\propto p(\mathbf{z}_k \mid \boldsymbol{\mu}_k^j, \boldsymbol{\theta}_k^{/j}, \boldsymbol{e}_k, m_k) \hat{p}(\boldsymbol{\mu}_k^j, \theta_k^{/j}, \boldsymbol{e}_k \mid \mathbf{z}_{1:k-1}) \end{aligned} \tag{2.33}$$

Equation (2.33) facilitates the application of a secondary MH sampling for drawing samples from (2.32). The overall refinement procedure is based on a Metropolis-within-Gibbs scheme for which the acceptance probability of a new candidate $\bar{\boldsymbol{\mu}}_k^j$ conditioned on the previously accepted one $\boldsymbol{\mu}_k^j$, is given by

$$\bar{\alpha} = \min \left\{ 1, \frac{p(\mathbf{z}_k \mid \bar{\boldsymbol{\mu}}_k^j, \boldsymbol{\theta}_k^{/j}, \boldsymbol{e}_k, m_k)}{p(\mathbf{z}_k \mid \boldsymbol{\mu}_k^j, \boldsymbol{\theta}_k^{/j}, \boldsymbol{e}_k, m_k)} \frac{\hat{p}(\bar{\boldsymbol{\mu}}_k^j, \boldsymbol{\theta}_k^{/j}, \boldsymbol{e}_k \mid \mathbf{z}_{1:k-1}) q(\boldsymbol{\mu}_k^j)}{\hat{p}(\boldsymbol{\mu}_k^j, \boldsymbol{\theta}_k^{/j}, \boldsymbol{e}_k \mid \mathbf{z}_{1:k-1}) q(\bar{\boldsymbol{\mu}}_k^j)} \right\} \tag{2.34}$$

Notice that the likelihood term in (2.34) is identical to $p(\mathbf{z}_k \mid \boldsymbol{\theta}_k, \boldsymbol{e}_k, m_k)$ of which an explicit expression is given in (2.18).

The instrumental density $q(\boldsymbol{\mu}_k^j)$ greatly affects the efficiency of the refinement stage. In this work we have used the observation set $\mathbf{z}_k = \{\mathbf{y}_k(i)\}_{i=1}^{m_k}$ to construct a smart proposal density. The underlying idea here is rather simple and is based on the fact that the observations themselves are typically concentrated in the vicinity of the unknown clusters' means. Following this our proposal density is composed as a regularized PDF

$$q(\boldsymbol{\mu}_k^j) \propto \sum_{i=1}^{m_k} \mathcal{N}(\boldsymbol{\mu}_k^j \mid \mathbf{y}_k(i), \boldsymbol{R}) \tag{2.35}$$

for which $\boldsymbol{R}$ is the regularization intensity covariance.

An alternative rather efficient construction of the proposal density $q(\boldsymbol{\mu}_k^j)$ consists of pruning out unlikely observations $\mathbf{y}_k(i)$ that might not originate from the jth cluster. This can be accomplished by defining a set of feasible indices in the sense

$$\mathcal{G}^j := \left\{ i \mid \| \mathbf{y}_k(i) - \boldsymbol{\mu}_k^l \|_2 \geq d, \quad \forall l \neq j, \quad i = 1, \ldots, m_k \right\} \tag{2.36}$$

where d is some threshold distance pertaining to the maximal allowable cluster size. The summation in (2.35) is then performed over the set $\mathcal{G}^j$ rather than over the entire m_k observations. We would like to point out that the primary purpose of this approach is to facilitate generation of distinguishable cluster mean candidates (i.e., properly separated) which thereby reduces the chance of having multiple detections of the same cluster. In this regard one could easily come up with other gating techniques for constructing various proposals.

In practice sampling from $q(\boldsymbol{\mu}_k^j)$ is performed by simply picking an observation index i uniformly at random either from $[1, m_k]$ or $\mathcal{G}^j$ and setting $\boldsymbol{\mu}_k^j \sim \mathcal{N}(\cdot \mid \mathbf{y}_k(i), \boldsymbol{R})$.

Computing Point Estimates and Posterior Moments

As distinct from random vectors, random sets cannot be averaged owing to their property of being invariant to element permutations. This fact in turn implies that the conventional approach of computing the empirical moments of $p(\boldsymbol{\theta}_k, \boldsymbol{e}_k \mid \mathbf{z}_{1:k})$ based on the MCMC output $\{\boldsymbol{\theta}_k^{(i)}, \boldsymbol{e}_k^{(i)}\}_{i=1}^N$ is strictly inadequate (this is otherwise known as the label switching problem [41]). For that reason, in what follows we propose an approach for obtaining valid estimates using the empirical approximation.

The proposed technique relies on reordering the elements of each and every particle $(\boldsymbol{\theta}_k^{(i)}, \boldsymbol{e}_k^{(i)})$ with respect to some reference point estimate $(\boldsymbol{\theta}_{ref}, \boldsymbol{e}_{ref})$. The permuted particles thus obtained are regarded as vectors for which the entries at fixed locations correspond to the same cluster. Consequently the realizations of say, the mean of the jth cluster at time k, $\boldsymbol{\mu}_k^j$, are given by all particles $\{\boldsymbol{\mu}_k^{j,(i)}\}_{i=1}^N$. This approach facilitates the application of conventional methods for computing statistical moments (e.g., the sample mean and covariance).

The reference point which plays a vital role here might be a mode of $p(\boldsymbol{\theta}_k, \boldsymbol{e}_k \mid \mathbf{z}_{1:k}) = p(\boldsymbol{\theta}_k, \mid \mathbf{z}_{1:k}, \boldsymbol{e}_k) p(\boldsymbol{e}_k \mid \mathbf{z}_{1:k})$, or analogously when using the empirical approximation $\hat{p}(\boldsymbol{\theta}_k, \boldsymbol{e}_k \mid \mathbf{z}_{1:k})$, some highly probable particle from $\{(\boldsymbol{\theta}_k^{(i)}, \boldsymbol{e}_k^{(i)}) \mid \sum_{j=1}^n \mathbf{e}_k^{j,(i)} = \sum_{j=1}^n \mathbf{e}_{ref}^j\}$, where $\mathbf{e}_{ref}$ is, similarly, a highly probable particle from $\hat{p}(\boldsymbol{e}_k \mid \mathbf{z}_{1:k})$. Note that as distinct from the importance sampling approach, which is in the heart of conventional particle filtering, MCMC essentially facilitates straightforward computation of proper estimates such as the maximum a posteriori (MAP), however, over some fixed dimension space (See [57] and Section 4 in [58]). Based on $(\boldsymbol{\theta}_{ref}, \boldsymbol{e}_{ref})$ we can now compose the following set of particles

$$T := \left\{ (\boldsymbol{\theta}_k^{(i)}, \boldsymbol{e}_k^{(i)}) \middle| \sum_{j=1}^{n} \mathbf{e}_k^{j,(i)} = \sum_{j=1}^{n} \mathbf{e}_{ref}^j, \quad i = 1, \ldots, N \right\} \tag{2.37}$$

that is, we pick all particles with the same number of active entities as in $\boldsymbol{e}_{ref}$, i.e., all elements in T have a fixed dimension $\sum_{j=1}^n \boldsymbol{e}_{ref}^j$. Further transforming the set particles in T into vectors, the entries of each individual particle $(\boldsymbol{\theta}_k, \boldsymbol{e}_k) \in T$ are reordered such that $\| \bar{\boldsymbol{\theta}}_k - \boldsymbol{\theta}_{ref} \|_2$ is minimized, where $\bar{\boldsymbol{\theta}}_k$ denotes the permuted particle. At this stage, the random vectors in T are of fixed dimension and as such can be used for computing empirical estimates. Thus, the sample mean and covariance over T are given by

$$\hat{\boldsymbol{\theta}}_k^{MSE} = \frac{1}{|T|} \sum_{\boldsymbol{\theta}_k \in T} \boldsymbol{\theta}_k \tag{2.38a}$$

$$\frac{1}{|T|} \sum_{\boldsymbol{\theta}_k \in T} \left[\boldsymbol{\theta}_k - \hat{\boldsymbol{\theta}}_k^{MSE} \right] \left[\boldsymbol{\theta}_k - \hat{\boldsymbol{\theta}}_k^{MSE} \right]^T \tag{2.38b}$$

where $|T|$ denotes the cardinality of T (i.e., the total number of particles in T).

MCMC Algorithm Summary

A single cycle of the basic MCMC filtering algorithm is summarized in Algorithms 3, 4 and 5.

2.5 Bayesian Causality Detection of Group Hierarchies

Hierarchies in Complex Systems

The field of *complex systems* and the study of *complexity* is a new interdisciplinary field of research. Complex systems are ubiquitous, ranging from particle physics to economies [59]. According to Bar-Yam [59], it is difficult to provide a succinct definition of complex systems. It is, however, possible to describe them. Thus, complex systems are characterized by the following properties: they contain a large number of mutually–interacting parts. They exhibit a purposive behavior – the dynamics of the system have a well-defined objective or function and often there is some method

Algorithm 3. Basic MCMC Filtering Algorithm for Cluster Tracking

```
Given previous time samples {θ_{k-1}^{(i)}, e_{k-1}^{(i)}}_{i=1}^{N} do
FOR i = 1,...,N + N_{Burn-in}
  FOR j = 1,...,n
    SIMULATE (μ̇_{k-1}^{j,(i)}, μ_{k-1}^{j,(i)}) ⟶ (μ̇_k^{j,(i)}, μ_k^{j,(i)}) using (2.21), (2.22) and (2.23).
    DRAW Σ_k^{j,(i)}, w_k^{j,(i)}, ρ_k^{j,(i)} according to (2.24), (2.25) and (2.26), respectively.
  END
  Perform MCMC move (Algorithm 4).
  DRAW e_k^{j,(i)}, j = 1,...,n for the accepted move according to (2.27).
  Perform Gibbs refinement (Algorithm 5).
END
```

Algorithm 4. MCMC Move

```
CALCULATE the MH acceptance probability α of the new move using (2.30).
DRAW u ~ U[0,1]
IF u < α
  ACCEPT s^{(i)} = (θ_k^{(i)}, e_k^{(i)}, θ_{k-1}^{(i)}, e_{k-1}^{(i)}) as the next sample of the realized chain.
ELSE
  RETAIN s^{(i)} = s^{(i-1)}.
END
```

Algorithm 5. Particles Refinement (Metropolis within Gibbs)

```
FOR j = 1,...,n
  IF e_k^{j,(i)} = 1
    FOR l = 1,...,N_{MH Steps}
      DRAW an observation ȳ uniformly at random from either {y_k(r)}_{r=1}^{m_k} or 𝒢^j in (2.36).
      PROPOSE a move μ̄_k^j ~ 𝒩(· | ȳ, R).
      CALCULATE the MH acceptance probability ᾱ of the new move using (2.34).
      DRAW u ~ U[0,1]
      IF u < ᾱ
        ACCEPT the new move by setting μ_k^{j,(i)} = μ̄_k^j.
      ELSE
        RETAIN previous μ_k^{j,(i)}.
      END
    END
  END
END
```

in the design of the systems, be it self-organization or careful engineering. Since analytic treatments do not yield complete theories of complex systems, computer simulations (cellular automata and Monte-Carlo methods) play a key role in the understanding of how these systems work.

The dynamics of complex systems are sometimes explained by chaos theory [59, 60]. In [60], adaptive behavior is described as a property that is self evident from the counter actions of simple components. It is of no importance if these components are neurons, amino-acids, ants, humans or strings of bits – adaptation can take place only if the collective behavior of the system is qualitatively different from the superposition of the separate components; this is exactly the definition of nonlinearity.

A well-studied example of a complex system is that of a *cooperative multi-agent system*. In general, multi-agent systems are inherently interdisciplinary, so different researchers use various terminologies. Mataric [4] uses the term multi-agent systems to describe systems of multiple robots, and the behavior that results from their interactions is referred to as a *group behavior*. Reif and Wang [61] refer to a system with multiple robots as a *very large scale robotic system*. Beni and Wang [62] use the term *swarm* to define a multi-robot system.

In multi-agent systems, the term *behavior* is used to describe two different phenomena: the actions and movement of one agent with respect to its environment, and the global actions and movement of all the agents. To disambiguate the two terms, the term *agent behavior* is used in reference to the individual agent behavior and the term *emergent behavior* describes the behavior of the whole system of agents.

Emergent behavior is an important characteristic of a complex system. Loosely speaking, emergent behavior means that one cannot predict the system properties from analyzing its individual components. In this context, Gaudiano [63] states that "traditional 'reductionist' approaches cannot cope with complex systems", because it is impossible to predict global emergent behavior from individual rules, and that "small changes in rules lead to dramatically different emergent behaviors". However, the "good news" according to Gaudiano is that it is possible to manipulate, or control, the behavior of the complex system by changing the individual rules of each component, making it possible to simulate the emergent behavior by using bottom-up techniques.

An important notion related to complex systems is that of *hierarchy*. Bar Yam in [59] claims that a complex system is formed out of a hierarchy of independent subsystems, but it is unclear whether bringing together complex systems to form a larger complex system must give rise to a collective complex system. As an example, consider the flocking of animals. Suppose we know the exact complexity of each individual animal. Does flocking of N animals requires to know all the information on each animal? The answer, of course, is negative, due to the fact that most of the information is either irrelevant or redundant (many replicas of the same information exist in the different individual animals). In conclusion, when observing migratory flocking of animals, the interest is in a small set of informative properties: size of the flock, density, grazing behavior activity, migration, and reproductive rates.

In general, therefore, the relationship between the complexity of the collective system and the complexities of the subsystems is crucially dependant on the existence of coherence and correlation between the individuals; this notion is what the current study is about.

Emergent Phenomena: The Manifestation of Causal Mechanisms

The current work essentially deals with the inverse problem in multi agent systems theory: how can patterns, individual agents, and similarity between components be identified? How patterns form and change in a given complex network? How can these patterns be categorized and predicted? To answer these questions, we utilize the notions of causality and causal interactions and treat the multi agent system as a dynamic causal network. We show that the language of causality is extremely efficient for identifying leaders and hierarchies in multi-agent systems. We illustrate the newly-suggested formalism by showing how leaders in a dynamic multi-agent system can be identified and how a human leader of a group of people can be detected by observing a video sequence of the group.

2.5.1 G-Causality and Causal Networks

Let us begin our discussion with the ideas laid out by both Wiener and Granger in [64] and [3], respectively. These works consider two time series that may or may not influence one another. The notion conceived by Wiener is that such an influence can be quantified upon attempting to predict one of the processes given some knowledge of the other. Thus, if the inclusion of some knowledge of process X improves the prediction accuracy of another process Y, then it is said that X has a *causal influence* on Y. The latter work by Granger has provided a formal mathematical interpretation of this concept by means of linear autoregressive models. Specifically, if the variance of the autoregressive prediction error of Y is decreased by the inclusion of some knowledge of X then it is said that X G-causes Y. In practice, the evaluation of the G-causality is carried out by performing an F-test of the null hypothesis, stating that the coefficients of the first series completely vanish in the autoregressive model describing the second series (which in turn implies that there is no causal influence in a certain direction).

The notion of G-causality can be extended to the multidimensional case where more than two time series are involved (cf. [65,66] and [67]). This approach, which is referred to as *Conditional* G-causality, relies on a series of F-tests for determining whether the causality influence is mediated through additional processes. The idea here is quite simple, and is based on the observation that if the causal influence of X on Y is entirely mediated by the Z variable (i. e., indirect causal influence), then one cannot expect any improvement in predicting $X \mid Z$ (X given the knowledge of Z) by including some knowledge of Y. The two connectivity patterns in which there are direct and indirect causal influences between X and Y are illustrated in Fig. 2.4.

Let us now assume that we are given a very large number of processes that influence one another. Such cases are frequently encountered in complex systems, where many components interact at the fundamental levels for producing a certain emergent behavior. The potential of the aforementioned approach for assessing the local casual influences in such a system can be readily recognized. Nevertheless, its

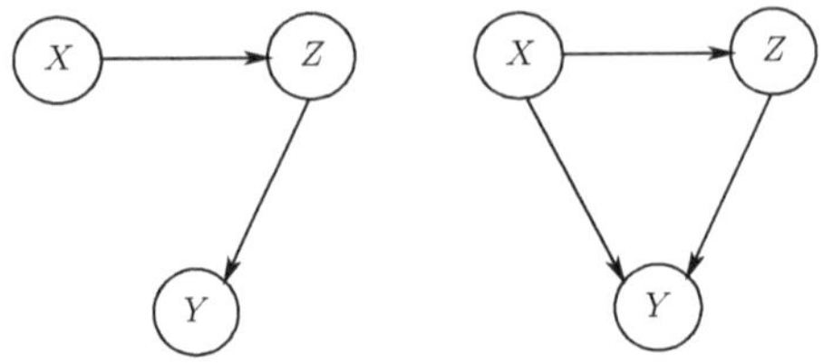

Fig. 2.4. Indirect (left) and direct (right) causal effects of X on Y, $X \to Y$. The conditional G-causality is able to distinguish between two such patterns whereas the plain G-causality cannot disambiguate them, because in both cases X G-causes Y.

practical implementation, which is aimed at answering a 'yes'/'no' type questions (i.e., testing a null hypothesis), turns out to be somewhat tedious for this specific purpose. The so called probabilistic approach to causality, which has reached maturity over the past two decades (see for example Pearl and Verma [68], Geffner [69], Shoam [70], and Pearl [2] for an extensive overview), establishes a much convenient framework for reasoning and inference of causal relations in complex structural models.

Many notions in probabilistic causality rely extensively on structural models and in particular on causal Bayesian networks which are normally referred to as simply causal networks (CN's). A CN is a directed acyclic graph compatible with a probability distribution that admits a Markovian factorization and certain structural restrictions. The following formal definition of a CN is due to Pearl [2].

Definition 1. *Let $p(v)$ be a probability distribution on a set V of v^i variables, and let $p_x(v)$ denote the distribution resulting from the intervention $X = x$ that sets a subset X of variables to constants x. Denote by p_* the set of all interventional distributions $p_x(v)$, $X \in V$, including $p(v)$, which represents no intervention. A directed acyclic graph G is said to be a causal Bayesian network compatible with p_* if and only if the following three conditions hold for every $p_x \in p_*$:*

1. *$p_x(v)$ is Markovian relative to G,*
2. *$p_x(v^i) = 1$ for all $v^i \in X$ whenever v^i is consistent with $X = x$,*
3. *$p_x(v^i \mid pa^i) = p(v^i \mid pa^i)$ for all $v^i \notin X$ whenever the set of Markovian parents pa^i of v^i, is consistent with $X = x$, i.e., each $p(v^i \mid pa^i)$ remains invariant to interventions not involving v^i.*

An alternative description of a CN, which will mostly occupy our attention in this work, is provided by structural equations [2]. In this framework the causal relations are formulated by a set of functions of the form

$$x^i = f^i(pa^i, \varepsilon^i), \tag{2.39}$$

for the Markovian parents pa^i of x^i and some (possibly random) disturbance ε^i, respectively. When the functions $f^i(\cdot)$ are linear, (2.39) reduces to the linear structural equation model that has become a standard tool in economics and social science. In detail,

$$x^i = \sum_{\{j|x^j \in pa^i\}} \alpha^{j \to i} x^j + \varepsilon^i, \tag{2.40}$$

where the coefficient $\alpha^{j \to i}$ represents a direct edge between x^j and x^i. It should be noted that the coefficients $\{\alpha^{j \to i}\}_{i \neq j}$ which may take the role of regressors in (2.40) are not necessarily identical to the path coefficients associated with edges in the underlying causal diagram G. The ability to correctly identify the path coefficients, which essentially signify the latent causal structure, requires that the underlying variables, say x^j and x^i, would be d-separated by a subset of variables in G^{ij}, a graph which lacks the direct edge between x^j and x^i but is otherwise identical to G. The reader is referred to Pearl [2] for insightful discussions on the identifiability of direct effects.

2.5.2 *Inferring Causal Relations from Empirical Data*

The problem of learning a latent causal structure based exclusively on observed data and possibly a limited prior knowledge has received much attention in many scientific fields. Over the past decade an enormous effort has been invested in coming up with computationally tractable and flexible schemes for inferring causal relations in a wide range of structural settings involving combinations of passive observations and interventional models. The frameworks suggest by Pearl [2], Sprites et al. [71] and Glymour et al. [72] (see also [73]) strongly rely on the Markovian property of causal networks. These methods essentially assess mutual statistical dependencies within the observed data and then exploit this knowledge for restraining the Markovian structure of the network. After enumerating all possible configurations that satisfy the underlying constraints, the one that is most consistent with the data is then chosen as the correct one. The primary limitation of these techniques is rooted in their inadequate, albeit efficient, manner of determining the underlying structural constraints which is based on assessing no more than two variables at a time.

Prevalent approaches for learning latent causal relations exploit direct Bayesian computations in which candidate structures are assigned likelihoods and priors reflecting the confidence in their propriety as generating mechanisms of the observed data. Notable works that devise such schemes are [74,75,76,77] and [78]. Another class of algorithms make use of active learning techniques for combining intervention selection and causal structure identification [79,80]. The primary idea underlying both these methods is to reformulate the intervention selection procedure as an information maximization problem. Hence, a specific intervention is chosen such that its effect is the most informative with respect to the latent causal structure.

Recently there has been an increased interest in information theoretic approaches to causality detection. The notions that are primarily used in this respect are these of conditional mutual information [81] and transfer entropy [82]. Whereas the former information metric is adequate for stationary signals, the latter is appropriate for quantifying causal relations among time series. In practice, the transfer entropy, which is normally expressed as the Kullback-Leibler divergence between transition densities, is approximated assuming the underlying processes are ergodic [82]. Other measures of causality for time series, which have a close relation to transfer entropy, are provided in Section 10.2.4 in [83].

The inference method derived in the ensuing is capable of quantifying causal interrelations in multi-channel time series data. As distinct from the previously mentioned approaches which rely on various (information) metrics for assessing and restraining direct links in the latent causal structure, our approach relaxes this computationally expensive task via using a juxtaposition of structural equation modeling and autoregressive processes. This in turn, renders our method viable in systems with prohibitively large number of components. It should be noted, however, that the recovered structure is prone to violate the assumptions underlying a CN.

Causal Hierarchies

In this work the term *causal hierarchies* refers to ranking of agents with respect to their cumulative effect on the actions of the remaining constituents in the system. The word "causal" here reflects the fact that our measure of distinction embodies the intensity of the causal relations between the agent under inspection and its counterparts. Adopting the information-theoretic standpoint, in which the links of a CN are regarded as information channels [81], one could readily deduce that the total effect of an agent is directly related to the local information flow entailed by its corresponding in and out degrees. To be more precise, the total effect of an agent is computed by summing up its associated path coefficients (obtained by any of the previously mentioned causality detection methods) of either inward or outward links. This concept is further illustrated in Fig. 2.5.

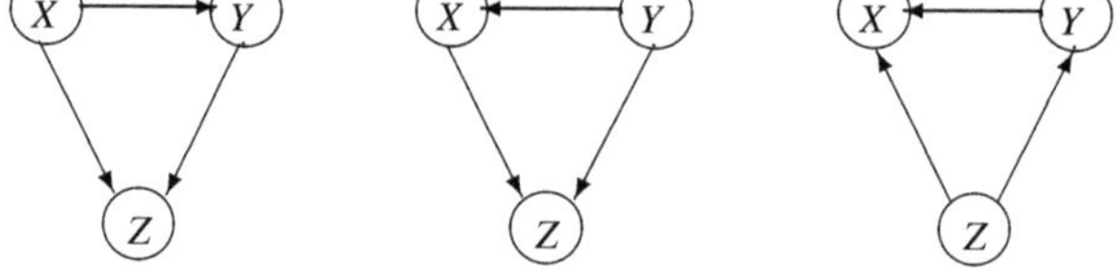

Fig. 2.5. From left to right: depiction of the causal hierarchies (based on out degrees) (X,Y,Z), (Y,X,Z), and (Z,Y,X). The most influential agents in the causal diagrams from left to right are X, Y and Z, respectively.

Inferring Causal Hierarchies by Other Methods

To some extent, causal hierarchies can be inferred using the class of principal component analysis (PCA)-based methods. Probably the most promising one in the context of our problem is the multi-channel singular spectrum analysis (M-SSA), which is otherwise known as extended empirical orthogonal function (EEOF) analysis [84]. The novel approach we suggest has some relations with M-SSA. The relevant details, however, are beyond the scope of this work. A performance evaluation of both our method and M-SSA is provided in the numerical study section in the ensuing.

2.5.3 Structural Dynamic Modeling

One cannot ignore the similarity shared by the structural linear model (2.40) and the autoregressive processes underlying G-causality. A unification of these two approaches results in a linear dynamical formulation of a CN

$$x_k^i = \sum_{j \neq i} \sum_{m=1}^{p} \alpha^{j \to i}(m) x_{k-m}^j + \varepsilon_k^i, \quad i = 1, .., n \tag{2.41}$$

where $\{x_k^i\}_{k=0}^{\infty}$ and $\{\varepsilon_k^i\}_{k=0}^{\infty}$ denote the ith random process and a corresponding white noise driving sequence, respectively. The coefficients $\{\alpha^{j \to i}(m)\}_{m=1}^{p}$ quantify the causal influence of the jth process on the ith process. Notice that the Markovian model (2.41) has a finite-time horizon of the order p (also referred to as the wake parameter). In the standard multivariate formulation, the coefficients $\alpha^{j \to i}(m)$ are square matrices of an appropriate dimension. For maintaining a reasonable level of coherency we assume that these coefficients are scalars irrespectively of the dimension of x_k^i. Nevertheless, our arguments throughout this Section can be readily extended to the standard multivariate case. Notice that as distinct from the conventional Granger models, the CN formulation (2.41) lacks the first few terms corresponding to the process x_k^i itself.

The methodology underlying Granger causality considers an F-test of the null hypothesis $\alpha^{j \to i}(m) = 0, \quad m = 1, \ldots, p$ for determining whether the jth process G-causes the ith process. The key idea here follows the simple intuitive wisdom that the more significant these coefficients the more likely they are to reflect a causal influence. In the framework of CNs the causal coefficients are related to the conditional dependencies within the probabilistic network, which in turn implies that their values can be learned based on the realizations of the time series $\{x_k^i\}_{k=0}^{\infty}$, $i = 1, \ldots, n$. In what follows, we demonstrate how the knowledge of these coefficients allows us to infer the fundamental role of individual agents within the system. Before proceeding, however, we shall define the following key quantity.

Definition 2 (Causation Matrix). *The causal influence of the process x^j on the process x^i can be quantified by*

$$A_{ij} = \sum_{m} \left[\alpha^{j \to i}(m)\right]^2 \geq 0. \tag{2.42}$$

In the above definition, A_{ij} denotes the coefficient relating the two processes x^j and x^i so as to suggest an overall matrix structure that would provide a comprehensive picture of the causal influences among the underlying processes. The matrix $A = [A_{ij}] \in \mathbb{R}^{n \times n}$, termed the *causation matrix*, essentially quantifies the intensity of all possible causal influences within the system (note that according to the definition of a CN, the diagonal entries in A vanish). It can be easily recognized that a single row in this matrix exclusively represents the causal interactions affecting each individual process. Similarly, a specific column in A is comprised of the causal influences of a single corresponding process on the entire system. This premise motivates us to introduce the notion of total causal influence.

Definition 3 (Total Causal Influence Measure). *The total causal influence (TCI) T_j of the process x_k^j is obtained as the l_1-norm of the jth column in the causation matrix A, that is*

$$T_j = \sum_{i=1}^{n} A_{ij}. \tag{2.43}$$

Having formulated the above concepts we are now ready to elucidate the primary contributions of this work, both of which rely on the TCI measure defined above.

2.5.4 Dominance and Similarity

A rather intuitive, but nonetheless striking, observation about the TCI is that it essentially reflects the dominance of each individual process in producing the underlying emergent behavior. This allows us to dissolve any complex act into its prominent behavioral building blocks (processes) using a hierarchical ordering of the form

$$\text{Least dominant } T_{j_1} \leq T_{j_2} \leq \ldots \leq T_{j_n} \text{ Most dominant} \tag{2.44}$$

Equation (2.44) is given an interesting interpretation in the application part of this work, where the underlying processes $\{x_k^j\}_{j=1}^n$ correspond to the motion of individual agents within a group. In the context of this example, the dominance of an agent is directly related to its leadership capabilities. By using the TCI measure it is therefore possible to distinguish between leaders and followers.

Another interesting implication of the TCI is exemplified in the following argument. Consider the two extreme processes in (2.44), one of which is the most dominant, $x_k^{j_n}$, while the other is the least dominant, $x_k^{j_1}$. Now, suppose we are given a new process $x_k^i, \ i \neq j_1, j_n$ and are asked to assess its dominance based exclusively on the two extremals, with respect to the entire system. Then, a common intuition would suggest to categorize x_k^i as a dominant process in the system whenever it resembles $x_k^{j_n}$ more than $x_k^{j_1}$ in the sense of $|T_{j_n} - T_i| < |T_{j_1} - T_i|$ and vice versa. This idea is summarized below.

Definition 4 (Causal Similarity). *A process x_k^j is said to resemble x_k^i more than x_k^l if and only if $|T_j - T_i| < |T_j - T_l|$.*

In the context of the previously–mentioned example, we expect that dominant agents with high leadership capabilities would possess similar TCIs that would distinguish them from the remaining agents, the followers.

2.5.5 Bayesian Estimation of $\alpha^{j\to i}$

In typical applications the coefficients $\alpha^{j\to i}(m)$, $m=1,\dots,p$ in (2.41) may be unknown. Providing that the realizations of the underlying processes are available it is fairly simple to estimate these coefficients by treating them as regressors. Such an approach by no means guarantees an adequate recovery of the underlying causal structure (see the discussion about the identifiability of path coefficients following (2.40), and a related assertion concerning non-parametric functional modeling in [2] pp. 156 – 157, both have a clear connotation to the "fundamental problem of causal inference" [85]). Nevertheless, it provides a computationally efficient framework for making inference in systems with exceptionally large number of components. This premise is evident by noting from (2.41) that while fixing i the coefficients $\alpha^{j\to i}(m)$, $\forall j\neq i$, $m=1,\dots,p$ are statistically independent of $\alpha^{j\to l}(m)$, $\forall l\neq i$.

In a Bayesian framework we confine the latent causal structure by imposing a prior on the coefficients $\alpha^{j\to i}(m)$. Let p^i_α be some prior of $\{\alpha^{j\to i}(m)\}$, and let also p^i_ε be some prescribed (not necessarily Gaussian) probability density of the white noise in (2.41). Then,

$$\begin{aligned} p(\alpha^{j\to i}(m)\mid \mathbf{x}^{1:n}_{0:k}) &\propto p^i_\alpha \prod_{t=p}^{k} p(\mathbf{x}^i_t \mid \alpha^{j\to i}(m), \mathbf{x}^j_{t-p:t-1}, j\neq i) \\ &= p^i_\alpha \prod_{t=p}^{k} p^i_\varepsilon(\mathbf{x}^i_t - \sum_{j\neq i}\sum_{m=1}^{p} \alpha^{j\to i}(m)\mathbf{x}^j_{t-m}), \quad i=1,\dots,n, \end{aligned} \tag{2.45}$$

where $\mathbf{x}^{1:n}_{0:k}=\{\mathbf{x}^1_0,\dots,\mathbf{x}^n_0,\dots,\mathbf{x}^1_k,\dots,\mathbf{x}^n_k\}$, and $\mathbf{x}^j_{t-p:t-1}=\{\mathbf{x}^j_{t-p},\dots,\mathbf{x}^j_{t-1}\}$. When the underlying densities are Gaussian, the Kalman filter provides an exact recursive computation of (2.45). A rather viable alternative which works well in most generalized settings is a Metropolis-within-Gibbs sampler that operates either sequentially or concurrently on the conditionals

$$p(\alpha^{j\to i}(m)\mid \mathbf{x}^{1:n}_{0:k}, \alpha^{l\to i}, l\neq j,i) \propto p^i_\alpha \prod_{t=p}^{k} p(\mathbf{x}^i_t \mid \alpha^{l\to i}(m), \mathbf{x}^l_{t-p:t-1}, l\neq j,i) \tag{2.46}$$

In either schemes the obtained estimates at time k are taken as the expected value of the posterior (2.45).

2.5.6 A Unified Causal Reasoning and Tracking Paradigm

In many practical applications the constituent underlying traits, which are represented here by the processes $\{\mathbf{x}^j_k\}^n_{j=1}$, may not be perfectly known (in the context of our work these could be the estimated object position and velocity, $\hat{\boldsymbol{\mu}}^j_k$, $\hat{\dot{\boldsymbol{\mu}}}^j_k$). Hence instead of the actual traits one would be forced to use approximations that might not be consistent estimates of the original quantities. As a consequence, the previously suggested structure might cease being an adequate representation of the latent causal mechanism. A plausible approach for alleviating this problem is to introduce a compensated causal structure that takes into account the exogenous disturbances induced by the possibly inconsistent estimates. Such a model can be readily formulated as a modified version of (2.41), that is

$$\hat{\boldsymbol{\mu}}_k^i = \sum_{j \neq i} \sum_{m=1}^{p} \alpha^{j \to i}(m) \hat{\boldsymbol{\mu}}_{k-m}^j + \boldsymbol{\varepsilon}_k^i + \boldsymbol{\zeta}_k^i, \quad i = 1,..,n, \tag{2.47}$$

where the additional factor $\boldsymbol{\zeta}_k^i$ denotes an exogenous bias. This premise further implies that one can use (2.47) to predict the effects of interventions in $\boldsymbol{\zeta}_k^i$ directly from passive observations (which are taken as an output of a tracking algorithm, e.g., $\hat{\boldsymbol{\mu}}_k^j$ or $\hat{\dot{\boldsymbol{\mu}}}_k^j$) without adjusting for confounding factors. See [2] (p. 166) for further elaborations on the subject.

2.6 Numerical Study

We demonstrate the performance of our suggested reasoning methodology and some of the previously mentioned concepts using both synthetic and realistic examples. All the scenarios considered here involve a group of dynamic agents, some of which are leaders that behave independently of all others. The leaders themselves may exhibit a highly nonlinear and non-predictive motion pattern which in turn affects the group's emergent behavior. We use a standard CN (2.41) with a predetermined time horizon p for disambiguating leaders from followers based exclusively on their instantaneous TCIs. In all cases the processes $\mathbf{x}_k^i$, $i = 1, \ldots, n$ are taken as either the increment $\dot{\boldsymbol{\mu}}_k^i$ or position $\boldsymbol{\mu}_k^i$ of each individual agent in the group. In addition, the unified tracking and reasoning paradigm is demonstrated by replacing the actual position and increment with the corresponding outputs of the MCMC cluster tracking algorithm, $\hat{\dot{\boldsymbol{\mu}}}_k^i$ and $\hat{\boldsymbol{\mu}}_k^i$.

The performance of the causality inference scheme, which employs a Kalman filter for estimating $\alpha^{j \to i}$, is directly related to its ability to classify leaders based on their TCI values. As leaders are, by definition, more dominant than followers in some measure space, essentially shaping the overall group behavior, we expect that their TCI values (computed based on either the out degree or in degree, in which case the transposed causation matrix, $\mathbf{A}^T$, is used in Definition 3) would reflect this fact. Furthermore, the hierarchy (2.44) should allow us to disambiguate them from the remaining agents according to the notion of causal similarity which was introduced in Section 2.5.4. Following this argument we define a rather distinctive performance measure which allows us to assess the aforementioned qualities.

Let G be a set containing the leaders indices, i.e.,

$$G = \{j \mid x_k^j \text{ is a leader's instantaneous position or velocity}\}.$$

Let also $\boldsymbol{v}$ be a vector containing the agents' ordered indices according to the instantaneous hierarchy at time k

$$T_{j_1} \leq \cdots \leq T_{j_n}, \tag{2.48}$$

i.e., $\boldsymbol{v} = [j_n, \ldots, j_1]^T$. Having stated this we can now define the following *performance index*

$$e = \arg\max_{i \in [1,n]} (\boldsymbol{v}_i \in G) . \tag{2.49}$$

The above quantity indicates the worst TCI ranking of a leader. As an example, consider a case with, say, 5 leaders. Then the best performance index we could expect would be 5, implying that all leaders have been identified and were properly ranked according to their TCIs. If the performance index yields a value greater than 5, say 10, it implies that all leaders are ranked among the top 10 agents according to their TCIs. The performance index cannot go below the total number of leaders and cannot exceed the total number of agents.

Example: Swarming of Multiple Interacting Agents (Boids)

Our first example pertains to identification of leaders and followers in a dynamical system of multiple interacting agents, collectively performing in a manner usually referred to as *swarming* or *flocking*.

In the current example, Reynolds-inspired flocking [1] is used to create a complex motion pattern of multiple agents. Among these agents, there are leaders, who independently determine their own position and velocity (based on the motion model (2.21)), and followers, who interact among themselves and follow the leader agents. Typical trajectories of the entire system along with two instantaneous snapshots are provided in Figs 2.6 and 2.7, respectively.

The inference scheme performance over 100 Monte Carlo runs, in which the agents initial state and velocity were randomly picked, is provided in Fig. 2.8. The synthetic scenario considered consists of 30 agents, 4 of which are actual leaders. The performance index cumulative distribution function (CDF) for this scenario, which is illustrated via the 50, 70 and 90 percentile lines, is shown over the entire time interval in the left panel in this figure. The percentiles indicate how many runs out of 100 yielded a performance index below a certain value. Thus, 50 percent of the runs yielded a performance index below the 50 percentile, 70 percent of the runs attained values below the 70 percentile, and so on. Following this, it can be readily recognized that from around $k = 150$ the inference scheme is able to accurately identify the actual leaders in 50 percent of the runs. A further examination of this figure reveals that the 4 actual leaders are ranked among the top 6 from around $k = 180$ in 90 percent of the runs.

A comparison of leaders ranking capabilities of the proposed approach with that of the M-SSA method is provided in the right panel in Fig. 2.8. Hence, the instantaneous CDFs of both techniques are shown when using either position or velocity time series data. This figure clearly demonstrates the superiority of the suggested approach with respect to the M-SSA.

Figure 2.9 illustrates the notion of causal similarity. In this scenario, the 4 actual leaders are assigned a unique bright color value (yellow) for distinguishing them from the remaining agents. At each time step k the agents, and therefore also their associated colors, are ordered according to their in degree and out degree TCIs. In this figure, the ordered color values are averaged over 100 runs, essentially unfolding a distinctive pattern in which all the actual leaders are closely ranked over the entire time interval. This example confirms the key idea conveyed by causal similarity, which generally states that *dominant agents are likely to resemble other dominant agents in the sense of their TCIs.*

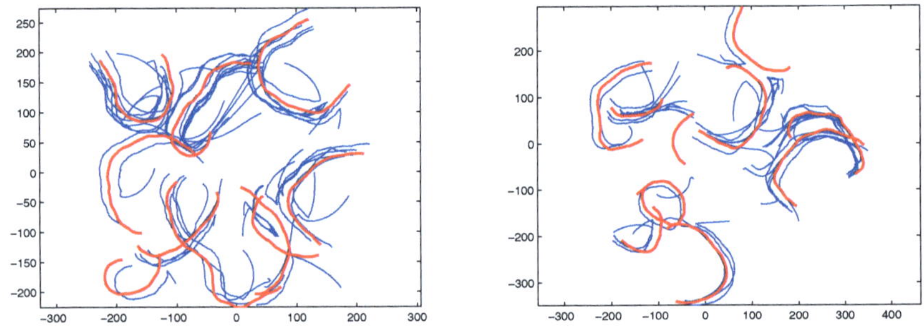

(a) Sample trajectories **(b) Sample trajectories**

Fig. 2.6. Sample trajectories of the Boids system with 50 agents (thin lines) and 10 leaders (thick lines).

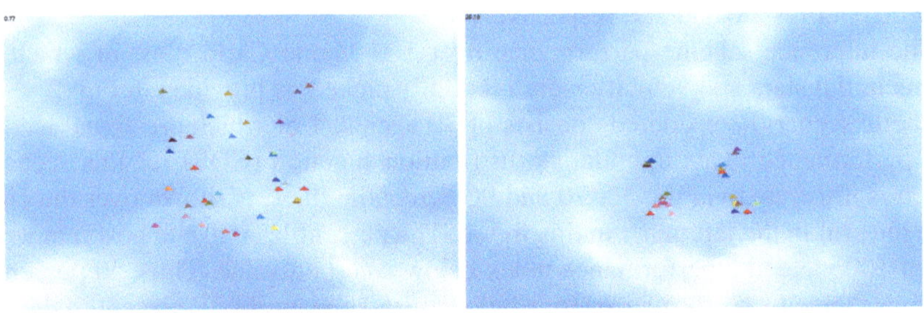

(a) Sample snapshot $k = 0$ **(b) Sample snapshot** $k = 260$

Fig. 2.7. Sample snapshots of the Boids system at various time instances.

A visualization of the instantaneous causation matrix is provided in Fig. 2.10. The out degree and in degree TCIs are illustrated by the horizontal and vertical indexed bars in this figure, respectively. Large TCI values correspond to bright colors and vice versa. Hence it can be readily recognized that the actual leaders (indexed 1, 3, 5 and 26) are assigned the largest out degree TCIs (brightest marks in the horizontal bar) and, respectively, the smallest in degree TCIs (darkest marks in the vertical bar). This further concurs with results in Fig. 2.9.

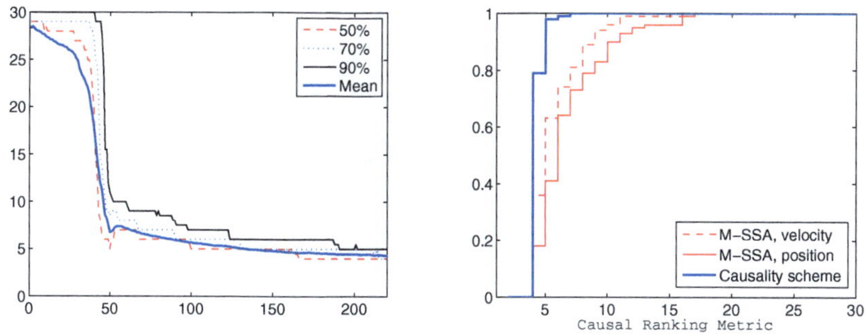

(a) Causal Ranking (out degree) **(b) Causal Ranking cdf**

Fig. 2.8. Identification performance over time of the causality scheme (left) and the ranking cdf at time $t = 220$ of both the causality scheme and the M-SSA method (using either velocity or position data) based on 100 Monte Carlo runs. Boids system with 30 agents and 4 leaders.

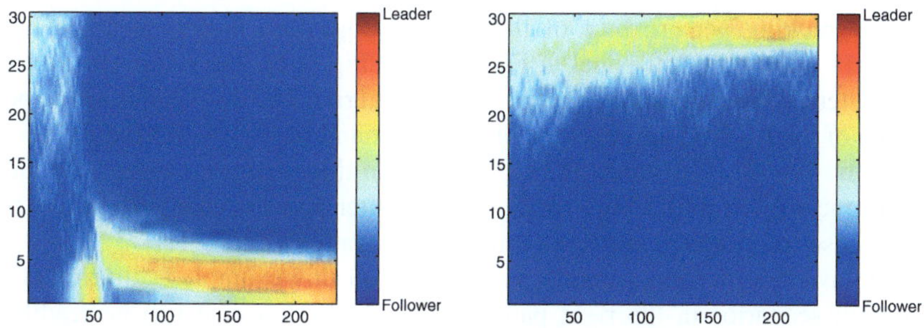

(a) Causal Ranking (out degree) **(b) Causal Ranking (in degree)**

Fig. 2.9. Two distinct causal rankings show how dominant agents share a similar behavioral pattern. Boids system with 30 agents and 4 leaders. Computed based on 100 Monte Carlo runs.

Imperfect Observations

In the following example the actual agent tracks are replaced by the output of the MCMC cluster tracking algorithm of Section 2.3. The scenario consists of 4 agents out of which 2 are leaders. As before we use the Boids system for simulating the entire system. This time, however, the produced trajectories are contaminated with clutter and additional points representing multiple emissions from possibly the same agent (i.e., agents are assumed to be extended objects). These observations are then used by the MCMC tracking algorithm of which the output is fed to the causality detection method, in a fashion similar to the one described in Section 2.5.6.

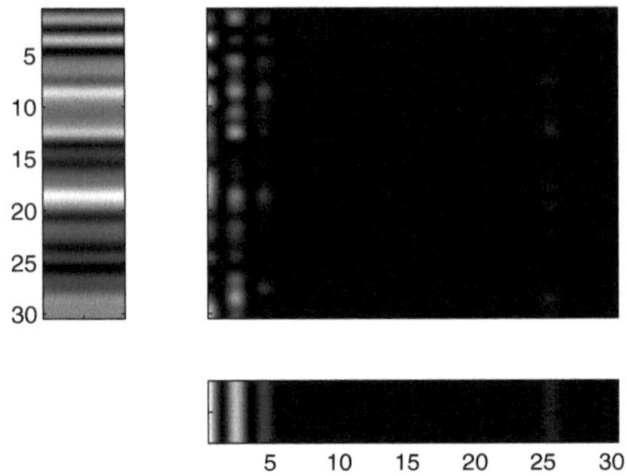

Fig. 2.10. Visualization of the causation matrix at time $t = 220$. The actual dominant agents are indexed $1, 3, 5$ and 26.

The tracking performance of the MCMC algorithm is demonstrated both in Fig. 2.11 and in the left panel in Fig. 2.12. In Fig. 2.11, the estimated tracks and the cluttered observations are shown for a typical run. The averaged tracking performance of the MCMC approach is further illustrated based on 20 Monte Carlo runs using the Hausdorff distance in Fig. 2.12. From this Figure it can be seen that the mean tracking errors become smaller than 1 after approximately 50 time steps in either cases of cluttered and non-cluttered observations.

The averaged leaders ranking performance in this example is illustrated for 3 different scenarios in the right panel in Fig. 2.12. Hence, it can be readily recognized that the two leaders are accurately identified after approximately 10 time steps when the agent positions are perfectly known. As expected, this performance is deteriorated in the presence of clutter and multiple emissions, essentially attaining an averaged ranking metric of nearly 2.5 after 60 time steps. Unexceptionally, the underlying leaders behavior validates the notion of causal similarity as illustrated in Fig. 2.13 by averaged color maps for both cases of cluttered and non-cluttered observations.

Example: Identifying Leaders in a Group of People from a Video Sequence

Our second, more practical example, deals with the following application. Consider a group of people, among which there are subgroups of leaders and followers. The followers coordinate their paths and motion with the leader. Using video observations only of the group, determine who the group leaders are. To that end, one must first develop a procedure for estimating the trajectories of n people from a given video sequence. The input to the described procedure is a movie with n moving people, where n is known. The objective is to track each person along the frame

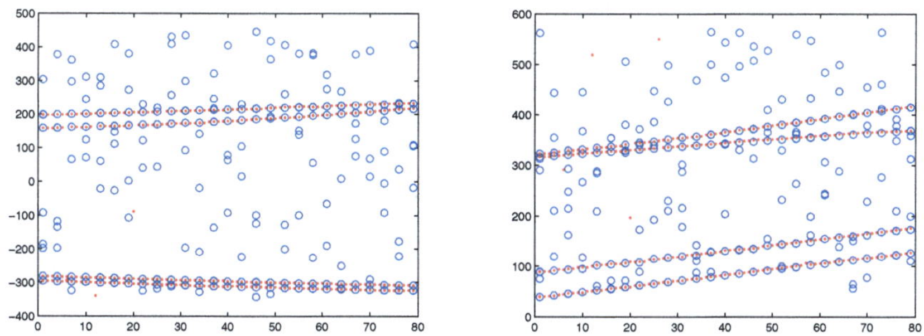

(a) X (b) Y

Fig. 2.11. Point observations and estimated tracks. Boids system with 4 agents and 2 leaders (extended objects).

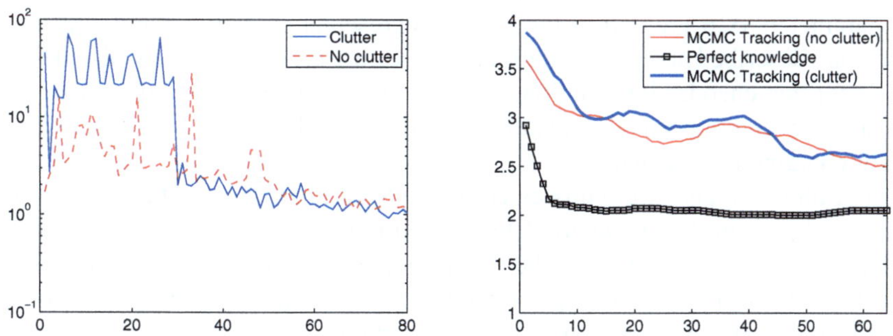

(a) Mean Hausdorff Distance **(b) Causality Ranking (out degree)**

Fig. 2.12. Tracking performance and causality ranking averaged over 20 Monte Carlo runs. Boids system with 4 agents and 2 leaders (extended objects).

sequence, and then feed this information into the CN mechanism for inferring the leaders and followers.

It is assumed that all the objects do not leave the camera field of view, and that there are no occlusions during the whole frame sequence. In addition, the movie is assumed to be captured with a fixed camera. This problem is tackled in two steps: calculation of stable points that represent the moving objects in each frame, and enforcing consistency among these points along the whole frame sequence, such that the ith stable point in all frames will belong to the same person ($i = 1, \dots, n$).

The implementation of the first phase is performed as follows. Each image is processed by the SIFT algorithm [86], providing a set of features $\{x, y, \sigma, \theta\}$, where x, y are the image coordinates of the area surrounding the object, σ is the scale parameter for which an extremum in a scale-space is found, and θ is the orientation parameter calculated based on local image properties for each feature. Although

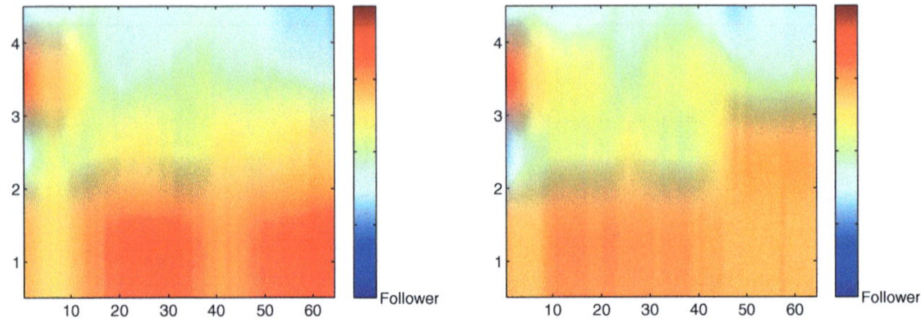

(a) No clutter **(b) With clutter**

Fig. 2.13. Causality ranking performance over 20 Monte Carlo runs. Boids system with 4 agents and 2 leaders (extended objects).

the parameters σ, θ are part of the feature representation, they are not used in the procedure described herein. In addition, each feature is accompanied with a descriptor vector that encodes information regarding the feature and is used to distinguish this specific feature from other features. In a default SIFT implementation [87], this vector contains 128 elements.

After all the images were processed by the SIFT algorithm, features that belong to static objects in the movie are detected and removed, yielding a set of dynamic features, i. e. features that describe the n people, since they are the only moving objects in the scene. This was performed as follows. Recalling that the camera is fixed throughout the frame sequence, all the static objects should have the same image coordinates in all the images. Yet, in practice, a different set of features is obtained for each image, although all the images capture a similar scene. In order to obtain the set of dynamic features along the frame sequence, we first performed matching between each pair of consecutive images, based on the descriptor vectors attached to the SIFT features. Only matched features proceed.

Next, the difference in image coordinates, $\Delta \in \mathbb{R}^2$, was computed for each matching pair of features. If (x^i, y^i) and (x^j, y^j) are the image coordinates of some matching pair of features in the ith and jth images, then $\Delta = (x^j, y^j)^T - (x^i, y^i)^T$. Matching pairs with very small $\|\Delta\|_2$ were classified as static features, and therefore all matching pairs with $\|\Delta\|_2$ below some threshold were removed. In addition, matching pairs of features with $\|\Delta\|_2$ larger than some (other) threshold were also removed, since these represent false matching.

The outcome of the above is a set of features in each image, that describe the n people, and only them. Ideally, no other objects should be represented by these features. From all these features, the objective now is to calculate a representative point for each human. The tracking task would be then accomplished if these points could be consistently re-arranged, such that the ith point in each image corresponds to the same person.

Utilizing the fact that the number of people in the movie is known, a clustering algorithm was applied from which n cluster centers were obtained. In the current implementation, the k-means algorithm [88] was used. Since it was assumed that throughout the whole frame sequence the objects do not leave the camera field of view, and that there are no occlusions, the calculated n cluster centers should represent each of the n people. Thus, in each image we have n cluster centers, and the goal now is to find how these cluster centers are related among the images, which is the second phase of processing.

In order to enforce the consistency of cluster centers along the image sequence, it is sufficient to enforce consistency of cluster centers between each adjacent pair of images. Assume that image i, denoted as I_i, was already processed, i. e. its n cluster centers are successfully associated with the n cluster centers of the previous image. The objective now is to associate the n cluster centers of the next image, I_{i+1}, to the cluster centers of I_i. This task was accomplished by associating the cluster center sets between the two images so that the distances between these two sets are minimized.

This technology was applied to two different video sequences in which there were 5 followers and 1 leader. Snapshots are shown in the upper panel in Fig. 2.14. In these videos, the actual leader (designated by a red shirt) performs a random trajectory, and the followers loosely follow its motion pattern. The clustering procedure described above is used to estimate the trajectories of the objects (the trajectories were filtered using a simple moving-average procedure to reduce the amount of noise contributed by the k-means clustering method). These trajectories were fed into the causality inference scheme.

The results of this procedure are shown in Fig. 2.15, which depicts the causality performance index for two values of the finite-time horizon (wake parameter), p. It is clearly seen that from a certain time point the algorithm identifies the *actual* leader in both videos irrespective to the value of p.

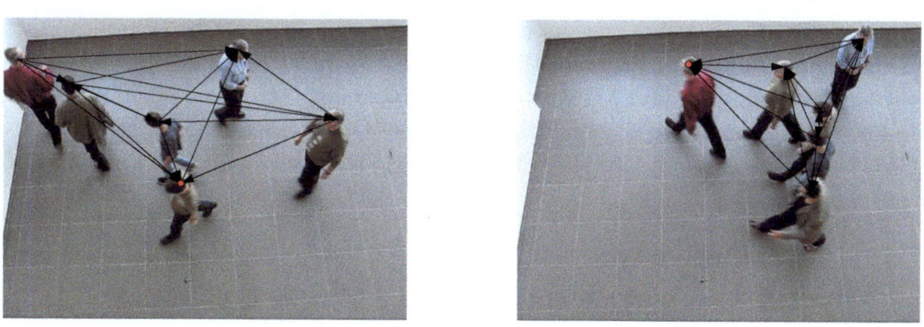

(a) Video 1 **(b) Video 2**

Fig. 2.14. Reconstructed instantaneous causal diagrams shown with the corresponding video frames.

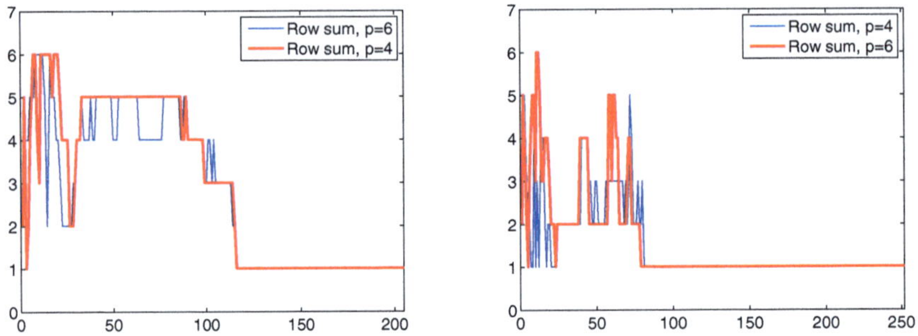

(a) Video 1 (b) Video 2

Fig. 2.15. Causality ranking performance over time.

2.7 Concluding Remarks

Several novel algorithms have been derived for tracking multiple groups of a potentially large number of coordinated objects in cluttered environments. The new methods either utilize sequential Monte Carlo techniques in conjunction with Markov chain Monte Carlo moves, or rely entirely on the latter technique for approximating the state statistics in high dimensional settings. In addition, a novel causal reasoning framework has been proposed for ranking agents with respect to their contribution in shaping the collective behavior of the system. This approach has been successfully applied for identifying leaders in groups in both synthetic and realistic scenarios.

As a concluding remark, we note that the causality inference scheme derived in this work is inadequate for tracking time-varying hierarchies, which is a highly probable venue in many applications. As part of future work this scheme would be made adaptive so as to accommodate this need.

Acknowledgments. L. Mihaylova acknowledges the support from the [European Community's] Seventh Framework Program [FP7/2007-2013] under grant agreement No 238710 (Monte Carlo based Innovative Management and Processing for an Unrivalled Leap in Sensor Exploitation). We appreciate also the help from QinetiQ, UK, including for providing us with the GMTI radar data, using RAISIN, QinetiQs GMTI Processor and the QinetiQ PodSAR radar.

2.8 Appendix: Algorithms for the Evolving Graphs

2.8.1 Evolving Graph Models

The aim is to determine an evolution model $\mathbf{G}_k = f(\mathbf{G}_{k-1}, \mathbf{X}_k)$ for the group structure, for time $k > 0$ and an initialization process $\mathbf{G}_0 = f(\mathbf{X}_0)$ for $k = 0$. The vector $\mathbf{X}_k = (\mathbf{x}_{k,1}, \ldots, \mathbf{x}_{k,n})$ comprises the state vectors of all the targets and f denotes the desired evolution model.

The system

$$\begin{cases} t = 0, \ \mathbf{G}_0 = f_I(\mathbf{X}_0), \\ t > 0, \ \mathbf{G}_k = f_{NS} \circ f_{NI} \circ f_{EU}(\mathbf{G}_{k-1}, \mathbf{X}_k), \end{cases} \tag{2.50}$$

shows the decomposition of the evolution model f according to the time k and according to three distinctive steps: edge update, node incorporation and node removal where $\circ$ denotes the composition operation; f_I is an **I**nitialization model that will be defined in Section 2.8.2; f_{EU} is the graph **E**dge **U**pdating model; f_{NI} is the graph **N**odes **I**ncorporation model that will be defined in Section 2.8.4; f_{NS} is the graph **N**odes **S**uppression model .

2.8.2 *Graph Initialization - Model* f_I

In this Section, we assume that, at time $t = 0$, the number of targets and their respective states are known, given by one of the detection techniques from [24]. Let us consider N targets constituting the set of vertices $\{\mathbf{v}_1, \ldots, \mathbf{v}_N\}$. Each vertex $\mathbf{v}_i$ is associated with the target state $\mathbf{x}_{0,i}$ at time $t = 0$, as well as the target state's corresponding variance matrix $\mathbf{P}_{0,i}$. Model 1, given below describes the proposed edge initialization method where E_0 is the set of edges linking the set of vertices $\{\mathbf{v}_1, \ldots, \mathbf{v}_N\}$. Initially E_0 is the empty set $\{\varnothing\}$. The Mahalanobis distance $d_{i,k}$ between vertices $\mathbf{v}_i$ and $\mathbf{v}_k$ is calculated and we evaluate whether it exceeds a chosen decision threshold ε. The edge between nodes $\mathbf{v}_i$ and $\mathbf{v}_k$ is denoted by (i,k). Using Model 1, the initial graph structure $\mathbf{G}_0 = (\{\mathbf{v}_1, \ldots, \mathbf{v}_N\}, E_0)$ is obtained.

Model 1.f_I-The Edge Creation Process.

```
E_0 = {∅}
FOR   i = 1,...,N−1
    FOR   k = i+1,...,N
        CALCULATE d_{i,k}
        IF d_{i,k} < ε, E_0 = E_0 ∪ {(i,k)}
        END
    END
END
```

2.8.3 *Edge Updating - Model* f_{EU}

The evolving graph of group of targets is more dynamic than those studied in the literature [21]. Existing edges should be updated at each time instant since the graph structure is related with the dynamic spatial configuration. In a straightforward way, Model 1 can recalculate the distance between any pair of nodes. However, the computational complexity can be reduced when some information about group centres (means, covariances and the distances between them) is used. For each group $\mathbf{g}$ we define its centre $\mathbf{O}^{\mathbf{g}} = \frac{1}{n_g} \sum_{\mathbf{v}_k \in \mathbf{g}} \mathbf{x}_k^{\mathbf{g}}$ and its corresponding average covariance matrix

$\mathbf{P^g} = \frac{1}{n_g} \sum_{v_k \varepsilon \mathbf{g}} \mathbf{P}_k^{\mathbf{g}}$ where n_g is defined as the number of targets in $\mathbf{g}$. The centre and covariance matrix of each group can be characterized differently, e.g., based on a mixture of Gaussian components.

Using the Mahalanobis distance criterion, an appropriate threshold $\varepsilon' >> \varepsilon$, and based on Model 1, a second graph $\mathbf{G}' = (\{\mathbf{v}'_1, \ldots, \mathbf{v}'_{n_G}\}, E')$ can be introduced with nodes $\mathbf{v}'_i$ being characterized by their position $\mathbf{O}^{\mathbf{g}_i}$. A couple of connected nodes in the set E' can be interpreted as two groups that can possibly have interactions (exchange of targets). Model 2 summarizes the edge updating process between neighboring groups. The graph $\mathbf{G}'$ will also be used in the node incorporation process.

Model 2. f_{EU}-Edges Updating Process.

```
FOR  i = 1,...,n_G - 1
    APPLY Model 1
    to the set of nodes in g_i and update E
    FOR  k = i+1,...,n_G
        IF edge (i,k) ∈ E'
          FOR each node in group g_i,
            CALCULATE the distance to each node in group g_k
            COMPARE with ε and update E
        END
    END
END
i = n_G APPLY Model 1
to the set of nodes in g_i and update E
```

Model 2 can be illustrated using the example from Figure 2.16. The considered graph contains 3 groups of 12 nodes. In Figure 2.16 (*a*), by introducing the centre of each group, the graph $\mathbf{G}'$ is represented: it contains 3 nodes, corresponding to the centre of each group, and one edge between $\mathbf{g}_1$ and $\mathbf{g}_2$. Figure 2.16 (*b*) and (*c*) illustrates the update of Model 2.

In each group, distances between any couple of nodes are calculated as shown in Figure 2.16 (*b*). Furthermore, in Figure 2.16 (*c*), for any couple of groups ($\mathbf{g}_i$, $\mathbf{g}_j$) connected in graph $\mathbf{G}'$ (in this example, only $\mathbf{g}_1$ and $\mathbf{g}_2$ are connected). The distances between any couple of nodes ($\mathbf{v}_i$, $\mathbf{v}_j$), chosen respectively in groups $\mathbf{g}_i$ and $\mathbf{g}_j$, are calculated. The use of Model 2, in this example, avoids calculations of distances between nodes in $\mathbf{g}_3$ and nodes in $\mathbf{g}_1$ and $\mathbf{g}_2$, respectively.

2.8.4 New Node Incorporation - Model f_{NI}

Classical approaches rely on either random or preferential approaches (the mixture of the two also exists) in order to assign edges to the new nodes. Additionally, in

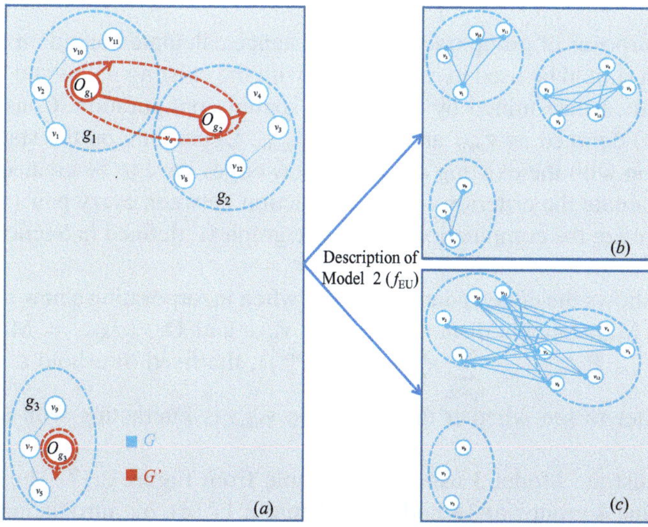

Fig. 2.16. Model 2: (*a*) use a graph structure $\mathbf{G}'$ and, for the edge updating process, (*b*) calculate distances between nodes in the same group and (*c*) calculate distances between nodes in groups that are connected through $\mathbf{G}'$.

Model 3. f_{NI}- Incorporation of new nodes.

```
Consider group i = 1
NodeNearGroup = false
DO
    CALCULATE d_{new,i}
    IF d_{new,i} < ε''
        NodeNearGroup = true
        FOR each node in g_i,
            CALCULATE the distance between v_new and each node in g_i
            COMPARE with ε and update E
        END
        FOR  k = i+1...n_G
        IF edge (i,k) ∈ E'
            CALCULATE the distance between v_new and each node in g_k
            COMPARE with ε and update E
        END
    END
    i = i+1
WHILE(i = n_G + 1 or NodeNearGroup = true)
```

classical graph techniques, the number of new edges assigned to each new node is fixed. The approach proposed in this paper differs from the above mentioned techniques.

For the purposes of group tracking, the distance calculated based on the interaction criterion should be used to create edges with the existing nodes and the number of edges is then determined by the nodes' spatial configuration. Consider a new node (vertex) denoted as $\mathbf{v}_{new}$ and its state $\mathbf{x}_{new}$. Depending on the state $\mathbf{x}_{new}$ and in comparison with the existing n_G nodes, new edges have to be created. A simple way is to evaluate the criterion for the interaction between every pair $(\mathbf{v}_{new}, \mathbf{v}_i)$. In order to optimize the computational time, the graph $\mathbf{G}'$ defined in Section 2.8.3 can be used.

Model 3 shows the edge updating process when incorporating a new node, where $d_{new,i}$ is the Mahalanobis distance between $\mathbf{v}_{new}$ and $\mathbf{O}_{g_i}$ ($d_{new,i} =$ Mahalanobis-distance $((\mathbf{x}_{new}, \mathbf{P}_{new}), (\frac{1}{n_{g_i}} \sum_{\mathbf{v}_k \varepsilon \mathbf{g_i}} \mathbf{x}_k^{\mathbf{g_i}}, \frac{1}{n_{g_i}} \sum_{\mathbf{v}_k \varepsilon \mathbf{g_i}} \mathbf{P}_k^{\mathbf{g_i}}))$; the fixed threshold $\varepsilon'' > \varepsilon$ introduced in order to see whether the new node $\mathbf{v}_{new}$ is interacting with a node in a group $\mathbf{g}$.

Let us illustrate Model 3 using the example from Figure 2.17. The considered graph contains 4 groups of 14 nodes. In Figure 2.17 (*a*), by introducing the centre of each group, the graph $\mathbf{G}'$ is represented: it contains 4 nodes, corresponding to the centre of each group, and two edges between, respectively, $\mathbf{g}_1$ and $\mathbf{g}_2$ and $\mathbf{g}_3$ and $\mathbf{g}_4$. Distances $d_{new,i}$ between the new node $\mathbf{v}_{new}$ and centres of groups $\mathbf{O}_i$ are computed. The principle of Model 3 is to calculate distances $d_{new,i}$ until finding one neighbor group of node $\mathbf{v}_{new}$ according to a threshold ε'' or until reaching the last index i ($i = n_G$). Note that ε'' is chosen such that $\varepsilon'' << \varepsilon'$ so that a new node close to one group $\mathbf{g}$ according the threshold ε'' is far from any group that is not connected with $\mathbf{g}$ according to the threshold ε'.

For the example presented on Figure 2.17, $\mathbf{g}_1$ and $\mathbf{g}_2$ are not neighbors of $\mathbf{v}_{new}$ according to the distance criterion. In contrast, $\mathbf{g}_3$ satisfies the distance criterion. Then, the calculated distances of Model 3, used to update graph $\mathbf{G}$, are illustrated. Distances between $\mathbf{v}_{new}$ and any node in $\mathbf{g}_3$ are calculated as shown in Figure 2.17 (*b*). Furthermore, since $\mathbf{g}_4$ is connected to group $\mathbf{g}_3$, in graph $\mathbf{G}'$, distances between $\mathbf{v}_{new}$ and any node in $\mathbf{g}_4$ are also calculated.

The use of Model 3, in this example, avoids the calculation of distances between $\mathbf{v}_{new}$ and nodes in $\mathbf{g}_1$ and $\mathbf{g}_2$, respectively.

2.8.5 *Old Node Suppression - Model f_{NS}*

This is the simplest graphical evolution modeling part and consists of removing death targets by removing corresponding nodes and their related edges. A target in the graph will be removed if the measurements do not contain any information about it after a certain period of time.

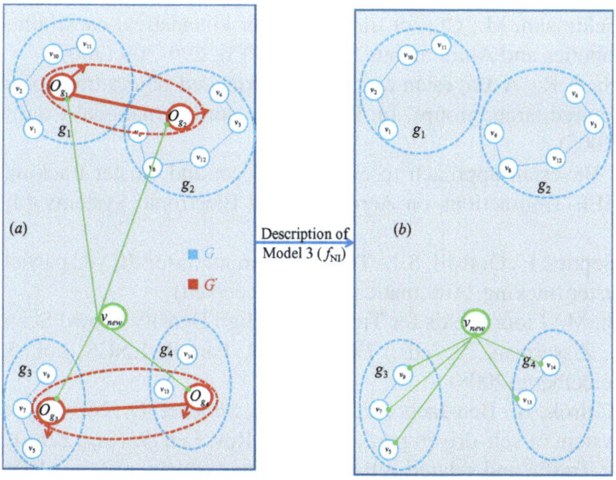

Fig. 2.17. Model 3: (*a*) use graph structure $\mathbf{G}'$ by calculating the distance from the new node $\mathbf{v}_{new}$ to the centres of all groups. Once one group $\mathbf{g_i}$ satisfies the distance threshold, (*b*) calculate the distances between $\mathbf{v}_{new}$ and any node in $\mathbf{g_i}$ in addition to the distance between $\mathbf{v}_{new}$ and any node in a group connected to $\mathbf{g_i}$ through $\mathbf{G}'$.

References

1. Reynolds, C.W.: Flocks, herds, and schools: A distributed behavioral model. Computer Graphics 21, 25–34 (1987)
2. Pearl, J.: Causality: Models, Reasoning, and Inference. Cambridge University Press (2000)
3. Granger, C.W.J.: Investigating causal relations by econometric models and cross-spectral methods. Econometrica 37, 424–438 (1969)
4. Mataric, M.J.: Designing and understanding adaptive group behaviors. Adaptive Behavior 4, 51–80 (1995)
5. Gurfil, P., Kivelevitch, E.: Flock properties effect on task assignment and formation flying of cooperating unmanned aerial vehicles. Proceedings of the Institution of Mechanical Engineers, Part G: Journal of Aerospace Engineering 221(3), 401–416 (2007)
6. Khan, Z., Balch, T., Dellaert, F.: Efficient particle filter-based tracking of multiple intercating targets using an MRF-based motion model. In: Proc. of the 2003 IEEE/RSJ Intl. Conf. on Intelligent Robots and Systems, USA, October 27-31 (2003)
7. Khan, Z., Balch, T., Dellaert, F.: A Rao-Blackwellized particle filter for eigentracking. In: Proc. of the IEEE Conf. on Computer Vision and Pattern Recognition (June 2004)
8. Khan, Z., Balch, T., Dellaert, F.: MCMC-based particle filtering for tracking a variable number of interacting targets. IEEE Transactions on Pattern Analysis and Machine Intelligence 27(11), 1805–1819 (2005)
9. Gning, A., Mihaylova, L., Maskell, S., Pang, S.K., Godsill, S.: Group object structure and state estimation with evolving networks and Monte Carlo methods. IEEE Transactions on Signal Processing 12(2), 523–536 (2011)
10. Pang, S.K., Li, J., Godsill, S.J.: Detection and tracking of coordinated groups. IEEE Transactions on Aerospace and Electronic Systems 47(1), 472–502 (2011)

11. Koch, W., Feldmann, M.: Cluster tracking under kinematical constraints using random matrices. Robotics and Autonomous Systems 57(3), 296–309 (2009)
12. Koch, W., Saul, R.: A Bayesian approach to extended object tracking and tracking of loosely structured target groups. In: Proc. of the 8th International Conf. on Inform. Fusion, ISIF (2005)
13. Koch, J.W.: Bayesian approach to extended object and cluster tracking using random matrices. IEEE Transactions on Aerospace and Electronic Systems 44(3), 1042–1059 (2008)
14. Carmi, A., Septier, F., Godsill, S.J.: The Gaussian mixture MCMC particle algorithm for dynamic cluster tracking. Automatica (2010) (accepted)
15. Ali, S., Shah, M.: Floor Fields for Tracking in High Density Crowd Scenes. In: Forsyth, D., Torr, P., Zisserman, A. (eds.) ECCV 2008, Part II. LNCS, vol. 5303, pp. 1–14. Springer, Heidelberg (2008)
16. Vicsek, T., Czirók, A., Ben-Jacob, E., Cohen, I., Shochet, O.: Novel type of phase transition in a system of self-driven particles. Phys. Rev. Lett. 75(6), 1226–1229 (1995)
17. Helbing, D.: Traffic and related self-driven many-particle systems. Review of Modern Physics 73, 1067–1141 (2002)
18. Waxman, M.J., Drummond, O.E.: A bibliography of cluster (group) tracking. In: Drummond, O.E. (ed.) Proceedings of the SPIE Signal and Data Processing of Small Targets, vol. 5428, pp. 551–560 (August 2004)
19. Celikkanat, H., Sahin, E.: Steering self-organized robot flocks through externally guided individuals. Neural Computing & and Applications 19, 849–865 (2010)
20. Mahler, R.: Statistical Multisource-multitarget Information Fusion. Artech House, Boston (2007)
21. Dorogovtsev, S.N., Mendes, J.F.F.: Evolution of networks. Advances in Physics 51, 1079–1187 (2002)
22. Albert, R., Barabási, A.-L.: Statistical mechanics of complex networks. Reviews of Modern Physics 74(1), 47–97 (2002)
23. Bar-Shalom, Y., Blair, W.: Multitarget-Multisensor Tracking: Applications and Advances, vol. III. Artech House, Boston (2000)
24. Blackman, S., Popoli, R.: Design and Analysis of Modern Tracking Systems. Artech House Radar Library (1999)
25. Clark, D., Godsill, S.: Group target tracking with the Gaussian mixture probability density filter. In: Proc. of the 3rd International Conf. on Intelligent Sensors, Sensor Networks and Information Processing (2007)
26. Pang, S.K., Li, J., Godsill, S.: Models and Algorithms for Detection and Tracking of Coordinated Groups. In: Proceedings of the IEEE Aerospace Conf. (March 2008)
27. Ristic, B., Clark, D., Vo, B.-N.: Improved SMC implementation of the PHD filter. In: Proceedings of the 13th Conference on Information Fusion (FUSION), July 2010, pp. 1–8 (2010)
28. Salmond, D.J., Gordon, N.J.: Group and extended object tracking. In: Proc. IEE Colloquium on Target Tracking: Algorithms and Applications, pp. 16/1–16/4 (1999)
29. Gilholm, K., Godsill, S., Maskell, S., Salmond, D.: Poisson models for extended target and group tracking. In: Proceedings of SPIE, vol. 5913 (2005)
30. Ristic, B., Arulampalam, S., Gordon, N.: Beyond the Kalman Filter: Particle Filters for Tracking Applications. Artech House, London (2004)
31. Koch, W.: On exploiting 'negative' sensor evidence for target tracking and sensor data fusion. Inf. Fusion 8(1), 28–39 (2007)
32. Ulmke, M., Koch, W.: Road-map assisted ground moving target tracking. IEEE Transactions on Aerospace and Electronic Systems 42(4), 1264–1274 (2006)

33. Mihaylova, L., Boel, R., Hegyi, A.: Freeway traffic estimation within recursive Bayesian framework. Automatica 43(2), 290–300 (2007)
34. Hegyi, A., Mihaylova, L., Boel, R., Lendek, Z.: Parallelized particle filtering for freeway traffic state tracking. In: Proceedings of the European Control Conference, Kos, Greece, July 2-5, pp. 2442–2449 (2007)
35. Arulampalam, M., Maskell, S., Gordon, N., Clapp, T.: A tutorial on particle filters for online nonlinear/non-Gaussian Bayesian tracking. IEEE Trans. on Signal Proc. 50(2), 174–188 (2002)
36. Angelova, D., Mihaylova, L.: Extended object tracking using Monte Carlo methods. IEEE Transactions on Signal Processing 56(2), 825–832 (2008)
37. Petrov, N., Mihaylova, L., Gning, A., Angelova, D.: A novel sequential monte carlo approach for extended object tracking based on border parameterization. In: Proceedings of the 14th International Conference on Information Fusion (Fusion 2011), Chicago, USA (2011)
38. Baum, M., Hanebeck, U.D.: Random hypersurface models for extended object tracking. In: Proc. of the IEEE International Symp. on Signal Processing and Information Technology (ISSPIT), pp. 178–183 (2009)
39. Baum, M., Feldmann, M., Fränken, D., Hanebeck, U.D., Koch, W.: Extended object and group tracking: A Comparison of Random Matrices and Random Hypersurface Models. LNCS (2010)
40. Jasra, A., Holmes, C.C., Stephens, D.A.: Markov chain Monte Carlo methods and the label switching problem in Bayesian mixture modeling. Statistical Science 1, 50–67 (2005)
41. Stephens, M.: Dealing with Label Switching in Mixture Models. Journal of the Royal Statistical Society (2000)
42. Vo, B., Singh, S., Doucet, A.: Sequential Monte Carlo Methods for Multi-Target Filtering with Random Finite Sets. IEEE Transactions on Aerospace and Electronic Systems 41(4), 1224–1245 (2005)
43. Goodman, I., Mahler, R., Nguyen, H.: Mathematics of Data Fusion. Kluwer Academic Publishing Co., Boston (1997)
44. Mahler, R.: Multi-Target Bayes Filtering via First-Order Multi-Target Moments. IEEE Transactions on Aerospace and Electronic Systems 39(4), 1152–1178 (2003)
45. Granström, K., Lundquist, C., Orguner, U.: A Gaussian Mixture PHD filter for extended target tracking. In: Proceedings of the International Conference on Information Fusion, Edinburgh, UK (2010)
46. Ng, W., Li, J., Godsill, S.J., Pang, S.K.: Multitarget Initiation, Tracking and Termination Using Bayesian Monte Carlo Methods. Computer Journal 50(6), 674–693 (2007)
47. Berzuini, C., Nicola, G., Gilks, W.R., Larizza, C.: Dynamic Conditional Independence Models and Markov Chain Monte Carlo Methods. Journal of the American Statistical Association 92(440), 1403–1412 (1997)
48. Green, P.J.: Reversible Jump Markov Chain Monte Carlo Computation and Bayesian Model Determination. Biometrika 82(4), 711–732 (1995)
49. Godsill, S.J.: On the relationship between Markov chain Monte Carlo methods for model uncertainty. Journal of Comp. Graph. Stats. 10(2), 230–248 (2001)
50. Li, X.R., Jilkov, V.: A survey of maneuveuvering target tracking. Part I: Dynamic models. IEEE Trans. on Aerosp. and Electr. Systems 39(4), 1333–1364 (2003)
51. Bar-Shalom, Y., Li, X.R.: Estimation and Tracking: Principles, Techniques and Software. Artech House (1993)
52. AlRashidi, M.R., El-Hawary, M.E.: A survey of particle swarm optimization applications in electric power systems. IEEE Transactions on Evolutionary Computation 13(4), 913–918 (2009)

53. Gning, A., Mihaylova, L., Maskell, S., Pang, S.K., Godsill, S.: Evolving networks for group object motion estimation. In: Proc. of IET Seminar on Target Tracking and Data Fusion: Algorithms and Applications, Birmingham, UK, pp. 99–106 (2008)
54. Gilks, W.R., Berzuini, C.: Following a moving target-monte carlo inference for dynamic bayesian models. Journal of the Royal Statistical Society. Series B (Statistical Methodology) 63(1), 127–146 (2001)
55. Gilholm, K., Godsill, S., Maskell, S., Salmond, D.: Poisson Models for Extended Target and Group Tracking. In: Proceedings of the SPIE Conference, pp. 230–241 (August 2005)
56. Rasmussen, C.E.: The Infinite Gaussian Mixture Model. In: Advances in Neural Information Processing Systems, vol. 12, pp. 554–560. MIT Press (2000)
57. Godsill, S., Doucet, A., West, M.: Maximum a posteriori inference sequence estimation using monte carlo particle filters. Annals of the Institute of Statistical Mathematics 53(1), 82–96 (2001)
58. Djuric, P.M., Chun, J.: An MCMC Sampling Approach to Estimation of Nonstationary Hidden Markov Models. IEEE Transactions on Signal Processing 50(5), 1113–1123 (2002)
59. Bar-Yam, Y.: Dynamics of Complex Systems, 1st edn. Addison-Wesley (1997)
60. Gleick, J.: Chaos – Making a New Science, Penguin USA, Paper (1988)
61. Reif, J.H., Wang, H.: Social potential fields: A distributed behavioral control for autonomous robots. In: Workshop on Algorithmic Foundations of Robotics, WAFR 1994 (1994)
62. Beni, G., Wang, J.: Swarm intelligence in cellular robotic systems. In: Proceedings of the NATO Advanced Workshop on Robotics and Biological Systems (1989)
63. Guadiano, P., Shargel, B., Bonabeau, E., Clough, B.: Control of UAV swarms: What the bugs can teach us. In: Proceedings of the 2nd AIAA Unmanned Unlimited Systems, Technologies, and Operations-Aerospace. AIAA, San Diego,; number AIAA 2003-6624 (2003)
64. Wiener, N.: The Theory of Prediction. In: Modern Mathematics for Engineers, McGraw-Hill, New York (1956)
65. Geweke, J.: Measurement of linear dependence and feedback between multiple time series. Journal of American Statistical Association 77, 304–313 (1982)
66. Chena, Y., Bressler, S.L., Ding, M.: Frequency decomposition of conditional granger causality and application to multivariate neural field potential data. Journal of Neuroscience Methods 150, 228–237 (2006)
67. Hosoya, Y.: Elimination of third-series effect and defining partial measures of causality. Journal of Time Series Analysis 22, 537–554 (2001)
68. Pearl, J., Verma, T.S.: A theory of inferred cauzation. In: Proceedings of the 2nd International Conference on Principles of Knowledge Representation and Reasoning, San Mateo, CA, pp. 441–452 (1991)
69. Geffner, H.: Default Reasoning: Causal and Conditional Theories. MIT Press (1992)
70. Shoam, Y.: Reasoning About Change: Time and Cauzation from the Standpoint of Artificial Intelligence. MIT Press (1988)
71. Sprites, P., Glymour, C., Scheines, R.: Cauzation, Prediction, and Search. MIT Press (2000)
72. Glymour, C., Cooper, G. (eds.): Computation, Cauzation, and Discovery. MIT Press (1999)
73. Friedman, N., Nachman, I., Peer, D.: Learning Bayesian network structure from massive datasets: The "sparse candidate" algorithm, pp. 206–215 (1999)

74. Cooper, G.F., Herskovits, E.: A Bayesian method for the induction of probabilistic networks from data. Machine Learning 9, 309–347 (1992)
75. Friedman, N., Koller, D.: Being Bayesian about network structure. A Bayesian approach to structure discovery in Bayesian networks. Machine Learning 50, 95–125 (2003)
76. Heckerman, D.: A Tutorial on Learning with Bayesian Networks. In: Learning in Graphical Models. MIT Press (1999)
77. Heckerman, D., Meek, C., Cooper, G.: A Bayesian Approach to Causal Discovery. In: Computation, Cauzation and Discovery. MIT Press (1999)
78. Tenenbaum, J.B., Griffiths, T.L.: Structure learning in human causal induction. In: Advances in Neural Information Processing Systems (2001)
79. Murphy, K.P.: Active learning of causal Bayes net structure. Tech. Rep., U.C. Berkeley (2001)
80. Tong, S., Koller, D.: Active learning for structure in Bayesian networks. In: Proceedings of the Seventeenth International Joint Conference on Artificial Intelligence, Seattle, WA, pp. 863–869 (2001)
81. Cheng, J., Greiner, R., Kelly, J., Bell, D., Liu, W.: Learning Bayesian network from data: An information-theory based approach. Artificial Intelligence 1-2, 43–90 (2002)
82. Hlavackova-Schindlera, K., Palusb, M., Vejmelkab, M., Bhattacharyaa, J.: Causality detection based on information-theoretic approaches in time series analysis. Physics Reports 441, 1–46 (2007)
83. Gourieroux, C., Monfort, A.: Time series and dynamic models. Cambridge Press (1997)
84. Golyandina, N., Nekrutkin, V., Zhigljavsky, A. (eds.): Analysis of Time Series Structure: SSA and related techniques. Chapman and Hall (2001)
85. Holland, P.W.: Statistics and causal inference. Journal of the American Statistical Association 81, 945–960 (1986)
86. Lowe, D.G.: Distinctive image features from scale-invariant keypoints. International Journal of Computer Vision 60(2), 91–110 (2004)
87. Vedaldi, A.: An open implementation of the SIFT detector and descriptor, Tech. Rep., Technical Report 070012, UCLA CSD. (February 2007)
88. MacQueen, J.B.: Some methods for classification and analysis of multivariate observations. In: Proceedings of 5th Berkeley Symposium on Mathematical Statistics and Probability, vol. 1, pp. 281–297 (1967)

Chapter 3
A Sequential Monte Carlo Method for Multi-target Tracking with the Intensity Filter

Marek Schikora, Wolfgang Koch, Roy Streit, and Daniel Cremers

Abstract. Multi-target tracking is a common problem with many applications. In most of these the expected number of targets is not known a priori, so that it has to be estimated from the measured data. Poisson point processes (PPPs) are a very useful theoretical model for diverse applications. One of those is multi-target tracking of an unknown number of targets, leading to the intensity filter (iFilter) and the probability hypothesis density (PHD) filter. This chapter presents a sequential Monte Carlo (SMC) implementation of the iFilter. In theory it was shown that the iFilter can estimate a clutter model from the measurements and thus does not need it as a priori knowledge, like the PHD filter does. Our studies show that this property holds not only in simulations but also in real world applications. In addition it can be shown that the performance of the PHD filter decreases substantially if the a priori knowledge of the clutter intensity is chosen incorrectly.

3.1 Introduction

This chapter presents a novel sequential Monte Carlo approach for multi-target tracking, called the intensity filter. In general, multi-target tracking involves the joint estimation of states and number of targets from a sequence of observations in the presence of detection uncertainty, association uncertainty and clutter [2]. Classical approaches such as the Joint Probabilistic Data Association filter (JPDAF) [9] and

Marek Schikora · Wolfgang Koch
Department of Sensor Data and Information Fusion,
Fraunhofer FKIE, Wachtberg, Germany
e-mail: {marek.schikora,wolfgang.koch}@fkie.fraunhofer.de,

Roy Streit
Metron Inc., Reston VA, USA
e-mail: r.streit@ieee.org

Daniel Cremers
Department of Computer Science, TU Munich, Garching Germany
e-mail: cremers@tum.de

P. Georgieva et al. (Eds.): Adv. in Intelligent Signal Processing and Data Mining, SCI 410, pp. 55–87.
springerlink.com

multi hypothesis tracking (MHT) [16] need in general the knowledge of the expected number of targets. Recently, the intensity filter (iFilter) [31, 28, 29] has been presented, which is similar to the well-known probability density hypothesis (PHD) filter [12]. Both filters give estimates to multi-target and multi-measurement states along with the number of targets. While the PHD filter was originally derived using finite set statistics, the iFilter was derived through Poisson point processes (PPPs). Furthermore it was shown that the PHD filter is a special case of the iFilter [29]. From an engineering point of view the main difference between the PHD and iFilter is the clutter rate model, which has to be known for the PHD filter a priori and is estimated by the iFilter. Many implementations of the PHD filter have been proposed either using sequential Monte Carlo methods [27, 37, 32], or with Gaussian mixtures [33]. An implementation and analysis of the iFilter was published in [24].

The main contribution of [24] is the first presentation of an implementation scheme for the iFilter using a sequential Monte Carlo method, also called particle filtering. The authors present a performance analysis of this new filter on simulated and real data. To obtain an objective judgement the PHD filter was also used for the same scenarios. This chapter however intends to improve the implementation proposed in [24] by adding a measurement steered birth model and a novel state extraction schema without the need of clustering techniques. This improvement is mainly based on the ideas presented by Ristic et al. [18]. The state extraction schema especially is well designed for scenarios with a high probability of detection. In this paper we will limit ourself to such experiments.

Moreover, we analyze the iFilter in demanding situations with various clutter rates and show the correct behavior of it. One example is the case, where no clutter measurements are present. The iFilter models its birth model through a clutter space $\mathscr{S}_\phi$, which is mainly responsible for the clutter rate estimation and will be explained in this chapter. If there is no clutter, the expected number of elements in this space will tend to zero and so one might think that target birth is no longer possible. However, this guess is not correct. In various experiments we show that the filter increases the intensity on $\mathscr{S}_\phi$, when a new target appears, and then can model target birth correctly, even in the case of no clutter measurements.

The remaining part of this book chapter is organized in the following way: Section 3.2 contains an introduction to Poisson Point Processes with a strong focus on multi-target tracking. The following Section 3.3 describes the intensity filter and how it can be implemented by a sequential Monte Carlo method. An intensive study of performance for simulated data is given in Section 3.4. In addition this section provides a comparison to the PHD filter and reveals the connection between both filters. Section 3.5 illustrates two applications in which the iFilter is used for the purpose of multi-target tracking. The first application is tracking of multiple objects with bearings as measurements, often called bearings-only tracking (BOT). The second presents a combination of high-accuracy optical flow and multi-target tracking algorithms for video tracking. Conclusions are drawn in the final Section 3.6.

3.2 Poisson Point Processes (PPPs)

This section gives a short introduction to basics of Poisson Point Processes (PPPs), which are used in the following. See [29] for further background. They are named after Siméon Denis Poisson (1781-1840).

Definition 1. ***(Poisson distribution)*** *A random variable* $x : \Omega \to \mathbb{N}_0$ *is said to be Pois(*λ*) distributed , with* $\lambda > 0$*, if*

$$p(x = n) = \exp(-\lambda)\frac{\lambda^n}{n!}, \tag{3.1}$$

for all $n \in \mathbb{N}_0$. $p(x = n)$ *denotes the probability for n occurrences of x.*

Every PPP is defined on a general set $\mathscr{S}$. In most of our applications this space will be considered as Euclidean, $\mathscr{S} \subseteq \mathbb{R}^d$, with $d \geq 1$ denoting the dimension, e.g. $d = 2$ for targets located in the (x,y)-plane. Other more complicated spaces are also possible. Realizations of PPPs on $\mathscr{R} \subset \mathscr{S}$ consist of $n > 0$ points $\mathbf{x}_1, \mathbf{x}_2, ..., \mathbf{x}_n \in \mathscr{R}$. We denote a realization as ordered pair

$$\xi = (n, \{\mathbf{x}_1, \mathbf{x}_2, ..., \mathbf{x}_n\}). \tag{3.2}$$

If n is equal to zero we set $\xi = (0, \emptyset)$, with $\emptyset$ the empty set. Through this notation we emphasize that the ordering of $\mathbf{x}_1, \mathbf{x}_2, ..., \mathbf{x}_n$ is irrelevant, but not that the points are necessarily distinct.

Definition 2. ***(Event Space)*** *The event space of a PPP is defined as*

$$\mathscr{E}(\mathscr{S}) := \{(0, \emptyset)\} \cup \{(n, \{\mathbf{x}_1, ..., \mathbf{x}_n\}) : \mathbf{x}_i \in \mathscr{S}, i = 1, ..., n\}_{n=1}^{\infty}. \tag{3.3}$$

Definition 3. ***(Intensity)*** *Every PPP is parameterized by only one function* $g : \mathscr{S} \to \mathbb{R}, s \mapsto g(s)$, $s \in \mathscr{S}$*, called the intensity. We call* $g(s)$ *the intensity at point s. For all* $s \in \mathscr{S}$*, if* $g(s) = c, c \geq 0$*, where c is constant, the PPP is called homogeneous; otherwise it is non-homogeneous.*

It is assumed that

$$0 \leq \int_{\mathscr{R}} g(s) \mathrm{d}s < \infty \tag{3.4}$$

holds for all bounded subsets $\mathscr{R}$ of $\mathscr{S}$, $\mathscr{R} \subset \mathscr{S}$. One realization of the PPP with intensity $g(s)$ comprises the number and the locations of points in $\mathscr{R}$.

3.2.1 PPP Sampling Procedure

A two step sampling procedure reveals the basic structure. Firstly, the number, $n \geq 0$, of points is determined by sampling the discrete Poisson variable with probability mass function given by

$$\Pr[n] = \exp\left(-\int_{\mathscr{R}} g(s)\mathrm{d}s\right) \frac{\left(\int_{\mathscr{R}} g(s)\mathrm{d}s\right)^n}{n!}, n = 0, 1, 2, ... \tag{3.5}$$

It follows from (3.5) that

$$\mathrm{E}[n] = \int_{\mathscr{R}} g(s)\mathrm{d}s. \tag{3.6}$$

Secondly, the n points in $\mathscr{R}$ are obtained as independent and identically distributed (i.i.d.) samples of the PDF (probability density function) given by $g(s)/\int_{\mathscr{R}} g(s)\mathrm{d}s$.

Two PPPs on $\mathscr{S}$ with intensities g and h are linearly superposed, if independent realizations of each are combined into one event. If $\xi_1 = (n, \{\mathbf{x}_1, ..., \mathbf{x}_n\})$ and $\xi_2 = (m, \{\mathbf{y}_1, ..., \mathbf{y}_m\})$ are two such realizations, the combined event is $\xi_3 = (n+m, \{\mathbf{x}_1, ..., \mathbf{x}_n, \mathbf{y}_1, ..., \mathbf{y}_m\})$. This event is probabilistically equivalent to a realization of a PPP whose intensity is $g+h$. In words, superposition of linearly independent PPPs yields another PPP whose intensity is the sum of the intensities of the superposed PPPs.

3.2.2 PPPs for Multi-target Tracking

In multi-target tracking applications two sequences of PPPs are usually used: one which corresponds to the multi-target state $\mathscr{X}_0, \mathscr{X}_1, ..., \mathscr{X}_k$ and one that corresponds to measurements $\mathscr{Z}_1, \mathscr{Z}_2, ..., \mathscr{Z}_k$. Both are bound to discrete time steps $t_0, t_1, ..., t_k$, with $t_{j-1} < t_j$ for $j = 1, ...k$. Measurements are assumed to be only available for time steps $j > 0$. An important but subtle point is hidden in this language. The multi-target process is not assumed to be a PPP, but it is approximated at every time step by a PPP. These PPP approximations are the $\mathscr{X}_k$. Similarly, measurement sets are not assumed to be PPPs. However, under the approximate PPP target models, the measurements are realizations of PPPs. These are the $\mathscr{Z}_k$.

We define now $\mathscr{S} \subseteq \mathbb{R}^{n_x}$, with $n_x \geq 1$, an n_x dimensional bounded single target state space. The multi-target state space is then an augmented space $\mathscr{S}^+ = \mathscr{S} \cup \mathscr{S}_\phi$, where $\mathscr{S}_\phi$ represents space of the "target absent" hypothesis ϕ. $\mathscr{S}^+$ is a discrete-continuous space. The main concepts of PPPs can be adapted to this space , but some modifications are needed. Here, $g(s)$ is a intensity defined for all $s \in \mathscr{S}^+$. Integrals of $g(s)$ over bounded subsets $\mathscr{R}$ of $\mathscr{S}^+$ must be finite, giving a discrete–continuous integral:

$$\begin{aligned} 0 &\leq \int_{\mathscr{R}^+} g(s)\mathrm{d}s \\ &\equiv g(\phi) + \int_{\mathscr{R}} g(\mathbf{x})\mathrm{d}\mathbf{x} < \infty, \end{aligned} \tag{3.7}$$

with $\mathscr{R}^+ \subset \mathscr{S}^+$, $\mathscr{R} \subset \mathscr{S}$, $\phi \in \mathscr{S}_\phi$ and $g(\phi)$ being a dimensionless intensity on $\mathscr{S}_\phi$. The number of copies of ϕ, or "clutter targets", in a realization is Poisson distributed with mean $g(\phi)$. In other words, $g(\phi)$ is the expected number of targets in ϕ. The augmented state space enables estimates of both target birth and measurement clutter. The integral $\int_{\mathscr{R}} g(\mathbf{x})d\mathbf{x}$ is the expected number of targets in $\mathscr{S}$. The measurement sequence is defined on the measurement space $\mathscr{Z} \subset \mathbb{R}^{n_z}$, with $n_z \geq 1$ being the dimension of the individual measurement.

Further discussion of PPPs defined on discrete-continuous and other spaces is given in the book by Streit [29, Section 2.12.].

3.3 The Intensity Filter

3.3.1 General Overview

The iFilter operates on the augmented space $\mathscr{S}^+$. In the same way as a standard single target filter consists of two main steps (Prediction and Update) also the iFilter predicts the intensity over $\mathscr{S}^+$ and then updates this intensity every time when new measurements arrive. Hidden here lies the main difference between a single target filter (e.g. a Kalman filter) and the iFilter. While a Bayesian filter can predict and update the posterior PDF, the iFilter only predicts and updates a first moment of this PDF. This first moment is called the intensity or probability hypothesis density (PHD). The only way to handle the posterior PDF in a multi-target case statistically correct would be by applying the multi-target Bayes filter, which is in practice not feasible.

In the following sections of this chapter an index $a|b$ in the intensity function $f_{a|b}(.)$ denotes that the intensity was updated in time step t_a with all the measurements up to time step t_b.

The intensity of the PPP $\mathscr{X}_k$ is $f_{k|k}(s), s \in \mathscr{S}^+$. We split the intensity $f_{k|k}(s)$ over $\mathscr{S}^+$ into two intensities $f_{k|k}(\mathbf{x})$ and $f_{k|k}(\phi)$. In general we can write

$$f_{k|k}(s) = \begin{cases} f_{k|k}(\mathbf{x}), & s = \mathbf{x} \in \mathscr{S} \\ f_{k|k}(\phi), & s = \phi \in \mathscr{S}_\phi \end{cases} \tag{3.8}$$

with $f_{k|k}(\mathbf{x})$ being the intensity over $\mathscr{S}$ and $f_{k|k}(\phi)$ the intensity for $\mathscr{S}_\phi$.

In order to describe the iFilter the following probabilities and PDFs have to be defined:

$\psi_k(\mathbf{x} \mid \phi)$	transition probability for new targets	(3.9)
$\psi_k(\mathbf{x} \mid \mathbf{y})$	target transition probability	(3.10)
$\psi_k(\phi \mid \phi)$	transition probability in $\mathscr{S}_\phi$	(3.11)
$\psi_k(\phi \mid \mathbf{x})$	transition probability for target death	(3.12)
$p_k(\mathbf{z} \mid \mathbf{x})$	measurement likelihood	(3.13)
$p_k(\mathbf{z} \mid \phi)$	likelihood for measurement fromϕ	(3.14)

$$p_k^D(\mathbf{x}) \qquad \text{detection probability for } \mathbf{x} \tag{3.15}$$

$$p_k^D(\phi) \qquad \text{detection probability for } \phi \tag{3.16}$$

with $\mathbf{x}, \mathbf{y} \in \mathscr{S}$ and $\mathbf{z} \in \mathscr{Z}$. Let us assume that we have the intensities $f_{k-1|k-1}(\mathbf{x})$ and $f_{k-1|k-1}(\phi)$, from the previous time step t_{k-1}. Similarly to most stochastic filtering techniques, the iFilter admits the Markovian assumption that the current state is only dependent of the last state.

In every time step t_k the likelihood (3.13) is set, according to:

$$p_k(\mathbf{z} \mid \phi) = \frac{(f_{k-1|k-1}(\phi))^{m_k}}{m_k!} e^{(-f_{k-1|k-1}(\phi))}, \tag{3.17}$$

with m_k the number of measurements in time step t_k.

In the prediction phase of the algorithm we have to predict the intensity on $\mathscr{S}^+$, denoted by $f_{k|k-1}(s), s \in \mathscr{S}^+$, as a convolution:

$$f_{k|k-1}(s) = \int_{\mathscr{S}+} \psi_k(s|y) f_{k-1|k-1}(y) \mathrm{d}y. \tag{3.18}$$

Using the definition (3.7) of a discrete–continuous integral gives the predicted intensity in the form

$$f_{k|k-1}(\mathbf{x}) = \psi_k(\mathbf{x} \mid \phi) f_{k-1|k-1}(\phi) + \int_{\mathscr{S}} \psi_k(\mathbf{x} \mid \mathbf{y}) f_{k-1|k-1}(\mathbf{y}) \mathrm{d}\mathbf{y} \tag{3.19}$$

and

$$f_{k|k-1}(\phi) = \psi_k(\phi \mid \phi) f_{k-1|k-1}(\phi) + \int_{\mathscr{S}} \psi_k(\phi \mid \mathbf{y}) f_{k-1|k-1}(\mathbf{y}) \mathrm{d}\mathbf{y}, \tag{3.20}$$

for all $\mathbf{x} \in \mathscr{S}$ and $\phi \in \mathscr{S}_\phi$, respectively. At time t_k we receive m_k measurements

$$\mathbf{Z}_k = \{\mathbf{z}_1, \ldots, \mathbf{z}_{m_k}\}, \tag{3.21}$$

with $\mathbf{z}_j \in \mathbb{R}^{n_z}$ and $j = 1, .., m_k$. The time step t_0 contains no measurements, so it is reserved for initialization. It may happen that the measurement set is empty for a given time step, in that case the update steps should be omitted. If $\mathbf{Z}_k$ is not the empty set, then the next step is to predict the measurement intensities. We do so by evaluating

$$\begin{aligned} \lambda_{k|k-1}(\mathbf{z}_j) &= \int_{\mathscr{S}+} p_k(\mathbf{z}_j|s) p_k^D(s) f_{k|k-1}(s) \mathrm{d}s \\ &= p_k(\mathbf{z}_j|\phi) p_k^D(\phi) f_{k|k-1}(\phi) + \int_{\mathscr{S}} p_k(\mathbf{z}_j|\mathbf{x}) p_k^D(\mathbf{x}) f_{k|k-1}(\mathbf{x}) \mathrm{d}\mathbf{x} \end{aligned} \tag{3.22}$$

for every measurement $\mathbf{z}_j$. The term $\lambda_{k|k-1}(\mathbf{z}_j)$ can also be called in the language of thermodynamics as a partition function evaluated at $\mathbf{z}_j$ for the space $\mathscr{S}^+$ [4].

Both intensities can be now updated at the time step t_k, giving:

$$f_{k|k}(\mathbf{x}) = (1 - p_k^D(\mathbf{x}))f_{k|k-1}(\mathbf{x}) + \left[\sum_{j=1}^{m_k} \frac{p_k(\mathbf{z}_j|\mathbf{x})p_k^D(\mathbf{x})}{\lambda_{k|k-1}(\mathbf{z}_j)}\right] f_{k|k-1}(\mathbf{x}) \tag{3.23}$$

and

$$f_{k|k}(\phi) = (1 - p_k^D(\phi))f_{k|k-1}(\phi) + \left[\sum_{j=1}^{m_k} \frac{p_k(\mathbf{z}_j|\phi)p_k^D(\phi)}{\lambda_{k|k-1}(\mathbf{z}_j)}\right] f_{k|k-1}(\phi). \tag{3.24}$$

Keep in mind these intensities are not the posterior PDF of $p_k(\mathbf{X}|\mathbf{Z}_k)$ but only a first moment of the posterior point process, i.e., in general:

$$\int_{\mathscr{S}} f_{k|k}(\mathbf{x})\mathrm{d}\mathbf{x} \neq 1. \tag{3.25}$$

In fact, it can be shown (cf. (3.6)) that the above integral is the expected number of targets for t_k, denoted by:

$$\eta_k = \int_{\mathscr{S}} f_{k|k}(\mathbf{x})\mathrm{d}\mathbf{x} \tag{3.26}$$

The main drawback of the above filter equations is that in general the involved integrals cannot be solved analytically. Therefore an appropriate numerical solution is needed. In the following we show a sequential Monte Carlo (SMC) version of the iFilter in which the intensity $f_{k|k}(\mathbf{x})$ will be approximated by particles (delta peaks) drawn from this intensity. Actually the particles approximate the involved integrals and the intensities. Another name for this kind of technique is particle based filtering [17].

3.3.2 The SMC-iFilter

The works of Vo et al. [32] and Ristic et al. [18] give efficient sequential Monte Carlo methods for the PHD filter. We present here a sequential Monte Carlo method for the iFilter. The following implementation is an improved version of our previously published work [24]. The main improvements are a measurement steered particle placement for target birth, and target state and covariance matrix estimation without the need of clustering.

The improved SMC-iFilter can be summarized in eight steps, which will be presented in the following. Here the particle set represents the target intensity of the PPP, which corresponds to the multi-target state $\mathscr{X}_k$. By analogy to the PHD-filter, the integral over this intensity (or sum, if using particles) is the estimated expected number of targets and it is not necessary equal to one. Given from the previous time step we have the particle set:

$$\{(\mathbf{x}_i, w_i)\}_{i=1}^{N_k}, \tag{3.27}$$

with $\mathbf{x}_i \in \mathbb{R}^{n_x}$, w_i the corresponding weight and N_k denoting the number of particles, estimated at time step t_{k-1}. This set represents the target intensity. In addition we have the intensity of the space $\mathscr{S}_\phi$ denoted by $f_{k-1|k-1}(\phi)$, c.f. (3.8). For the sake of simplicity, we will assume in the following uniformly distributed clutter. With this assumption the intensity $f_{k-1|k-1}(\phi)$ can be represented by a single number, called the number of ϕ hypotheses.

To initialize the particle cloud at time step t_0, $N_0 \in \mathbb{N}^+$ particles are distributed uniformly across the state space $\mathscr{S}$, e.g. $N_0 = 1000$. The weights are set to $w_i = 1/N_0$. $f_{0|0}(\phi)$ is set to a initial number, e.g. 2.

The implementation details using a particle representation are presented in the following. Steps 1 and 2 correspond to the prediction phase,steps 3-7 to the correction phase and step 8 to the resampling phase of a sequential Monte Carlo algorithm. A brief summary can be found in Algorithm 3.1.

1. **Predict target intensity**
   The resampled particle set gained from the previous step is denoted by $\{\mathbf{x}_i, w_i\}_{i=1}^{N_k}$, where N_k was estimated in time step t_{k-1}, c.f. Step 8. These particles represent the intensity over $\mathscr{S}$. Another interpretation is that every particle represents a possible target state (called microstates in the language of thermodynamics) in $\mathscr{S}$, so that the prediction of the whole set can be modeled by applying the Markovian transition model ψ_k to every particle. The weights are unchanged. In practical implementations this has the same effect as predicting the intensity distribution over $\mathscr{S}$ with a closed formula.

   Assuming a constant velocity model in two dimensions the prediction of the persistent particles can be modeled by:

$$\tilde{\mathbf{x}}_i = \begin{pmatrix} 1 & 0 & \Delta_t & 0 \\ 0 & 1 & 0 & \Delta_t \\ 0 & 0 & 1 & 0 \\ 0 & 0 & 0 & 1 \end{pmatrix} \mathbf{x}_i + \boldsymbol{\nu}, \tag{3.28}$$

   with $\Delta_t = t_k - t_{k-1}$ and $\boldsymbol{\nu} \sim \mathscr{N}(\mathbf{0}, \boldsymbol{\Sigma})$ a realization of a normal distributed random vector. The iFilter models the birth process by itself, so that the particle number has to be increased in order to represent newly born targets correctly. Then

$$N_{k,new} = \left\lceil \frac{N_k}{\eta_{k-1}} \cdot (1 - \psi_k(\phi \mid \phi)) \cdot f_{k-1|k-1}(\phi) \right\rceil \tag{3.29}$$

   denotes the additional number of particles. The term η_{k-1} denotes the estimated expected number of targets from the previous time step t_{k-1}, c.f. Step 8. The most general sampling for new born particles can be realized through a uniform sampling procedure in $\mathscr{S}$. The new born particles must cover the whole state space. In order to avoid a high number of additional particles in scenarios with

Algorithm 3.1. The SMC-iFilter

In: $\{(\mathbf{x}_i, w_i)\}_{i=1}^{N_k}, f_{k-1|k-1}(\phi), \mathbf{Z}_k, \mathbf{Z}_{k-1}$
Out: $\{(\mathbf{x}_i, w_i)\}_{i=1}^{N_k+1}, f_{k|k}(\phi), \{\hat{\mathbf{y}}_j, \hat{\mathbf{P}}_j\}$

1. Predict target intensity
   - For $i = 1, ..., N_k$ apply (3.28) to get $\tilde{\mathbf{x}}_i$.
   - Sample $N_{k,new}$ (3.29) many new particles; measurement steered according to $\mathbf{Z}_{k-1}$ and (3.30)
   - Weights for new particles are w_i (3.31)
2. Predict ϕ intensity
   - $f_{k|k-1}(\phi) = \hat{b}_k(\phi) + \hat{d}_k(\phi)$ (3.34)
3. Predict measurement intensity
   - $\lambda_{k|k-1}(\mathbf{z}_j) = \tilde{\lambda}_k(\mathbf{z}_j) + \nu_k(\mathbf{z}_j)$ (3.37)
4. Compute target states
   - Compute the set $\mathscr{J}$ (3.40)
   - For all $j \in \mathscr{J}$:
     $\hat{\mathbf{y}}_j = \frac{1}{W_j} \sum_{i=1}^{N_k} w_{j,i} \tilde{\mathbf{x}}_i$ (3.41)
5. Compute covariance matrices
   - For all $j \in \mathscr{J}$:
     $\mathbf{P}_j = \frac{1}{W_j} \sum_{i=1}^{N_k} w_{j,i} (\tilde{\mathbf{x}}_i - \hat{\mathbf{y}}_j)(\tilde{\mathbf{x}}_i - \hat{\mathbf{y}}_j)^T$ (3.42).
6. Update target intensity
   - For every particle $(\tilde{\mathbf{x}}_i, w_i)$, with $i = 1, .., N_k + N_{k,new}$ set the new weight according to (3.43).
7. Update ϕ intensity
   - Set $f_{k|k}(\phi)$ according to (3.44).
8. Resample
   - Compute $N_{k+1} = (N_k + N_{k,new}) \cdot p_S$ (3.47).
   - Use some standard resampling strategy to get $\{(\mathbf{x}_i, w_i)\}_{i=1}^{N_k+1}$

a high probability of detection, the authors in [18] proposed to sample new born particles according to the measurements from the previous time step $\mathbf{Z}_{k-1}$. Let m_{k-1} denote the number of measurements in time step t_{k-1}, then for each of these we sample

$$N_{k,new}^{j} = \lceil N_{k,new} / m_{k-1} \rceil, \quad j = 1, ..., m_{k-1} \tag{3.30}$$

many particles $\tilde{\mathbf{x}}_i$ drawn from a distribution proportional to the distribution $p_k(\mathbf{z}_j^{k-1}|\mathbf{x})$ centered around an old measurement $\mathbf{z}_j^{k-1}$. The operator $\lceil . \rceil$ rounds to the next bigger integer.

The weights of the new born particles are set to

$$w_i = \frac{\psi_k(\mathbf{x}_i \mid \phi) \cdot f_{k-1|k-1}(\phi)}{N_{k,new}}, \quad i = 1, ..., N_{k,new}. \tag{3.31}$$

This sampling is an approximation of the transition model $\psi_k(\mathbf{x} \mid \phi)$, which has proven very stable in experiments. We define $\{\tilde{\mathbf{x}}_i, w_i\}_{i=1}^{N_k + N_{k,new}}$ as the predicted particle set containing the newly created and the shifted particles.

2. **Predict hypothesis intensity**
   In order to predict the number of ϕ hypotheses, compute the predicted number of persistently absent $\tilde{b}_k$ and newly absent targets $\tilde{d}_k$ as

$$\tilde{b}_k(\phi) = \psi_k(\phi \mid \phi) \cdot f_{k-1|k-1}(\phi), \tag{3.32}$$

$$\tilde{d}_k(\phi) = \sum_{i=1}^{N_k} \psi_k(\phi \mid \tilde{\mathbf{x}}_i) \cdot w_i. \tag{3.33}$$

   The predicted number is then:

$$f_{k|k-1}(\phi) = \tilde{b}_k(\phi) + \tilde{d}_k(\phi). \tag{3.34}$$

3. **Predict measurement intensity**
   For all new measurements $\mathbf{z}_j$, with $j = 1, ..., m_k$ compute, according to (3.22), the partition functions evaluated at $\mathbf{z}_j$ for the state space and ϕ:

$$\nu_k(\mathbf{z}_j) = \sum_{i=1}^{N_k + N_{k,new}} p_k(\mathbf{z}_j \mid \tilde{\mathbf{x}}_i) p_k^D(\tilde{\mathbf{x}}_i) w_i \tag{3.35}$$

$$\tilde{\lambda}_k(\mathbf{z}_j) = p_k(\mathbf{z}_j \mid \phi) p_k^D(\phi) f_{k|k-1}(\phi). \tag{3.36}$$

   The sum of both is the predicted measurement intensity for $\mathbf{z}_j$

$$\lambda_{k|k-1}(\mathbf{z}_j) = \tilde{\lambda}_k(\mathbf{z}_j) + \nu_k(\mathbf{z}_j) \tag{3.37}$$

4. **Estimate target states**
   To avoid a clustering step the methodology presented in [18] is used and adopted to the iFilter. First, compute the following weights for all new measurements $\mathbf{z}_j, j = 1, ..., m_k$ and all persistent particles, i.e. not the new born, $\mathbf{x}_i, i = 1, ..., N_k$.

$$w_{j,i} = \frac{p_k(\mathbf{z}_j|\tilde{\mathbf{x}}_i)p_k^D(\tilde{\mathbf{x}}_i)}{\lambda_{k|k-1}(\mathbf{z}_j)} \cdot w_i \tag{3.38}$$

Then compute the following sum

$$W_j = \sum_{i=1}^{N_k} w_{j,i}, \tag{3.39}$$

which can be seen as a probability of existence for target j, similarly to the multi-target multi-Bernoulli filter. For further analysis only those j are considered for which W_j is above a specified threshold τ, i.e.

$$\mathscr{J} = \{j|W_j > \tau, j = 1,...,m_k\} \tag{3.40}$$

For all $j \in \mathscr{J}$ the estimated states are then:

$$\hat{\mathbf{y}}_j = \frac{1}{W_j}\sum_{i=1}^{N_k} w_{j,i}\tilde{\mathbf{x}}_i. \tag{3.41}$$

In Equation (3.41) we added, in contrast to [18], the normalization term $\frac{1}{W_j}$ to receive more accurate state estimates when W_j is not practically one. Note that only targets that have been detected at time step t_k can be reported as present. This follows the lack of "memory" of a PHD filter. The iFilter still suffers from this effect. The full characteristics are discussed in [8]. In experiments τ is usually set as $\tau = 0.75$.

5. **Estimate covariance matrices**
   For each estimated state $\hat{\mathbf{y}}_j$ compute its covariance matrix:

$$\mathbf{P}_j = \frac{1}{W_j}\sum_{i=1}^{N_k} w_{j,i}(\tilde{\mathbf{x}}_i - \hat{\mathbf{y}}_j)(\tilde{\mathbf{x}}_i - \hat{\mathbf{y}}_j)^T. \tag{3.42}$$

In Equation (3.42) we added, in contrast to [18], the normalization term $\frac{1}{W_j}$ to receive more accurate covariance matrix estimates when W_j is not practically one. The matrix $\mathbf{P}_j$ is not an error covariance matrix in the sense of single target Bayes filtering, but it characterizes the particle distribution of state $\hat{\mathbf{y}}_j$.

6. **Update target intensity**
   Given m_k new measurements the update of the state intensity is realized through a correction of the individual particle weights. For every particle $(\mathbf{x}_i, w_i)$, with $i = 1,..,N_k + N_{k,new}$ set:

$$\hat{w}_i = \left[(1 - p_k^D(\tilde{\mathbf{x}}_i)) + \sum_{j=1}^{m_k} \frac{p_k(\mathbf{z}_j \mid \tilde{\mathbf{x}}_i)p_k^D(\tilde{\mathbf{x}}_i)}{\lambda_{k|k-1}(\mathbf{z}_j)}\right] \cdot w_i \tag{3.43}$$

7. **Update hypothesis intensity**
Adjust also the number of ϕ hypotheses:

$$f_{k|k}(\phi) = \left[(1 - p_k^D(\phi)) + \sum_{j=1}^{m_k} \frac{p_k(\mathbf{z}_j \mid \phi) p_k^D(\phi)}{\lambda_{k|k-1}(\mathbf{z}_j)} \right] \cdot f_{k|k-1}(\phi) \tag{3.44}$$

8. **Resampling**
The number of particles in the state space may and should vary over time in order to represent the current situation better, e.g. more targets need more particles, so that the particle approximation accuracy is still sufficient. To estimate the correct number of particles resampled for the next time step compute first the estimated expected number of targets

$$\eta_k = \sum_{i=1}^{N_k + N_{k,new}} \hat{w}_i. \tag{3.45}$$

Then compute the following probability:

$$p_S = \frac{\eta_k}{\eta_k + f_{k|k}(\phi)}. \tag{3.46}$$

The number of resampled particles N_{k+1} is then the expectation of a binomial distribution with the probability p_S and samples equal to $N_k + N_{k,new}$, i.e.

$$N_{k+1} = \left(N_k + N_{k,new}\right) \cdot p_S. \tag{3.47}$$

The estimation of N_{k+1} at every time step prevails the particle number from growing against infinity. Given N_{k+1} any standard resampling technique for particle filtering can be used, e.g. the following:
Initialize the cumulative probability with $c_1 = 0$ and set

$$c_i = c_{i-1} + \frac{\hat{w}_i}{\eta_k}, \text{ for } i = 2, \ldots, N_k + N_{k,new}. \tag{3.48}$$

Draw a uniformly distributed starting point a_1 from the interval $[0, N_{k+1}^{-1}]$.
For $j = 1, \ldots, N_{k+1}$,

$$a_j = a_1 + N_{k+1}^{-1} \cdot (j-1) \tag{3.49}$$

$$\text{while } a_j > c_i,$$

$$i = i + 1.$$

$$\text{end while.} \tag{3.50}$$

$$\mathbf{x}_j = \tilde{\mathbf{x}}_i \tag{3.51}$$

$$w_j = N_{k+1}^{-1} \tag{3.52}$$

Rescale the weights by η_k to get a new particle set $\left\{\mathbf{x}_j, \eta_k / N_{k+1}\right\}_{j=1}^{N_{k+1}}$.

Remark 1. In practical implementations, we found it useful to limit the number of particles to a maximum number in the prediction step. In addition we use a minimum number of particles per target in the resampling step, to ensure a good approximation.

Remark 2. After every time step t_k we generate a particle cloud, which represents the PPP over $\mathscr{S}$. The estimation of target states can also be done by applying a clustering method. To make this section self-contained we briefly present the main idea of how to use a clustering technique. The iFilter filter estimates the number of objects for every time step, so it is possible to use a clustering technique, which requires the number of clusters, e.g. k-means clustering [10]. However, the estimated object number has a high variance. This behavior was already shown for the PHD filter [13]. The iFilter still suffers from this problem. To deal with it, one can use the adaptive resonance theory (ART) clustering [5], which estimates the number of clusters automatically, with a distance parameter as predefined user input. ART can be used to estimate the target count and the individual target states from the particle cloud. In fact, best results are achieved if one only uses a subset

$$S \subset \{\mathbf{x}_j, w_j\}_{j=1}^{N_{k+1}}, \tag{3.53}$$

with

$$(\mathbf{x}_j, w_j) \in S \text{ if } w_j \geq \tau. \tag{3.54}$$

In general we recommend the usage of the proposed state estimation scheme (steps 4 and 5) and not a clustering approach.

In situations with a low probability of detection (e.g. $p_k^D(\mathbf{x}) = 0.3$ or lower) the hybrid intensity and likelihood ratio (iLRT) filter [30] is better suited.

3.3.3 Relationship to the PHD Filter

The proposed derivation of the iFilter is very general and specializations for different applications are possible. The most known is the PHD filter, which was originally derived using the random finite set theory. Nevertheless, it can also be derived using PPPs, analog to the iFilter. Details on this topic can be found in [28] and [29].

The main differences are reducible to the augmented state space $\mathscr{S}^+$. While the iFilter uses $\mathscr{S}^+ = \mathscr{S} \cup \mathscr{S}_\phi$, the PHD filter only uses $\mathscr{S}$. The basis for the on-line estimation of the intensities of the target birth and measurement clutter PPPs is the state ϕ. If, however, the birth and clutter rates are known a priori then the state ϕ can be omitted, giving the PHD filter. This requires some care. In order to discard targets, so that the target count does not balloon out of control, the PHD filter uses a death probability before predicting the multi-target intensity $f_{k|k}(\mathbf{x})$. The iFilter models this through $\psi_k(\phi \mid \mathbf{x})$, because transition of a target into ϕ can be seen as target "death". A given a priori clutter rate can replace $\tilde{\lambda}_k(\mathbf{z}_j)$ in (3.22). An a priori birth model has to be considered in step 1 of the algorithm, see [32, 21] for details on this step of the PHD filter.

3.4 Numerical Studies

In order to analyze the performance of the iFilter we test it against its specialization the PHD filter using the OSPA-metric [26]. Some words on the strong relationship between both filters can be found in 3.3.3. In all experiments we use a sequential Monte Carlo implementation of both filters. A general description of the SMC-PHD implementation can be found in [32]. The implementation used here was published in [21]. To have a fair comparison we modified the implementation by the ideas presented in [18]. This method works well when the probability of detection $p_k^D(\mathbf{x})$ is high, as in the examples and applications presented in this paper. We used the PHD as it was described in those papers for a matched clutter rate. In the following we present results from simulated scenarios.

3.4.1 Scenario - 1

First, we analyze the behavior of the iFilter in a demanding linear scenario. Herein six inertial moved targets are placed in an area $A = [-500,500] \times [-500,500]$. The unit is assumed to be meters. The state space is $\mathscr{S} \subset \mathbb{R}^4$, where the first two components correspond to the x and y coordinates and the third and fourth their velocities. The measurement space consists of x and y measurements, so $\mathscr{Z} \subset \mathbb{R}^2$. New measurements occur for the sake of simplicity every second. The measurement noise is white gaussian noise with a standard deviation $\sigma_x = \sigma_y = 15$. The probability of detection is set equal for all states to $p_k^D(\mathbf{x}) = 0.95,\ \mathbf{x} \in \mathscr{S}$. Target placement and direction of movement is visualized in Figure 3.1. Targets 1 - 3 are present for all time steps. Target 4 is presented between time step 15 and 90. Target 5 and 6 are present between time step 30 and 75. The whole scenario has a length of 100 time steps (seconds). The transition probabilities were set to $\psi_k(\mathbf{x} \mid \phi) = 0.2$, $\psi_k(\phi \mid \phi) = 0.01$ and $\psi_k(\phi \mid \mathbf{x}) = 0.1$ and the probabilities of detection were set to $p_k^D(\mathbf{x}) = 0.95$ and $p_k^D(\phi) = 0.3$. The number of targets in the following experiments are the result of (3.45). The number of states is the number of states, which were extracted in Step 4 of the SMC-iFilter algorithm. Both numbers are average results after 500 Monte Carlo trials.

In a first experiment the iFilter is tested in a case where no clutter measurements are present. The iFilter models its birth process through the state ϕ, so if no clutter measurements are present $f_{k|k}(\phi)$ should tend to be zero. The question that then arises is: Will the iFilter be able to deal with this and produce a reliable birth model? We run 500 Monte Carlo simulations on the above scenario. As a modification we set $p_k^D(\phi) = 0$ for this experiment. Figure 3.2 demonstrates that the iFilter is capable to produce a good target birth model if no clutter is present. In time step 15 a new target appears, because of this $f_{k|k}(\phi)$ reaches a value above one in this time step (this is denoted as 'no. Phi" in Figure 3.2). From this increase the filter can model a new target state in the next time step. Similar behavior can be observed in time step 30, where two new targets appear. $f_{k|k}(\phi)$ reaches a value of above two and from

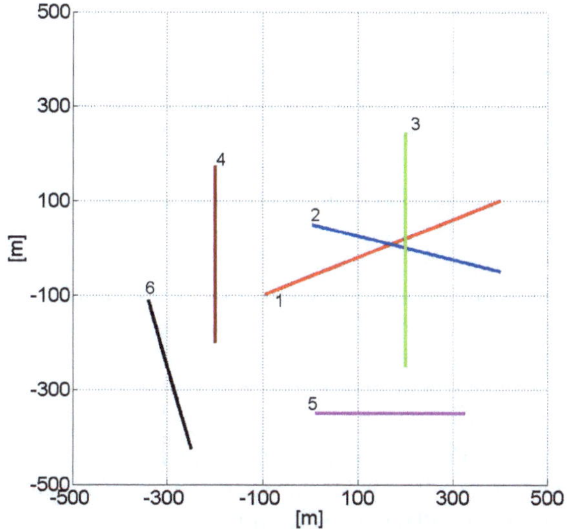

Fig. 3.1. Linear scenario used for performance evaluation. Six targets move inertially. The individual starting points of each target correspond to the denoted target number. Targets 1 - 3 are present for all time steps. Target 4 is presented between time step 15 and 90. Target 5 and 6 are present between time step 30 and 75.

this it can produce new target states. Even when the targets disappear the correct number targets is estimated.

In the second experiment, we will investigate how a estimated clutter rate can improve the results from a multi-target tracker in comparison to results gained from a multi-target tracker that needs a priori knowledge about the clutter rate. The later will be an SMC implementation of the PHD filter. In its standard formulation, the PHD needs exact knowledge about the underlying mean clutter rate for a given scenario. However, the clutter rate is hard to know in advance for a given real scenario and sensor setup. In total we perform three experiments with low, middle and high clutter rates. For each we evaluate the mean results after 500 Monte Carlo trials on the linear scenario, see Figure 3.1. The number of clutter measurements is estimated following a Poisson distribution with the mean value $|A| \cdot \rho_A$:

$$p(n_c) = \frac{1}{n_c!}(A \cdot \rho_A)^{n_c} \exp(-|A| \cdot \rho_A), \tag{3.55}$$

with $|A|$ denoting the volume of a observed area and ρ_A a parameter describing the clutter rate. For the low clutter scenario we used $\rho_A = 4 \cdot 10^{-6}$, $\rho_A = 9 \cdot 10^{-6}$ for middle and $\rho_A = 9 \cdot 10^{-5}$ for high clutter rates. Clutter measurements are generated by a i.i.d. process. We will match the PHD filter to a middle clutter rate to see its behavior in situations with less clutter on the one hand and more clutter on the other hand. In addition we will use the iFilter, which does not need any prior knowledge about the clutter rate.

Figure 3.4 displays the mean OSPA values for both filters after 500 Monte Carlo trials on a middle clutter rate. It can be observed that the achieved results are similar through the whole scenario. This is an obvious result, since the PHD is matched to the used middle clutter rate and the iFilter successfully learned the clutter rate. More important is the fact that the iFilter, although it has a more complicated estimation problem to solve (additional estimation of the clutter rate), reaches the same results as the PHD filter, which has the advantage of exact a priori knowledge. A close look on Figure 3.3 reveals another interesting effect of the PHD filter. The estimated expected number of targets (which is the integral of the intensity over the state space) is biased in comparison to the ground truth. The number of state estimates, which was produced by the strategy proposed in [18] (c.f. step 4 of the iFilter algorithm), improves the result of the PHD filter. Therefore we can claim that if not using this state estimation strategy the iFilter will outperform the PHD filter even in a case where it knows the exact clutter rate. Proofs for this statement were presented in [24].

This misbehavior of the PHD arises from the fact that it assumes a mean clutter rate for all time steps. Let us assume that we have as mean value one clutter measurement per scan. If we use now a realization of a Poisson distribution with this mean value, the probability to get zero clutter measurement or more than one clutter measurement is exactly $1 - e^{-1} = 0.63$, and so these events can occur frequently leading to the errors in the PHD. The iFilter on the other hand estimates the clutter rate for every time step individually and is, because of this, robust against a changing number of clutter measurements. This can be seen in Figure 3.3, where the number of targets and the number of states are very close to the true value for

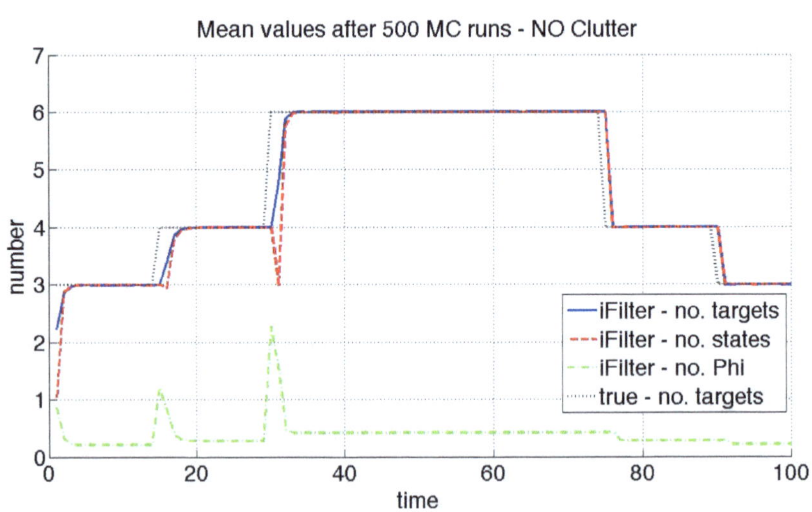

Fig. 3.2. Mean estimated target and state number after 500 Monte Carlo trials on the linear scenario.

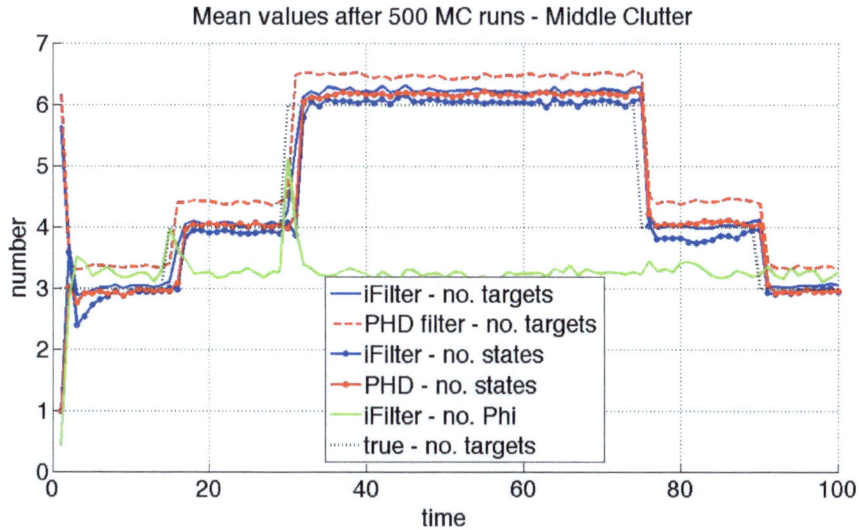

Fig. 3.3. Mean estimated target and state number after 500 Monte Carlo trials on the linear scenario with middle clutter.

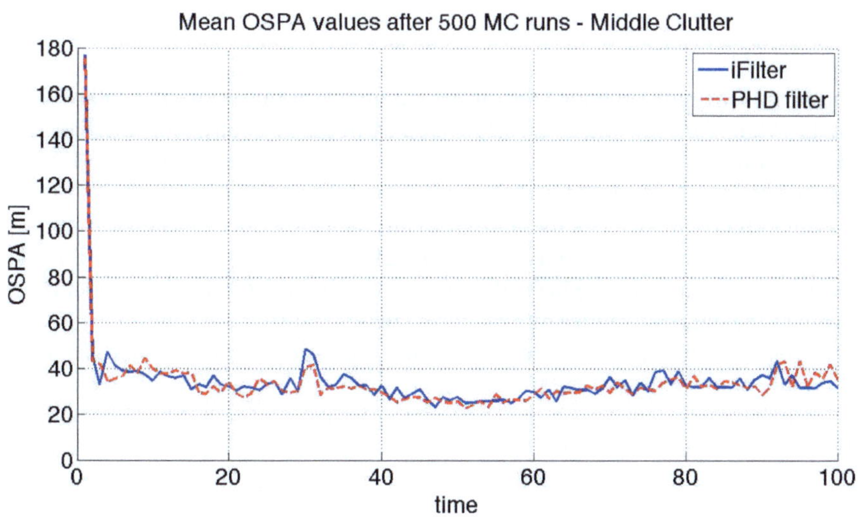

Fig. 3.4. Mean OSPA values over time after 500 Monte Carlo trials on the linear scenario with middle clutter.

all time steps. Further we can read from this figure that the mean number of clutter measurements for 500 Monte Carlo trials was estimated as 3.3 ('no. Phi" in Figure 3.3), which corresponds to the ground truth. Again we can observe that the intensity $f_{k|k}(\phi)$ increases, when new targets appear. So even in a clutter scenario the birth model of the iFilter works well.

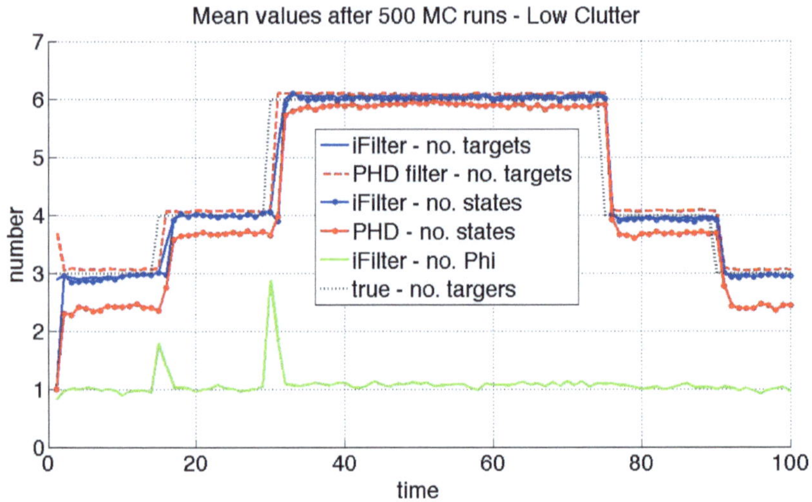

Fig. 3.5. Mean estimated target and state number after 500 Monte Carlo trials on the linear scenario with low clutter.

In the following we changed the clutter parameter of this scenario to a low value. Results obtain for both filters are illustrated in figures 3.5 and 3.6. The first thing to notice is the underestimation of the correct number of states for the PHD. This effect can be mainly observed in those time steps, where only few targets are present. Here the PHD filter matches the more measurements to clutter then the truth is. In the moment, where more targets appear, the effect is reduced but still visible. The impact can be seen in the mean OSPA values, c.f. Figure 3.6. The iFilter adapts very reliably to the changed clutter rate and gives better results.

The final experiment will use a high clutter rate. Again the PHD filter is matched to a middle clutter rate and the iFilter has no information about the clutter scenario. The corresponding results are depicted in the figures 3.7 and 3.8. Especially the PHD filter overestimates the number of targets and thus its OSPA values are higher than the corresponding values of the iFilter. Also worth mentioning is that estimated number of targets of the iFilter is slightly biased in comparison to the ground truth. But the estimated number of states keeps close to the true values. The PHD on the other hand has a strongly biased estimated number of targets.

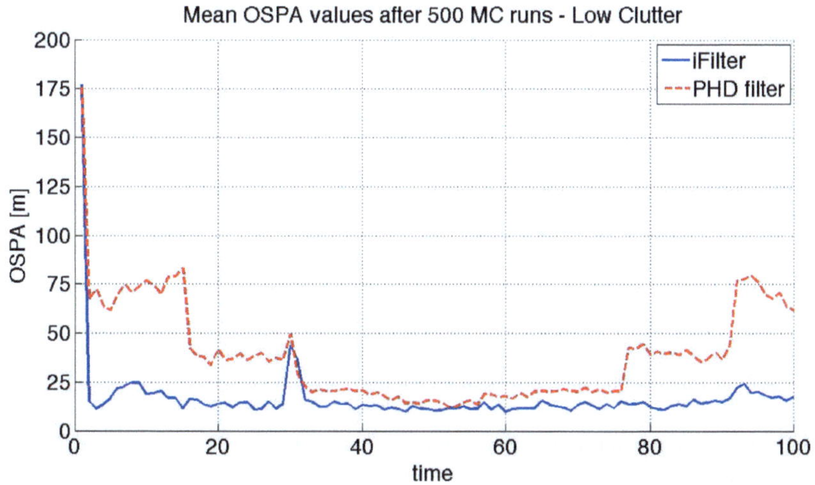

Fig. 3.6. Mean OSPA values over time after 500 Monte Carlo trials on the linear scenario with low clutter.

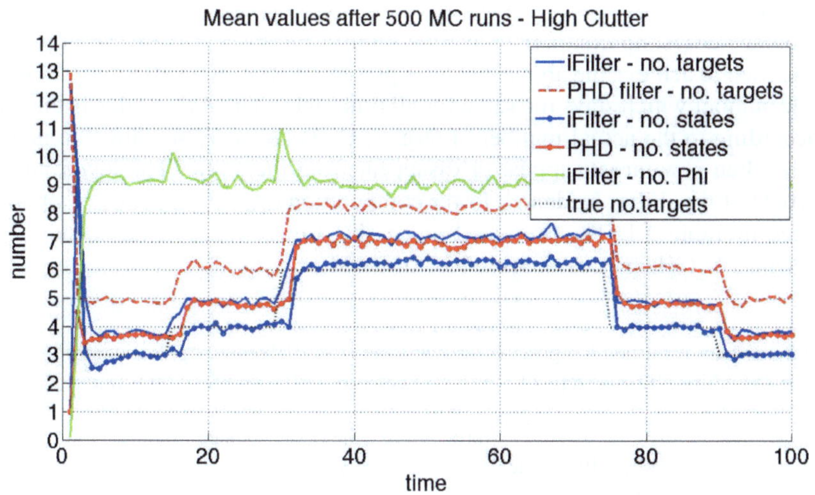

Fig. 3.7. Mean estimated target and state number after 500 Monte Carlo trials on the linear scenario with high clutter.

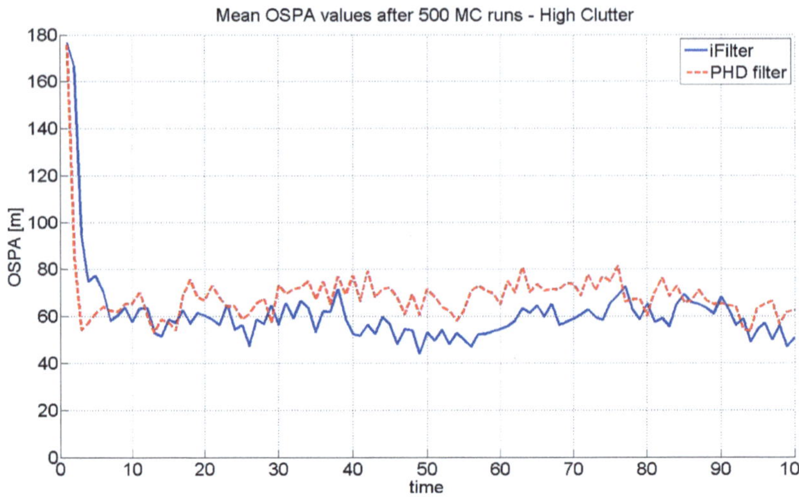

Fig. 3.8. Mean OSPA values over time after 500 Monte Carlo trials on the linear scenario with high clutter.

In general we can say that for both filters the used state estimation improves the results and makes both filters more robust against clutter. The iFilter however performs better, even in cases where the PHD filter is matched to the actual clutter rate.

Another interesting comparison can be seen in Figure 3.9. Here the number of particle used to achieve the above results is presented. The PHD filter has a constant particle number which stays the same for all time steps. The iFilter adjusts the number according to the actual number of targets. One can easily see how the number adjusts, when new targets appear or disappear from the scene. The lower number of particle of the iFilter leads to a lower computation time. A run time comparison can seen in Table 3.1. The presented processing times where achieved on a Intel Core2Duo 2.53GHz processor with 4GB of RAM.

Table 3.1. Mean runtimes for processing one time step. Values computed over 500 Monte Carlo trials and for all time steps of the linear scenario with a middle clutter rate.

	processing time (msec)	speedup
SMC-PHD filter	12.645	1.0
SMC-iFilter	7.46467	1.7

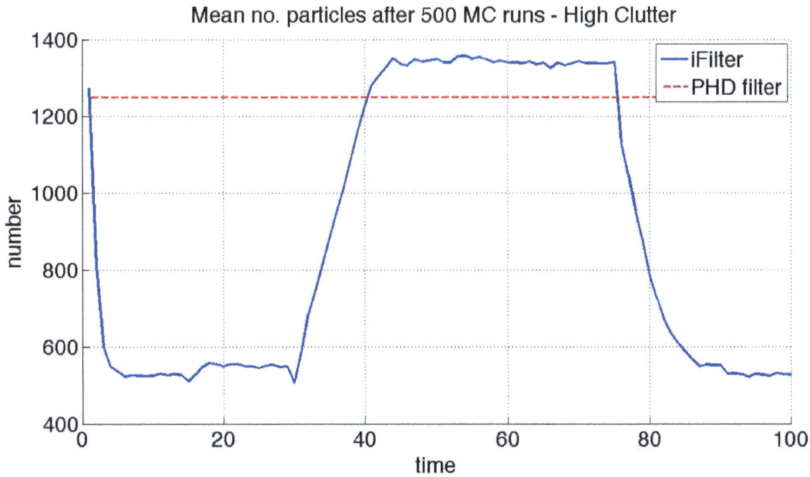

Fig. 3.9. Mean number of particles used over time after 500 Monte Carlo trials on the linear scenario with high clutter. New targets appear in time steps 15 and 30 and disappear in time step 75 and 90.

3.4.2 Scenario - 2

In the following both filters are tested on a non-linear scenario, c.f. Figure 3.10. Here we use bearing measurements (azimuth and elevation) to estimate position and velocity of multiple targets. The measurement likelihood is defined through:

$$p(\mathbf{z}|\mathbf{x}) = \frac{1}{2\pi|\mathbf{\Sigma}|^{\frac{1}{2}}} \exp\left(-\frac{1}{2}(\mathbf{z}-h(\mathbf{x}))^T\mathbf{\Sigma}^{-1}(\mathbf{z}-h(\mathbf{x}))\right), \tag{3.56}$$

with $\mathbf{\Sigma}$ the covariance matrix of the measurement noise and

$$h(\mathbf{x}) = \begin{pmatrix} \arctan\left(\frac{\mathbf{x}(1)-\mathbf{x}_{\text{obs}}(1)}{\mathbf{x}(2)-\mathbf{x}_{\text{obs}}(2)}\right) \\ \frac{\pi}{2} + \arctan\left(\frac{\mathbf{x}(3)-\mathbf{x}_{\text{obs}}(3)}{\sqrt{(\mathbf{x}(1)-\mathbf{x}_{\text{obs}}(1))^2+(\mathbf{x}(2)-\mathbf{x}_{\text{obs}}(2))^2}}\right) \end{pmatrix}. \tag{3.57}$$

An observer performs a half circle flight over a region of interest. For discrete time steps we obtain bearing measurements from three targets and additionally some clutter measurements. Details on this scenario can be found in [21]. The covariance matrix Σ was chosen according to sensor models for small antenna arrays, i.e. high angular error. The transition probabilities where set to $\psi_k(\mathbf{x} \mid \phi) = 0.2$, $\psi_k(\phi \mid \phi) = 0.01$ and $\psi_k(\phi \mid \mathbf{x}) = 0.1$ and the probabilities of detection where set to $p_k^D(\mathbf{x}) = 0.95$ and $p_k^D(\phi) = 0.3$. Again, we performed a Monte Carlo simulation with 500 trials, c.f. Figure 3.11. It can be easily seen that also in the nonlinear scenario the iFilter produces lower OSPA values compared to the PHD filter.

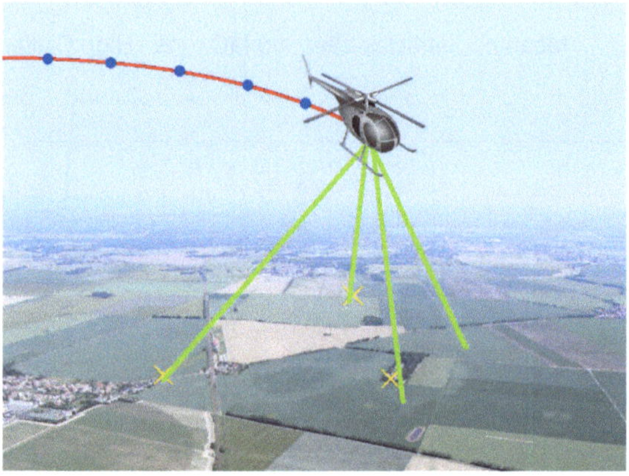

Fig. 3.10. Non linear scenario used in our simulations and real world experiments. A possible observer path is illustrated by the dotted curve, whereby the dots represent the discrete measurement positions. The crosses are objects of interest in this scene, which should be localized and tracked from the algorithm by bearing data. The latter is represented by rays which point in the directions resulting from processing the sensor output.

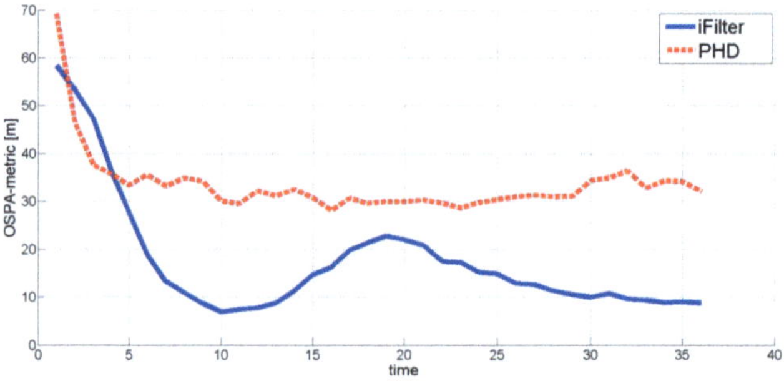

Fig. 3.11. OSPA-metric for 500 Monte Carlo runs on a non-linear scenario.

3.5 Applications

In this section, we present two example application for the usage of the iFilter for multi-target tracking. The first one is bearings-only tracking, where the states of individual targets have to be estimated from azimuth and elevation measurements. In practice, this problem involves highly non-linear measurement equations and high sensor errors with geometry depended bias. The second application is tracking in video sequences with a broad space of following application, e.g. security, surveillance and behavior analysis. Here we combine the methodology of the iFilter with a high-accuracy optical flow algorithm. In addition, we show how target labeling can be introduced to the iFilter.

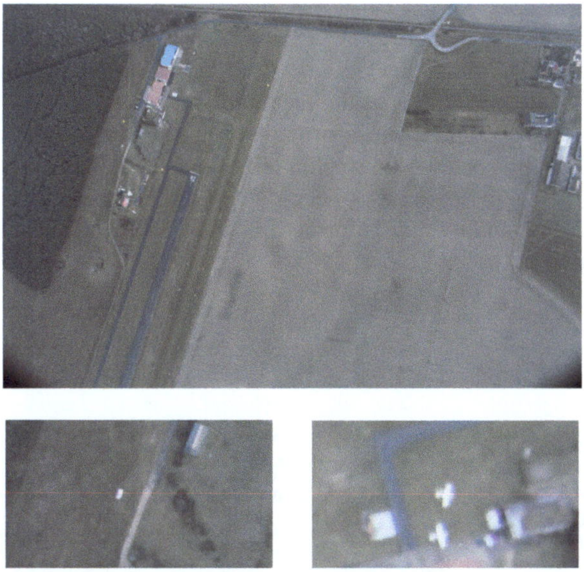

Fig. 3.12. Typical input images produced by our camera system. Top row: scaled image used for processing. Bottom row: cut in original size on a car and two airplanes.

3.5.1 Bearings-Only Tracking

In this subsection we present localization and tracking results achieved with real data. As sensor platform we used an unmanned aerial system (UAS). The UAS was equipped with a Global Positioning System (GPS) and an Inertial Navigation System (INS), so that at every time step the position and attitude information of the observer is available. The UAS was flying at a height of about 1000 meters above ground level. As measurement we used here again bearings (azimuth and elevation), like in the simulated non-linear scenario. As sensors for bearing measurements we used:

1. Antenna Array
   A three-element antenna array was mounted beneath a UAS. This small array is able to detect and compute bearing data for satellite telephone uplink communication. In order to obtain data from the received signal we used the strategy proposed in [25]. The challenge for a filter lies in a non-Gaussian error distribution and additional grating lobe effects, which leads to high errors in the estimated bearings. The errors here have a systematic and statistical component. In the filter, we only modeled the statistical errors.

2. Optical System
   In addition to the antenna array we used a fixed down-looking high resolution camera system. The field of view was 114 degree horizontal and 88 vertical. To detect possible object we use the technique presented in [19]. This detection procedure uses shape and color information to find objects in color images. For the experiments presented here we limited ourselves to cars and airplanes (c.f. Figure 3.12). Once an object has been detected bearing data can be computed using the position and attitude information of the UAS. The necessary formulas can be found in [22].

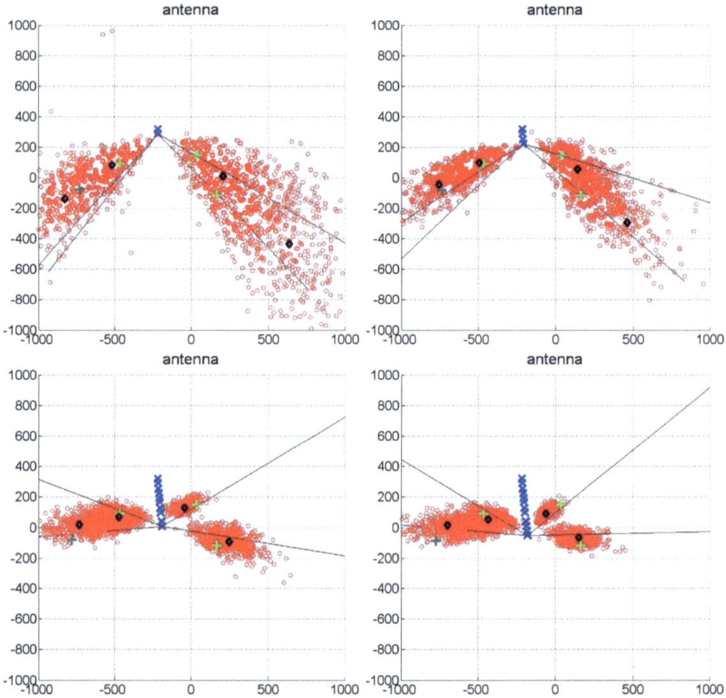

Fig. 3.13. Particle set evolution for an antenna array system at different time steps. The x's represent the observer position where bearing measurement were produced. The corresponding bearings are represented by rays. The individual particles are illustrated as circles, whereby the estimated localizations are displayed as diamonds and the ground truth is represented by crosses. The leftmost object is moving to the left and the three others are stationary targets. For a better perspective this Figure only shows the top view of the 3D scenario.

The transition probabilities of the iFilter were set to $\psi_k(\mathbf{x} \mid \phi) = 0.2$, $\psi_k(\phi \mid \phi) = 0.01$ and $\psi_k(\phi \mid \mathbf{x}) = 0.1$ and the probabilities of detection were set to $p_k^D(\mathbf{x}) = 0.95$ and $p_k^D(\phi) = 0.3$.

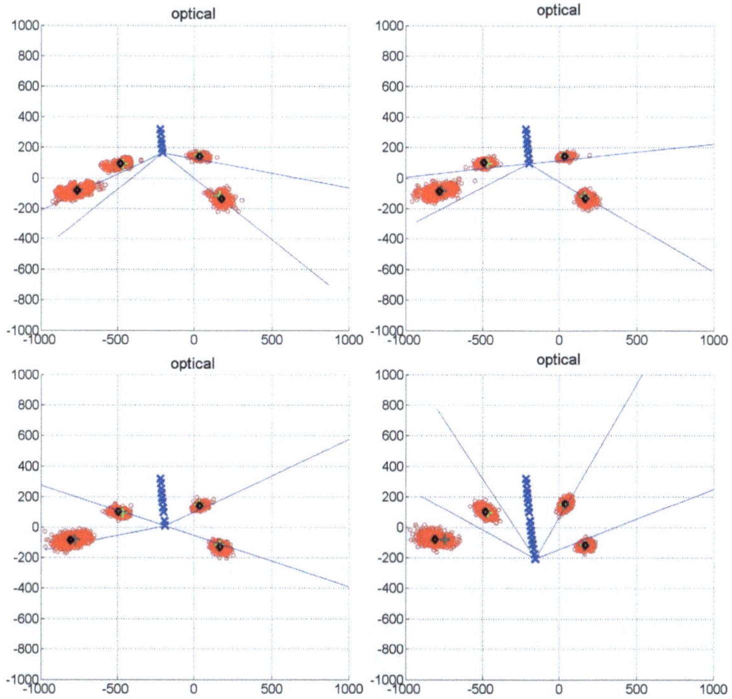

Fig. 3.14. Particle set evolution for an optical system at different time steps. The x's represent the observer position where bearing measurement were produced. The corresponding bearings are represented by rays. The individual particles are illustrated as circles, whereby the estimated localizations are displayed as diamonds and the ground truth is represented by crosses. The leftmost object is moving to the left and the three others are stationary targets. For a better perspective this Figure only shows the top view of the 3D scenario.

The results for the optical and antenna system are visualized in Figures 3.14 and 3.13, respectively. As it can be easily seen the performance for the optical sensor is much better in comparison to the antenna system. This relies on the fact that the bearing errors of the antenna system are very high and have additionally a strong systematic component. Nevertheless, for both sensor types the iFilter produces good results and estimates the expected number of targets correctly for all time steps. In addition we can observe an increase of the localization confidence given more measurements. These results state that the performance we achieved with simulated data (c.f. Figure 3.11) also holds for real data. In [21] comparable results with a PHD filter for this data was shown.

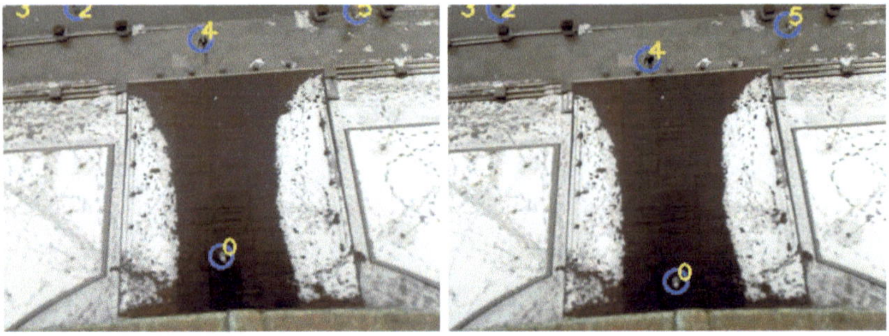

Fig. 3.15. Tracking an unknown number of people using high-accuracy optical flow. Tracking results as blue circles and person ID in yellow.

3.5.2 Video Tracking

Multi-object tracking using a monocular system is a challenging but very important problem in many computer vision applications. The aim is to estimate the number and the state information of every object for each time step in an image sequence. The problem becomes challenging when the number of objects is unknown and variable. Here we will use the proposed SMC-iFilter to estimate the number of targets and their states.

Algorithms based on the Joint Probabilistic Data Association filter (JPDAF) [3] tend to merge tracking results produced by closely spaced objects. This drawback cannot be observed when using an iFilter.

In [34, 11] the authors use a PHD filter for multi object tracking, with the drawback that only position information, gained from a detector, is used, so that the filter must estimate the velocity indirectly, which reduces the robustness of the filter. In this work we extend the classical iFilter to deal with image data and velocity information gained from optical flow. For every object tracking task some kind of measurement is needed. Using optical flow we directly obtain velocity measurements for every pixel. Unfortunately, this does not provide any information about the positions of objects in the scene. Fortunately, there has been a rapid progress in the field of object detection strategies [6, 20]. Since not all of these strategies are able to run in real-time we will present a fast strategy for moving object detection based on real-time optical flow. The main idea of our tracking algorithm is to run an object detector and additionally compute the optical flow information. This gives us two kinds of measurements, which can be used for a stable and robust multi-object tracking, sample results can be seen in Figure 3.15. Parts of this work have been previously published [23].

Optical Flow

The estimation of the optical flow between two images is a well-studied problem in low-level vision. A diverse range of optical flow estimation techniques have been developed and we refer to the survey [35] for a detailed review. Taking into account the so-called Middlebury dataset [1] the discontinuity-preserving variational models based on Total Variation (TV) regularization and L^1 data terms are among the most accurate flow estimation techniques. Because of this fact, we will use in this context the estimation technique proposed by Werlberger et. al. [36]. To make this chapter self-contained we briefly reflect their work.

For two input images $I_0, I_1 : \Omega \subset \mathbb{R}^2 \to [0,1]$ the optical flow model can be stated as

$$\min_{\mathbf{u}} \left\{ \int_{\Omega} \sum_{d=1}^{2} |\nabla u_d| + \lambda |\rho(\mathbf{u}(\mathbf{x}))| \mathrm{d}\mathbf{x} \right\}, \tag{3.58}$$

with $\mathbf{u}(\mathbf{x}) = (u_1(\mathbf{x}), u_2(\mathbf{x}))^T$, $u_d : \Omega \to \mathbb{R}$, the free parameter λ to balance the relative weight of data and regularization term and $\rho(\mathbf{u}(\mathbf{x})) = \mathbf{u}(\mathbf{x})^T \nabla I_1(\mathbf{x}) + I_1(\mathbf{x}) - I_0(\mathbf{x})$ the optical flow constrained equation. To improve the results and the accuracy the authors extend this approach using anisotropic Huber regularization. To still be able to perform a minimization a Legendre-Fenchel (LF) dual transform is needed. The final energy functional is then

$$\min_{\mathbf{u},\mathbf{v}} \sup_{|\mathbf{p}_d| \leq 1} = \left\{ \int_{\Omega} \sum_{d=1}^{2} \left[\left(D^{\frac{1}{2}} \nabla u_d \right) \mathbf{p}_d - \varepsilon \frac{|\mathbf{p}_d|^2}{2} + \frac{1}{2\Theta}(u_d - v_d)^2 \right] + \lambda |\rho(\mathbf{v}(\mathbf{x}))| dx \right\}. \tag{3.59}$$

This approach has several benefits: firstly, the energy functional is convex, which leads to globally optimal solutions, and, secondly, the minimization can be scheduled in parallel leading to a real-time computation. use the results without massive time drawbacks. Using this approach we can compute for every pixel $\mathbf{x} \in \Omega$ and each time step k the velocity $\mathbf{u}_k(\mathbf{x}) = (u_1, u_2)^T$ of this pixel.

Moving Object Detector

In this subsection we present a fast object detector, which is designed to detect moving objects. Given the flow field $\mathbf{u}_k$ at a given time step k, we can compute the probability, individually for every pixel that it belongs to a moving object:

$$p_m(\mathbf{x}) = 1.0 - \exp\left(-\frac{1}{2} \frac{(\|\mathbf{u}((x))\|_2 - \mu)^2}{\sigma^2} \right). \tag{3.60}$$

Here μ and σ correspond to a normal distribution indicating that a pixel does not move. Using a stationary camera μ would be zero. Using a flying platform with downward-looking camera μ would correspond to the actual velocity of the platform. A typical value for σ in our experiments is 0.5. Given this probability image $p_m : \Omega \to [0,1]$ we can compute the center of gravity for every region with a high probability of movement. Examples of this detector can be seen in Figure 3.16.

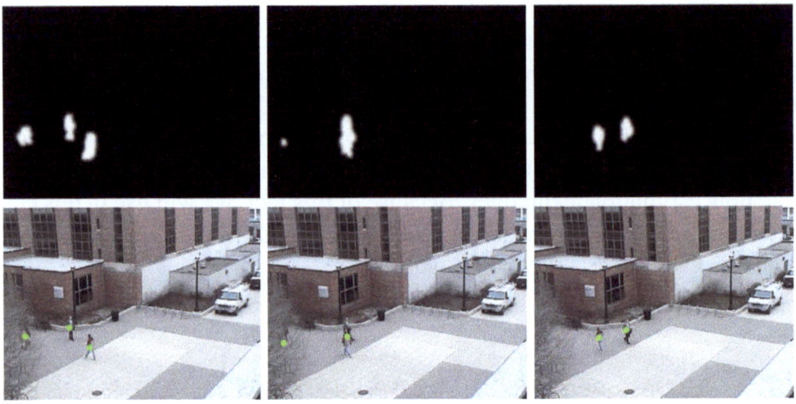

Fig. 3.16. Moving object detection for different frames in the image sequence. Top row: smoothed probability image for pixel movement; bottom row: position measurement displayed as black and white circles.

Muti-target Tracking and Labeling

We implemented the iFilter according to Section 3.3.2. In the following the state of an individual object will be represented by $\mathbf{x}_k \in \mathbb{R}^4$, with two random entries for the position and two random entries for the velocities. Each measurement $\mathbf{z}_k \in \mathbb{R}^4$ is represented analogous. The measurement and state unit is pixel. For the sake of simplicity we assume that the object motion model of each target is linear with a constant velocity. Since we use a high-accuracy optical flow with a high frame rate (e.g. 30 frames per second) we do not need a more complicated motion model in our experiments. With this the object state prediction can be written as:

$$\mathbf{x}_k = \begin{pmatrix} \mathbf{I}_2 & \Delta T \mathbf{I}_2 \\ \mathbf{0}_2 & \mathbf{I}_2 \end{pmatrix} \mathbf{x}_{k-1} + \mathbf{s}_k, \tag{3.61}$$

with $\mathbf{s}_k$ a zero mean Gaussian white process noise, ΔT the time difference between step k and $k-1$. $\mathbf{I}_2$ denotes the identity matrix for two dimensions and $\mathbf{0}_2$ a 2x2 matrix with zeros. The likelihood function is given by:

$$p(\mathbf{z}|\mathbf{x}) = \exp\left(-\frac{1}{2}(\mathbf{z}-\mathbf{x})^T \boldsymbol{\Sigma}(\mathbf{z}-\mathbf{x})\right), \tag{3.62}$$

with $\boldsymbol{\Sigma}$ the covariance matrix of the measurement noise.

At every time step k we have $\{\mathbf{x}_k^i, w_k^i\}_{i=1}^{N_k}$ as a particle-based approximation of the intensity over $\mathscr{S}^+$. The prediction, update and resampling for every time step and new measurements is done following the work in [24], see Section 3.3.2.

To establish an individual object trajectory we have to label each object correctly. We use two kinds of information: object state and the color distribution of the object.

For both information, we will use a likelihood-type function measuring the confidence that two objects from consecutive time steps $k-1$ and k are identical. The values of these likelihoods will be in the range of 0 (not identical) and 1 (identical). Let us assume that m is an object from the time step $k-1$ and n is a object from the time step k: then we can predict the object state of m using (3.61), so that we get $\tilde{m}$. The distance is defined as $d(\tilde{m},n) = \|\mathbf{x}^{\tilde{m}} - \mathbf{x}^n\|_2$, with $\mathbf{x}^{\tilde{m}}$ and $\mathbf{x}^n$ the state vectors of the objects $\tilde{m}$ and n. The likelihood function is then

$$L_{\text{state}}(m,n) = \exp\left(-\frac{(d(\tilde{m},n))^2}{2\sigma_d^2}\right), \tag{3.63}$$

with σ_d the standard deviation of the distance information.

The likelihood function for the color measurement is based on the idea of similarity measures on color histograms, which has the benefit to be robust against non-rigidity, rotation and partial occlusions [14]. Suppose that the distribution is discretized into η bins. The color histogram $\mathbf{p}(\mathbf{x}) = \{p(\mathbf{x}^{(c)})\}_{c=1,\dots,\eta}$ at position $\mathbf{x}$ is calculated as

$$p(\mathbf{x}^{(c)}) = f \sum_{\mathbf{x}_j \in \mathcal{N}(\mathbf{x})} g\left(\frac{\|\mathbf{x}-\mathbf{x}_j\|}{\alpha}\right) \delta(h(\mathbf{x}_j) - c). \tag{3.64}$$

In (3.64) f is a normalization factor, α is the scaling factor, $\mathcal{N}(\mathbf{x})$ denotes the neighborhood of pixel $\mathbf{x}$, δ is the Kronecker delta function and $g(.)$ is a weighting function given by

$$g(r) = \begin{cases} 1-r^2, & r<1 \\ 0, & \text{otherwise} \end{cases}. \tag{3.65}$$

$h(\mathbf{x})$ is a function, which assigns the color at location $\mathbf{x}$ to the corresponding bin. To measure the similarity between two color distributions, which are denoted by $\mathbf{p}(\mathbf{x}) = \{p(\mathbf{x}^{(c)})\}_{c=1,\dots,\eta}$ and $\mathbf{q}(\mathbf{x}) = \{q(\mathbf{x}^{(c)})\}_{c=1,\dots,\eta}$, we use the Bhattacharyya coefficient.

Let $\mathbf{p}_{\tilde{m}}$ and $\mathbf{q}_n$ be the color distribution of the objects $\tilde{m}$ and n, then the likelihood is:

$$L_{\text{color}}(m,n) = \sum_{c=1}^{\eta} \sqrt{p_{\tilde{m}}^{(c)} q_n^{(c)}}. \tag{3.66}$$

The likelihood that the objects m and n are identical, is a weighted sum over both likelihoods

$$L(m,n) = w_p L_{\text{state}}(m,n) + w_c L_{\text{color}}(m,n). \tag{3.67}$$

Using this measurement (3.67) we can compute the similarity between every object from the time step $k-1$ and every object from the time step k. If the measurement exceeds a threshold value, then the objects are labeled as identical. If a new object does not match any other object from the previous time step, then a new object is added to the database. Objects that are not supported by new measurements over the time are deleted from the database, assuming that the object has left the scene.

Results

In this section, we present experimental results of our tracking algorithm. The transition probabilities of the iFilter where set to $\psi_k(\mathbf{x} \mid \phi) = 0.2$, $\psi_k(\phi \mid \phi) = 0.01$ and $\psi_k(\phi \mid \mathbf{x}) = 0.1$ and the probabilities of detection where set to $p_k^D(\mathbf{x}) = 0.95$ and $p_k^D(\phi) = 0.3$. The image sequence used in Figure 3.17 was published in [7]. The top row of it shows the position measurement. In frame 61 the motion field of two persons merges (c.f. Figure 3.16), so that the detector measures only one moving object for a couple of frames. Because of the proposed labeling and the iFilter, we are able to track and label both persons correctly when the motion fields splits again. This can be seen in the bottom row. The center of the blue circle corresponds to the position information gained though the state estimation step. The radius of this circle is fixed and only used for presentation. The yellow number is the ID of a person. In frame 86, the ID of person 2 is still displayed to indicate the last known position of this person. For this scene, we had a hand-labeled ground truth. The mean position error between the proposed algorithm and the ground through lies by 2.68 pixel with a standard deviation of 1.5 pixel.

frame: 41 61 86

Fig. 3.17. Tracking result. Top row: position measurement displayed as black and white cirles in the image sequence. Bottom row: Position estimation of every object plotted as blue circle with fixed radius and ID-number of every object in yellow.

Tracking and labeling results for a different scene can be seen in Figure 3.15. The image sequence used here was published in [15]. The challenge in this scene lies in the high false alarm rate of about 5-10%, which comes from additional noise produced through snowfall. Nevertheless, the proposed strategy could track and label all persons in this scene correctly. We computed the achieved runtime as frame rate means from the individual runtimes for each frame in the scenes from the Figures 3.17 and 3.15: 500 particles 61.74fps, 1000 particles 54.43fps and 1500 particles 31.85fps. The results were computed on an Intel Q8220 Qaud Core CPU with 4GB RAM using a single core implementation.

3.6 Conclusions

This chapter presented an elementary introduction to PPPs. Their application for the multi-target tracking problem with an unknown number of targets was illustrated, which lead to the iFilter. For this concept, an improved sequential Monte Carlo implementation was derived. To verify the theory and analyze the filter behavior, various numerical studies were performed. It was demonstrated in linear and non-linear scenarios that iFilter has in general a better performance than the PHD filter, especially, if the clutter model for the PHD filter is not known perfectly. Even in situations where the clutter model of the PHD was matched to the clutter rates in the scenario, the iFilter outperformed the PHD. The implementation and usage of the PHD filter was done according to the references in the literature, which are published up to now. The only drawback of the iFilter is its slight slower initial convergence in comparison to the PHD filter.

In addition two applications were presented. They demonstrate the good performance of the iFilter in real world problems. The first presented the usability of the iFilter in a demanding estimation problem for real bearings-only data with high systematic and statistical errors. The second showed that the filter can be well used in time critical estimation problems like multi-target tracking in video sequence with a high frame rate.

References

1. Baker, S., Schastein, D., Lewis, J., Roth, S., Black, M., Szeliski, R.: A database and evaluation methodology for optical flow. In: ICCV (2007)
2. Bar-Shalom, Y., Fortmann, T.: Tracking and Data Association. Academic, San Diego (1988)
3. Bar-Shalom, Y., Fortmann, T., Scheffe, M.: Joint probabilistic data association for multiple targets in clutter. In: Conf. on Information Sciences and Systems (1980)
4. Callen, H.: Thermodynamics and an Introduction to Thermostatistics, 2nd edn. Wiley, New York (1985)
5. Carpenter, G., Grossberg, S.: ART 2: Self-organizing stable category recognition codes for analog input patterns. Applied Optics 26(23), 4919–4930 (1987)
6. Dalal, N., Triggs, B.: Histograms of oriented gradients for human detection. In: CVPR (2005)
7. Davis, J., Sharma, V.: Background-subtraction using contour-based fusion of thermal and visible imagery. Computer Vision and Understanding 106(2-3), 162–182 (2007)
8. Erdinc, O., Willett, P., Bar-Shalom, Y.: Probability hypothesis density filter for multitarget multisensor tracking. In: Proc. of the 8th International Conference on Information Fusion (FUSION), Philadelphia, PA, USA (July 2005)
9. Fortmann, T., Bar-Shalom, Y., Scheffe, M.: Sonar tracking of multiple targets using joint probabilistic data association. IEEE J. Oceanic Eng. 8, 173–184 (1983)
10. Lloyd, S.: Least squares quantziation in pcm. IEEE Trans. Inform. Theory 28(2), 129–137 (1982)
11. Maggio, E., Taj, M., Cavallaro, A.: Efficient multi-target visual tracking unsing random finite sets. IEEE Trans. on TCSVT 18(8), 1016–1027 (2008)

12. Mahler, R.: Multitarget Bayes filtering via first-order multitargets moments. IEEE Trans. Aerosp. Electron. Syst. 39(4), 1152–1178 (2003)
13. Mahler, R.: PHD filters of higher order in target number. IEEE Trans. Aerosp. Electron. Syst. 43(4), 1523–1543 (2007)
14. Nummiaro, K., Koller-Meier, E., Gool, L.V.: An adaptive color-based particle filter. Image and Vision Computing 21(1), 99–110 (2002)
15. Pellegrini, S., Ess, A., Schindler, K., van Gool, L.: You'll never walk alone: Modeling social behaviour for multi-target tracking. In: ICCV (2009)
16. Reid, D.B.: An algorithm for tracking multiple targets. IEEE Trans. Autom. Control 24(6), 843–854 (1979)
17. Ristic, B., Arulampalam, S., Gordon, N.: Beyond the Kalman filter: Particle filters for tracking applications. Artech House (2004)
18. Ristic, B., Clark, D., Vo, B.-N.: Improved SMC implementation of the PHD filter. In: Proc. of the 13th International Conference on Information Fusion, Edinburgh, UK (July 2010)
19. Schikora, M.: Global optimal multiple object detection using the fusion of shape and color information. In: Energy Minimization Methods in Computer Vision and Pattern Recognition, EMMCVPR (August 2009)
20. Schikora, M.: Global optimal multiple object detection using the fusion of shape and color information. In: EMMCVPR (2009)
21. Schikora, M., Bender, D., Cremers, D., Koch, W.: Passive multi-object localization and tracking using bearing data. In: 13th International Conference on Information Fusion, Edinburgh, UK (July 2010)
22. Schikora, M., Bender, D., Koch, W., Cremers, D.: Multitarget multisensor localization and tracking using passive antennas and optical sensors. In: Proc. SPIE, Security + Defense, vol. 7833 (2010)
23. Schikora, M., Koch, W., Cremers, D.: Multi-object tracking via high accuracy optical flow and finite set statistics. In: Proc. of the 36th International Conference on Acoustics, Speech and Signal Processing (ICASSP), Prag, Czech Republic. IEEE (May 2011)
24. Schikora, M., Koch, W., Streit, R., Cremers, D.: Sequential Monte Carlo method for the iFilter. In: Proc. of the 14th International Conference on Information Fusion (FUSION), Chicago, IL, USA (July 2011)
25. Schmidt, R.O.: Multiple emitter location and signal parameter estimation. In: Proc. RADC Spectrum Estimation Workshop, Griffith AFB, pp. 243–258 (1979)
26. Schumacher, D., Vo, B.-T., Vo, B.-N.: A consistent metric for performance evaluation of multi-object filters. IEEE Trans. Signal Processing 58(8), 3447–3457 (2008)
27. Sidenbladh, H.: Multi-target particle filtering for probability hypothesis density. In: International Conference on Information Fusion, Cairns, Australia, pp. 800–806 (2003)
28. Streit, R.: Multisensor multitarget intensity filter. In: 11th International Conference on Information Fusion (2008)
29. Streit, R.: Poisson Point Processes: Imaging, Tracking, and Sensing. Springer (2010)
30. Streit, R., Osborn, B., Orlov, K.: Hybrid intensity and likelihood ratio tracking (iLRT) filter for multitarget detection. In: Proceedings of the 14th International Conference on Information Fusion (FUSION), Chicago, IL, USA (July 2011)
31. Streit, R., Stone, L.: Bayes derivation of multitarget intensity filters. In: 11th International Conference on Information Fusion (2008)
32. Vo, B.-N., Singh, S., Doucet, A.: Sequential Monte Carlo methods for multi-target filtering with random finite sets. IEEE Trans. Aerosp. Electron. Syst. 41(4), 1224–1245 (2005)

33. Vo, B.-T., Ma, W.-K.: The Gaussian mixture probability hypothesis density filter. IEEE Trans. Signal Processing 55(11), 4091–4104 (2006)
34. Wang, Y., Wu, J., Kassim, A., Huang, W.: Tracking a variable number of human groups in video using probability hypothesis density. In: ICPR (2006)
35. Weickert, J., Bruhn, A., Brox, T., Papenberg, N.: A survey on variational optic flow methods for small displacements. Mathematical Models for Registration and Applications to Medical Images, 103–136 (2006)
36. Werlberger, M., Trobin, W., Pock, T., Wedel, A., Cremers, D., Bischof, H.: Anisotropic Huber-L1 optical flow. In: BMVC, London, UK (September 2009)
37. Zajic, T., Mahler, R.: A particle-systems implementation of the PHD multi-target tracking filter. In: Proc. SPIE, Signal Processing, Sensor Fusion Target Recognition XII, vol. 5096(4), pp. 291–299 (2003)

Chapter 4
Sequential Monte Carlo Methods for Localization in Wireless Networks

Lyudmila Mihaylova, Donka Angelova, and Anna Zvikhachevskaya

Abstract. Wireless indoor and outdoor localization systems received a great deal of attention in recent years. This chapter surveys first the current state-of-the-art of localization techniques. Next, it formulates the problem of localization within Bayesian framework and presents sequential Monte Carlo methods for localization based on received signal strength indicators (RSSIs). Multiple model particle filters are developed and their performance is evaluated with RSSIs by accounting for and without considering the measurement noise time correlation. A Gibbs sampling algorithm is presented for estimating the unknown parameters of the measurement noise which highly increases the accuracy of the localization process. Two approaches to deal with the correlated measurement noises are proposed in the framework of auxiliary particle filtering: with a noise augmented state vector and the second approach implements noise decorrelation. The performance of the two proposed multi-model auxiliary particle filters (MM AUX-PFs) is validated over simulated and real RSSIs and high localization accuracy is demonstrated.

Keywords: sequential Monte Carlo methods, auxiliary particle filtering, localization, wireless networks, correlated measurement noises, multiple model estimation, parameter estimation, Gibbs sampling.

4.1 Motivation

In many every day applications there is a need of finding the position of a moving staff of mobile transportation systems or a vehicle and to know the coordinates

Lyudmila Mihaylova · Anna Zvikhachevskaya
School of Computing and Communications,
Lancaster University, InfoLab21, South Drive,
Lancaster LA1 4WA, United Kingdom
e-mail: {mila.mihaylova,a.zvikhachevskaya}@lancaster.ac.uk,

Donka Angelova
the Institute of Information and Communication Technologies,
Bulgarian Academy of Sciences, Sofia, Bulgaria
e-mail: donka@bas.bg

P. Georgieva et al. (Eds.): Adv. in Intelligent Signal Processing and Data Mining, SCI 410, pp. 89–118.
springerlink.com

within a certain geographical region. Other motivating applications include monitoring of large geographical areas, such as for wildlife tracking [1], monitoring of buildings, production processes, warehouses [2] and E-Health systems.

In wireless networks if direct physical connections exist between the mobile unit and the base stations, the channels are considered to be in line-of-sight (LOS) and accurate location estimates can be obtained. However, in urban or indoor environments, the attenuation of the signals is high. Obstacles such as buildings, trees and walls cause obscurations. Reflections and diffractions occur due to shadowing and all these effects cause non-line-of-sight (NLOS). Hence, the movement patterns of mobile users vary significantly in different environments and the location systems should be able to cope with various challenges, such as NLOS, a changeable infrastructure of the wireless network and robustness to sensor failures. It is essential when nodes perform jointly certain tasks, such as decision making, sensor data fusion, object tracking [3–7] to localize the node positions and movement [2, 8, 9], as the transmitter range is generally fairly small with respect to the size of the area. Apart from the changeable network topology, the need of communications between the nodes under limited resources (energy, bandwidth), the need of processing noisy data and of overcoming losses pose additional challenges.

4.1.1 Methods for Localization

There is a great deal of methods for localization (see, e.g., the surveys [2, 8, 10–12]) between which the *range-based* [2, 13, 14] methods are widely used. They rely on the distances between nodes and are usually evaluated using received signal strengths, signal time-of-arrivals, time difference of arrivals or angle-of-arrivals and they vary in their complexity and accuracy. The range-based techniques can be divided into *radio frequency* (RF) ranging and *acoustic ranging*. The radio frequency ranging relies on the premise that by measuring the received signal strength a receiver can determine its distance to a transmitter. Another class of ranging schemes measures the time difference of arrival of acoustic and ultrasonic signals [13, 15].

While range-based algorithms need point-to-point distance estimation or angle estimation for positioning, *range-free* [16] algorithms do not require this information. Another classification subdivides the approaches for mobile nodes localization in wireless networks to approaches for *indoor* and *outdoor* environment [12]. In [17–19] indoor localization sensing systems are surveyed. Recently, another approach for indoor localization has been proposed, called *fingerprinting localization* [20, 21] which uses power maps, usually created offline. The main idea is to match the observed RSSIs with the map of previously measured RSSIs. Outdoor, Global Positioning Systems (GPSs) and differential GPSs [22] are the most successful systems.

From the point of view of the methods employed, a number of localization techniques rely on Extended Kalman filters [23, 24], Hidden Markov models [25, 26] (for coping with NLOS), Monte Carlo methods [10, 27–29], including nonparametric belief propagation [30], and on the knowledge of the connectivity between the nodes. Communications between nodes during the localization process are reduced to minimum due to energy and bandwidth constraints.

In [31] and [32] multiple model particle filtering techniques for mobility tracking of users in cellular networks are developed. A particle filter, a Rao-Blackwellized particle filter are presented and their performance is compared with an Extended Kalman filter over simulated and real data from base stations. In [33, 34] an auxiliary multiple model particle filter (PF) is proposed for bearings-only tracking problems. In [33] a deterministic splitting of each particle into several offsprings is performed for maneuvering target tracking, each offspring representing a different target maneuver. In [35] an auxiliary PF is designed for target tracking in binary sensor networks. When the measurements are co-linear a simulated annealing approach, such as the proposed in [36] can cope with these ambiguities.

Most of the above mentioned works, however, do not take into account the correlation of the measurement noise which can deteriorate significantly the localization accuracy. One of the most common correlation models for the slow shadow fading component is proposed by Gudmunsson [37]. The model of Gudmunsson consists of a decreasing autocorrelation function. The same first-order autoregressive correlation model is used in [38], jointly with a Kalman filter aimed to estimate the correlation coefficient of the shadowing component of the measurements. The shadow fading correlation properties are studied also in [39].

In our work we propose a solution to the self-localization problem of mobile nodes by taking into account the temporal correlation in the measurement noise. The methods that we present are general, although the applications considered are for outdoor localization. They can be applied also in indoor environment if a map is available. This chapter generalizes several sequential Monte Carlo techniques based on [28, 31, 32, 40, 41]. Multiple model auxiliary particle filtering techniques are proposed for localization of a single or several mobile nodes in wireless sensor networks. Their performance is validated on simulated and real RSSIs, over scenarios with several nodes and with a single mobile node, respectively. Similarly to [37], the correlated noise is modeled by a first order autoregressive model. Two approaches that can deal with the correlated measurement noises are described. The first one augments the state vector with the measurement noise, the second approach implements a noise decorrelation by introducing the so called "differenced measurement" [42]. The two approaches for localization under measurement noise uncertainty are implemented in the multiple model auxiliary particle filter framework and their performance is validated.

In the considered formulation of the problem, node mobility is modeled as a linear system driven by a discrete-time command Markov process whereas the measurement models are nonlinear and necessitate a reliable nonlinear estimation method. Due to the fact that the control process of the mobile nodes is unknown, node mobility is modeled with multiple acceleration modes (regimes). The proposed nonlinear estimation techniques can incorporate physical constraints and possibly communications among frequently maneuvering mobile nodes in the form of additional measurements.

The remaining part of the chapter is organized as follows. Section 4.2 formulates the considered problem for localization of mobile nodes and describes the motion model and the observation model. A multiple model auxiliary particle filter for mobile nodes self-localization is described in Section 4.3. The unknown measurement noise parameters can be estimated with a Gibbs sampling algorithm presented in Sections 4.4 and 4.5. Performance evaluation of the proposed Gibbs sammpling combined with the multiple model auxiliary particle filter is performed in Sections 4.6 and 4.7, over simulated and real data. Finally, Section 4.8 discusses the results and outlines open issues for future research.

4.2 Localization of Mobile Nodes

Consider the two-dimensional problem of simultaneous localization of n mobile nodes as formulated in [2]. The vector $\{(x_1, y_1), (x_2, y_2), \ldots, (x_n, y_n)\}$ of positions of the mobile nodes is estimated given n_r reference (anchor) nodes with known coordinates $\{(x_{n+1}, y_{n+1}), (x_{n+2}, y_{n+2}), \ldots, (x_{n+n_r}, y_{n+n_r})\}$ and pairwise measurements $\{z_{ij}\}$, where z_{ij} is a measurement between devices i and j. The reference nodes can obtain their coordinates in the network (through some external means such as a GPS). Apart from their positions, the mobile nodes estimate their speeds and accelerations. This includes applications in which each sensor is equipped with a wireless transceiver and the distance between sensor locations is estimated by RSSI measurements or time delay of arrivals between sensor locations.

4.2.1 Motion Model of the Mobile Nodes

Different mobility models have been developed for localization in wireless networks, such as random walk and pursue mobility models [43] and Singer-type models [8, 44–46]. Many of the models suggested in the target tracking literature [8, 47, 48] are also applicable for mobile nodes localization. In this work we adopt the discrete-time Singer-type model [49] because it characterizes the correlated accelerations of the mobile as a time correlated process and allows for accurate prediction of the position, speed and acceleration of mobile nodes. This higher-order model affords decreasing the estimation error [8], although mobility models that do not comprise the acceleration can be used too.

The state of each moving node at time instant k is defined by the vector $\boldsymbol{x}_k = (x_k, \dot{x}_k, \ddot{x}_k, y_k, \dot{y}_k, \ddot{y}_k)'$ where x_k and y_k specify the position, $\dot{x}_k$ and $\dot{y}_k$ specify the speed, $\ddot{x}_k$ and $\ddot{y}_k$ are, respectively, the acceleration in x and y directions in the two-dimensional plane; $'$ denotes the transpose operation. The motion of each mobile node can be described by the following Singer model [49, 50]

$$\boldsymbol{x}_k = \boldsymbol{A}(T, \alpha)\boldsymbol{x}_{k-1} + \boldsymbol{B}_u(T)\boldsymbol{u}_k + \boldsymbol{B}_w(T)\boldsymbol{w}_k, \tag{4.1}$$

where $\boldsymbol{u}_k = (u_{x,k}, u_{y,k})'$ is a discrete-time command process, and the respective matrices from (4.1) are in the form

$$\boldsymbol{A}(T,\alpha) = \begin{pmatrix} \tilde{\boldsymbol{A}} & \boldsymbol{0}_{3\times 3} \\ \boldsymbol{0}_{3\times 3} & \tilde{\boldsymbol{A}} \end{pmatrix}, \boldsymbol{B}_i(T) = \begin{pmatrix} \tilde{\boldsymbol{B}}_i & \boldsymbol{0}_{3\times 1} \\ \boldsymbol{0}_{3\times 1} & \tilde{\boldsymbol{B}}_i \end{pmatrix}, \tag{4.2}$$

$$\tilde{\boldsymbol{A}} = \begin{pmatrix} 1 & T & T^2/2 \\ 0 & 1 & T \\ 0 & 0 & \alpha \end{pmatrix}, \tilde{\boldsymbol{B}}_u = \begin{pmatrix} T^2/2 \\ T \\ 0 \end{pmatrix}, \tilde{\boldsymbol{B}}_w = \begin{pmatrix} T^2/2 \\ T \\ 1 \end{pmatrix}. \tag{4.3}$$

The subscript i in the matrix $\boldsymbol{B}(T)$ in (4.2) stands for u or w, respectively. The random process $\boldsymbol{w}_k$ is a 2×1 vector and T is the discretization period. The parameter α is the reciprocal of the maneuver time constant and thus depends on how long the maneuver lasts. The process noise $\boldsymbol{w}_k$ is a white sequence, with covariance matrix $E[\boldsymbol{w}\boldsymbol{w}'] = \boldsymbol{Q} = \sigma_w^2 \boldsymbol{I}$, where $E[.]$ is the mathematical expectation operation, $\boldsymbol{I}$ denotes the unit matrix and σ_w is the standard deviation.

The unknown command processes $u_{x,k}$ and $u_{y,k}$ take values from a set of acceleration levels $\mathcal{M}_x$ and $\mathcal{M}_y$. The process $\boldsymbol{u}_k$ takes values from the set

$$\mathrm{M} = \mathcal{M}_{\mathrm{x}} \times \mathcal{M}_{\mathrm{y}} \triangleq \{\boldsymbol{u}_1, \ldots, \boldsymbol{u}_{\mathrm{r}}\}. \tag{4.4}$$

Let $\mathbb{S} \triangleq \{1, 2, \ldots, r\}$ denote the set of models and let $m_k \in \mathbb{S}$ be the regime variable, modelled as a first-order Markov chain with transition probabilities $\Pi_{ij} = P(m_k = j | m_{k-1} = i)$, $i, j = 1, \ldots, r$ and initial probability distribution $\tilde{\mu}_{i,0} = P_0\{m = m_i\}$ for $m_i \in \mathbb{S}$ such that $\tilde{\mu}_{i,0} \geq 0$ and $\sum_{i=1}^{r} \tilde{\mu}_{i,0} = 1$.

The next section describes the observation model used in the sequential Monte Carlo methods for localization.

4.2.2 Observation Model

In wireless networks, the distance between mobile and reference (anchor) nodes can be inferred from RSSIs or pilot signals of nodes. The RSSI $z_{\ell j,k}$ received at the mobile node N_ℓ with coordinates $(x_{\ell,k}, y_{\ell,k})$ at time k, after being transmitted from the node N_j with coordinates $(x_{j,k}, y_{j,k})$, propagates as follows [8, 37, 51]

$$z_{\ell j,k} = \kappa_j - 10\gamma \log_{10}(d_{\ell j,k}) + v_{\ell j,k}, \tag{4.5}$$

where κ_j is a constant depending on the transmission power, wavelength, and gain of node N_j, γ is the slope index (according to [37] $\gamma = 3.3$ for suburban environment and $\gamma = 5$ for microcells in urban environment), $v_{\ell j,k}$ is the logarithm of the shadowing component, which is usually correlated in time, and $d_{\ell j,k}$ is the distance between nodes N_ℓ and N_j

$$d_{\ell j,k} = \sqrt{(x_{\ell,k} - x_{j,k})^2 + (y_{\ell,k} - y_{j,k})^2}. \tag{4.6}$$

All mobile nodes in the group send their pilot signals to the reference nodes. In order to locate a single mobile node in the two-dimensional plane, the three largest RSSIs to reference neighboring nodes are necessary to enable triangulation. The measurement equation can be written in the form

$$\boldsymbol{z}_k \triangleq (z_{\ell 1,k}, z_{\ell 2,k}, z_{\ell 3,k})' = \boldsymbol{h}(\boldsymbol{x}_k) + \boldsymbol{v}_k, \tag{4.7}$$

where $\boldsymbol{h}(\boldsymbol{x}_k) = (h_{\ell 1,k}, h_{\ell 2,k}, h_{\ell 3,k})'$, is a nonlinear function, with components $h_{\ell j,k} = \kappa_j - 10\gamma \log_{10}(d_{\ell j,k}), j = 1,2,3, \boldsymbol{z}_k \in \mathbb{R}^{n_z}$ and $n_z = 3$. The noise $\boldsymbol{v}_k = (v_{1,k}, v_{2,k}, v_{3,k})'$, $\boldsymbol{v}_k \in \mathbb{R}^{n_z}$ with covariance $E[\boldsymbol{v}_k \boldsymbol{v}_k'] = \boldsymbol{R}_k$, characterizes the shadowing component.

In the general case with $\ell = 1, \ldots n$ mobile nodes and $j = 1, \ldots n_r$ reference nodes, the overall observation vector will contain $L = n * n_r$ number of measurements.

In the next section we describe a technique to model the temporal correlation of the measurement noise from (4.5) and after that we link this model with the sequential Monte Carlo techniques for localization.

4.2.3 *Correlated in Time Measurement Noise*

In urban and suburban environment the autocorrelation function of the measurement noise (shadowing component) $v_{\ell j,k}$ from (4.5) can be modeled with the relation [37, 38]

$$c_v(\tau) = \sigma_v^2 \exp\{-\mathrm{v}|\tau|/D_c\}, \tag{4.8}$$

where τ is the time lag, σ_v denotes the standard deviation of the shadowing process, D_c is the effective correlation distance, which is of key importance in a wireless environment and v is the velocity of the mobile node. In [37] is shown that

$$D_c = -\frac{d_{\ell j,k}}{\ln(\varepsilon_D)} \geq 0, \tag{4.9}$$

where ε_D is the correlation coefficient of the shadow process between two mobile nodes separated by the distance $d_{\ell j,k}$. Usually D_c is in the range between 10 meters in urban environment and 500 meters in suburban environment. The value of the standard deviation σ_v varies dependent on the environment and in suburban areas is typically 8 dB [37, 38], whereas in urban environments it is roughly 4 dB.

The shadowing process can be modeled by a first-order autoregressive model (AR) [37, 38]

$$v_{\ell j,k} = a v_{\ell j,k-1} + \phi_k, \tag{4.10}$$

where ϕ_k is a zero mean white Gaussian process with variance $\sigma_\phi^2 = (1 - a^2)\sigma_v^2$. The coefficient a is given by

$$a = \exp(-\mathrm{v}T/D_c), \tag{4.11}$$

where T is the measurement sampling period.

It is assumed that the AR model parameters (correlation coefficient and variance) are known and have typical values for urban and suburban environments. These parameters are experimentally obtained by different authors for urban and suburban environment, e.g. in [37, 38, 52].

4.2.4 Motion and Observation Models for Simultaneous Localization of Multiple Mobile Nodes

A combined state vector $\boldsymbol{X}_k = \{\boldsymbol{x}'_{1,k}, \dots, \boldsymbol{x}'_{n,k},\}$ is composed and all states of multiple mobile nodes can be simultaneously estimated. The motion model (4.1)-(4.3) can be generalized to the form

$$\boldsymbol{X}_k = \boldsymbol{f}(\boldsymbol{X}_{k-1}, \mathrm{M}_\mathrm{k}, \boldsymbol{U}_\mathrm{k}, \boldsymbol{W}_\mathrm{k}), \tag{4.12}$$

where $\boldsymbol{X}_k \in \mathbb{R}^{n*n_x}$ is the combined system *base* state vector, $\boldsymbol{U}_k \in \mathbb{R}^{n*n_u}$ specifies the command processes for all mobile nodes, and the *modal (discrete) state* $\mathrm{M}_\mathrm{k} \in \mathbb{S}$ of the different system modes (regimes). The dimension of the combined system noise vector is $\boldsymbol{W}_k \in \mathbb{R}^{n*n_w}$.

The measurement equation (4.7) can be generalized to

$$\boldsymbol{Z}_k = \boldsymbol{h}(\boldsymbol{X}_k) + \boldsymbol{V}_k, \tag{4.13}$$

$$\boldsymbol{V}_k = a\boldsymbol{V}_{k-1} + \boldsymbol{\Phi}_k, \tag{4.14}$$

where $\boldsymbol{Z}_k \in \mathbb{R}^{n*n_z}$ is a generalized measurement vector, and the generalized noise vector $\boldsymbol{V}_k \in \mathbb{R}^{n*n_z}$ characterizes the correlated in time shadowing components; $\boldsymbol{\Phi}_k$ is a $(n*n_z)$-dimensional white noise with covariance matrix $E[\boldsymbol{\Phi}_k \boldsymbol{\Phi}_k^{'}] = \sigma_\phi^2 \boldsymbol{I}$ and $\boldsymbol{I}$ denotes the identity matrix.

Equations (4.12)-(4.14) constitute the whole model for the motion of the mobile nodes and observations with a correlated in time noise.

4.3 Sequential Bayesian Framework

4.3.1 General Filtering Framework

Within the sequential Bayesian framework the localization of mobile nodes reduces to approximation of the state probability density function (PDF) given a sequence of measurements. According to the Bayes' rule the filtering PDF $p(\boldsymbol{X}_k|\boldsymbol{Z}_{1:k})$ of the state vector $\boldsymbol{X}_k \in \mathbb{R}^{n*n_x}$ given a sequence of sensor measurements $\boldsymbol{Z}_{1:k}$ up to time k can be written as

$$p(\boldsymbol{X}_k|\boldsymbol{Z}_{1:k}) = \frac{p(\boldsymbol{Z}_k|\boldsymbol{X}_k)p(\boldsymbol{X}_k|\boldsymbol{Z}_{1:k-1})}{p(\boldsymbol{Z}_k|\boldsymbol{Z}_{1:k-1},)}, \tag{4.15}$$

where $(\boldsymbol{Z}_k|\boldsymbol{Z}_{1:k-1})$ is the normalizing constant. The state *predictive* distribution is given by the Chapman-Kolmogorov equation

$$p(\boldsymbol{X}_k|\boldsymbol{Z}_{1:k-1}) = \int_{\mathbb{R}^{n*n_x}} p(\boldsymbol{X}_k|\boldsymbol{X}_{k-1})p(\boldsymbol{X}_{k-1}|\boldsymbol{Z}_{1:k-1})d\boldsymbol{X}_{k-1}. \tag{4.16}$$

The evaluation of the right hand side of (4.15) involves integration which can be avoided in the particle filtering approach [53] by approximating the posterior PDF $p(\boldsymbol{X}_k|\boldsymbol{Z}_{1:k})$ with a set of particles $\boldsymbol{X}_{0:k}^{(i)}$, $i = 1, \ldots, N$ and their corresponding weights $w_k^{(i)}$. Then the posterior density can be written as follows

$$p(\boldsymbol{X}_{0:k}|\boldsymbol{Z}_{1:k}) = \sum_{i=1}^{N} w_k^{(i)} \delta(\boldsymbol{X}_{0:k} - \boldsymbol{X}_{0:k}^{(i)}), \tag{4.17}$$

where $\delta(.)$ is the Dirac delta function, and the weights are normalized such that $\sum_i w_k^{(i)} = 1$.

Each pair $\{\boldsymbol{X}_{0:k}^{(i)}, w_k^{(i)}\}$ characterizes the belief that the object is in state $\boldsymbol{X}_{0:k}^{(i)}$. An estimate of the variable of interest is obtained by the weighted sum of particles. Two major stages can be distinguished: *prediction* and *update*. During prediction, each particle is modified according to the state model, including the addition of random noise in order to simulate the effect of the noise on the state. In the update stage, each particle's weight is re-evaluated based on the new data. A *resampling* procedure introduces variety in the particles by eliminating those with small weights and replicating the particles with larger weights such that the approximation in (4.17) still holds. The residual resampling algorithm [54, 55] is applied here. This is a two step process making use of sampling-importance-resampling scheme.

Since the command process of the mobile nodes is unknown, a multiple model auxiliary particle filter (MM AUX-PF) for localization of the mobile nodes is designed and presented in the next section. Given the set M covering well the possible acceleration values, the unknown accelerations are supposed to evolve as a first-order Markov chain with transition probability matrix $\boldsymbol{\Pi}$. The particles for the base state are generated from the transition prior, according to (4.12)-(4.13) (where the motion model for each mobile is given by (4.1)-(4.3).

4.3.2 Auxiliary Multiple Model Particle Filtering for Localization

The auxiliary Sampling Importance Resampling (SIR) PF was introduced by Pitt and Shephard [56]. The auxiliary PF draws particles from an importance function which is close as possible to the optimal one. The auxiliary PF introduces an importance function $q(\boldsymbol{X}_k, i^{(j)})_{i=1}^N$ where $i^{(j)}$ refers to the index of the particle at $k-1$. The filter obtains samples from the joint density $p(\boldsymbol{X}_k, i|\boldsymbol{Z}_{1:k})$ and then omits the index i in the pair $(\boldsymbol{X}_k, i)$ to produce a sample $\{\boldsymbol{X}_k^{(i)}\}_{i=1}^N$ from the marginalized density

$p(\boldsymbol{X}_k|\boldsymbol{Z}_{1:k})$. The importance density that generates the sample $\{\boldsymbol{X}_k^{(i)}\}_{i=1}^N$ is defined to satisfy the relation [53]

$$q(\boldsymbol{X}_k, i|\boldsymbol{Z}_{1:k}) \propto p(\boldsymbol{Z}_k|\boldsymbol{\mu}_k^{(i)})p(\boldsymbol{X}_k|\boldsymbol{X}_{k-1}^{(i)})w_{k-1}^{(i)}, \tag{4.18}$$

where $\boldsymbol{\mu}_k^{(i)}$ is some characteristic of $\boldsymbol{X}_k$ given $\boldsymbol{X}_{k-1}^{(i)}$ (e.g., the mean or median).

The selection of the most promising particles is carried out by sampling from a multinomial distribution where the number of possible outcomes is N^{out}. The auxiliary PF [56] resamples the predicted particles to select which particles to use in the prediction and measurement update.

For the purposes of mobile node localization we propose an auxiliary MM PF. The MM AUX-PF represents the PDF $p(\boldsymbol{X}_k, i, \mathrm{M}_k|\boldsymbol{Z}_{1:k})$ where i refers to the i-th particle at $k-1$ and $\mathrm{M_k}$ is the set of acceleration levels, defined as in (4.4). After marginalization, the representation of $p(\boldsymbol{X}_k|\boldsymbol{Z}_{1:k})$ can be obtained.

Similarly to [34], the joint probability density function $p(\boldsymbol{X}_k, i, \mathrm{M}_k|\boldsymbol{Z}_{1:k})$ can be written using the Bayesian rule as:

$$\begin{aligned} p(\boldsymbol{X}_k, i, \mathrm{M_k}|\boldsymbol{Z}_{1:k}) &\propto p(\boldsymbol{Z}_k|\boldsymbol{X}_k)p(\boldsymbol{X}_k, i, \mathrm{M}_k|\boldsymbol{Z}_{1:k-1}) \\ &= p(\boldsymbol{Z}_k|\boldsymbol{X}_k)p(\boldsymbol{X}_k|\boldsymbol{X}_{k-1}^{(i)}, \mathrm{M}_k)p(\mathrm{M}_k|\mathrm{M}_{k-1}^{(i)})w_{k-1}^{(i)}, \end{aligned} \tag{4.19}$$

where $p(\mathrm{M}_k|\mathrm{M}_{k-1})$ is an element of the transition probability matrix $\boldsymbol{\Pi}$. Since sampling directly from $p(\boldsymbol{X}_k, i, \mathrm{M}_k|\boldsymbol{Z}_{1:k})$ is difficult, the following importance function $q(\boldsymbol{X}_k, i, \mathrm{M}_k|\boldsymbol{Z}_{1:k})$ is introduced

$$q(\boldsymbol{X}_k, i, \mathrm{M}_k|\boldsymbol{Z}_{1:k}) \propto p(\boldsymbol{Z}_k|\boldsymbol{\mu}_k^{(i)}(\mathrm{M}_k))p(\boldsymbol{X}_k|\boldsymbol{X}_{k-1}^{(i)}, \mathrm{M}_k)p(\mathrm{M}_k|\mathrm{M}_{k-1}^{(i)})w_{k-1}^{(i)}, \tag{4.20}$$

where

$$\boldsymbol{\mu}_k^{(i)}(\mathrm{M}_k) = E\left(\boldsymbol{X}_k|\boldsymbol{X}_{k-1}^{(i)}, \mathrm{M}_k\right). \tag{4.21}$$

The importance density $q(\boldsymbol{X}_k, i, \mathrm{M}_k|\boldsymbol{Z}_{1:k})$ differs from (4.19) only in the first factor. Marginalization over $\boldsymbol{X}_k$ yields

$$q(i, \mathrm{M}_k|\boldsymbol{Z}_k) \propto p(\boldsymbol{Z}_k|\boldsymbol{\mu}_k^{(i)}(\mathrm{M}_k))p(\mathrm{M}_k|\mathrm{M}_{k-1}^{(i)}).w_{k-1}^{(i)}. \tag{4.22}$$

By using (4.22) a random sample from the density $q(\boldsymbol{X}_k, i, \mathrm{M}_k|\boldsymbol{Z}_{1:k})$ can be obtained as follows. First, a sample $\left\{i^{(j)}, \mathrm{M}_k^{(j)}\right\}_{j=1}^N$ is drawn from the multinomial distribution $q(i, \mathrm{M}_k|\boldsymbol{Z}_{1:k})$, (4.22), by splitting each of the N particles at $k-1$ into r groups. Each of the $N * r$ particles is assigned a weight proportional to the right hand site of (4.22). Next, a sample $\left\{\boldsymbol{X}_k^{(j)}\right\}_{j=1}^N$ from the joint density $q(\boldsymbol{X}_k, i, \mathrm{M}_k|\boldsymbol{Z}_{1:k})$ is generated from $p(\boldsymbol{X}_k|\boldsymbol{X}_{k-1}^{(i^j)}, \mathrm{M}_k^{(j)})$. To use the samples $\left\{\boldsymbol{X}_k^{(j)}, i^j, \mathrm{M}_k^{(j)}\right\}_{j=1}^N$ to characterize the density $p(\boldsymbol{X}_k, i, \mathrm{M}_k|\boldsymbol{Z}_{1:k})$, we attach to each particle the weight

$$w_k^{(j)} = \frac{p(\boldsymbol{Z}_k|\boldsymbol{X}_k^{(j)})}{p(\boldsymbol{Z}_k|\mu_k^{(ij)}(\mathrm{M}_k))}, \tag{4.23}$$

which represents the ratio of (4.20) and (4.19). By omitting the $\left\{i^{(j)}, \mathrm{M}_k^{(j)}\right\}$ components from the triplet sample $\left\{\boldsymbol{X}_k^{(j)}, i^{(j)}, \mathrm{M}_k^{(j)}\right\}_{j=1}^{N}$, we have representation of the marginalized density $p(\boldsymbol{X}_k|\boldsymbol{Z}_{1:k})$, i.e.

$$p(\boldsymbol{X}_k|\boldsymbol{Z}_{1:k}) \approx \sum_{j=1}^{N} w_k^{(j)} \delta(\boldsymbol{X}_k - \boldsymbol{X}_k^{(j)}). \tag{4.24}$$

The conditional mean $\boldsymbol{\mu}_k^{(i)}(\mathrm{M}_k^{(i)})$ for each particle in the MM AUX-PF comprises the mean vectors of all mobile nodes. The following deterministic mobility equation is used to calculate the mean for each mobile node:

$$\boldsymbol{x}_k = \boldsymbol{A}(T, \alpha)\boldsymbol{x}_{k-1} + \boldsymbol{B}_u(T)\boldsymbol{u}_k. \tag{4.25}$$

The whole MM AUX-PF for mobile nodes localization is presented as Algorithm 2. The MM AUX-PF takes into account speed constraints, i.e., the speed of each mobile node cannot exceed the maximum value $V_{\max}$. Finally, resampling is performed only when the efficient number of particles, N_{eff} is smaller than a given threshold N_{thresh}.

4.3.3 Approaches to Deal with the Time Correlated Measurement Noise

The time correlation in the measurements can be taken with the account with the following two approaches described below [40].

Approach 1. One approach to overcome the problem with correlated measurement noise is to augment the mobile state $\boldsymbol{x}_k$ with the noise $\boldsymbol{v}_k$. Then the localization algorithm (MM AUX-PF) described in Subsection 4.3.2 can be applied to the whole augmented state vector of size $n*(n_x+n_z)$ (comprising the state vectors of the mobile nodes and measurement noise). This algorithm with the state vector augmented with the correlated noise is referred to as a MM-AUX PF with augmented state (AS).

Approach 2. Another decorrelation technique introduces the following "artificial measurement": $\bar{\boldsymbol{z}}_k = \boldsymbol{z}_k - a\boldsymbol{z}_{k-1}$. The measurement equation can then be written in the form

$$\begin{aligned} \bar{\boldsymbol{z}}_k &= \boldsymbol{z}_k - a\boldsymbol{z}_{k-1} && (4.26)\\ &= \boldsymbol{h}(\boldsymbol{x}_k) + \boldsymbol{v}_k - a\left[\boldsymbol{h}(\boldsymbol{x}_{k-1}) + \boldsymbol{v}_{k-1}\right] && (4.27)\\ &= \boldsymbol{h}(\boldsymbol{x}_k) - a\boldsymbol{h}(\boldsymbol{x}_{k-1}) + \bar{\boldsymbol{v}}_k. && (4.28) \end{aligned}$$

Algorithm 1. A Multiple Model Auxiliary PF for Mobile Nodes Localization

Initialization

I. $k = 0$, for $i = 1, \ldots, N$,
Generate samples $\{\boldsymbol{X}_0^{(i)} \sim p(\boldsymbol{X}_0), \mathrm{M}_0^{(i)} \sim P_0(\mathrm{M})\}$,
and set initial weights $w_0^{(i)} = 1/N$.

II. For $k = 1, 2, \ldots,$
(1) For $i = 1, \ldots, N * r$,
- Calculate the conditional mean:
$$\boldsymbol{\mu}_k^{(i)}(\mathrm{M}_k) = E\left(\boldsymbol{X}_k | \boldsymbol{X}_{k-1}^{(i)}, \mathrm{M}_k\right) \text{ for every } \mathrm{M}_k \in \mathbb{S}$$
(2) Generate $\left\{i^j, \mathrm{M}_k^{(j)}\right\}, j = 1, \ldots, N$ by sampling from $q(i, \mathrm{M}_k | \boldsymbol{Z}_{1:k})$,
where $q(i, \mathrm{M}_k | \boldsymbol{Z}_{1:k}) \propto p(\boldsymbol{Z}_k | \boldsymbol{\mu}_k^{(i)}(\mathrm{M}_k)) p(\mathrm{M}_k | \mathrm{M}_{k-1}^{(i)}) w_{k-1}^{(i)}$.
(3) ***Prediction Step***
For $j = 1, \ldots, N$, predict the particles according to
$$\boldsymbol{X}_k^{(j)} = f(\boldsymbol{X}_{k-1}^{(i^j)}, \mathrm{M}_k^{(j)}, \boldsymbol{w}_k^{(j)})$$
with noise realizations $\boldsymbol{w}_k^{(j)} \sim \mathcal{N}(0, \boldsymbol{Q})$. Impose the speed constraints.

(4) ***Measurement update***
For $j = 1, \ldots, N$ compute the weights
$$w_k^{(j)} = p(\boldsymbol{Z}_k | \boldsymbol{X}_k^{(j)}) / p(\boldsymbol{Z}_k | \boldsymbol{\mu}_k^{(i^j)}(\mathrm{M}_k)).$$
Normalize the weights: $\widetilde{w}_k^{(j)} = w_k^{(j)} / \sum_{j=1}^N w_k^{(j)}$

(5) ***Output estimate***
The posterior mean $E[\boldsymbol{X}_k | z_k]$
$$\hat{\boldsymbol{X}}_k = \sum_{j=1}^N \widetilde{w}_k^{(j)} \boldsymbol{X}_k^{(j)}.$$

(6) ***Resampling step***:
Compute the effective sample size
$N_{eff} = 1 / \sum_{j=1}^N (\widetilde{w}_k^{(j)})^2$,

Resample if $N_{eff} < N_{thresh}$
* For $i = 1, \ldots, N$, set $w_k^{(i)} = 1/N$.

The noise $\bar{\boldsymbol{v}}_k = \boldsymbol{v}_k - a\boldsymbol{v}_{k-1}$ in the new measurement equation is white but correlated with the process noise. The cross-correlation between two noise sequences can be eliminated by a procedure, given in [47]. In most practical algorithms this cross-correlation is omitted due to the little performance degradation. Thus the measurement equation in the case of one mobile node can be modified to

$$\bar{z}_k = \bar{\boldsymbol{h}}(\boldsymbol{x}_k) + \bar{\boldsymbol{v}}_k, \tag{4.29}$$

where $\bar{\boldsymbol{h}}(\boldsymbol{x}_k) = \boldsymbol{h}(\boldsymbol{x}_k) - a\boldsymbol{h}(\boldsymbol{x}_{k-1})$.

The MM-AUX PF algorithm with decorrelation is referred to as with an artificial measurement (AM).

4.4 Estimation of the Measurement Noise Parameters

The parameters of the RSSI measurement noise vary significantly depending on the environment: urban, suburban or rural. This causes difficulties to most of localization techniques since their accuracy depends to a high extent on the noise parameters. In practice, the noise characteristics of the received signal strengths can vary in a large range, e.g., between 3dB and 24dB depending on the environment: urban or semi-urban, different meteorological conditions (snow, rain), presence of obscuration or attenuation of the signals and are typically correlated [37]. A relatively small number of works consider the noise parameter estimation. Between the related works we have to mention [41] and [57]. In [57] the noise parameter estimation is performed with a single Dirichlet process and in [58] based on a mixture of Dirichlet processes. In [41] a Gibbs sampling approach is proposed for estimating the unknown noise parameters of the measurement noise based on batch measurements and next the estimated parameters are fed to the localization techniques.

The next section presents a Gibbs sampling algorithm for estimating the unknown measurement noise parameters. The localization of mobile nodes is performed after that with these noise parameter estimates embedded in the MM AUX-PF. The proposed approach deals successfully with the highly nonlinear measurement models with non-Gaussian measurement errors, can incorporate physical constraints and possibly communications among frequently maneuvering mobile nodes in the form of additional measurements.

The noise $\boldsymbol{v}_k$ characterizes the shadowing components, assumed to be uncorrelated in time and with unknown parameters. One simple, but effective solution to the localization problem with unknown noise parameters is to model the measurement noise as a n_{mix}-component Gaussian mixture

$$\boldsymbol{v}_k \sim \sum_{i=1}^{n_{mix}} \pi_i \mathcal{N}(\mu_i, \sigma_i^2), \tag{4.30}$$

where μ_i and σ_i^2 are the mean and variance of the mixture component i and $\boldsymbol{\pi} = (\pi_1, \ldots, \pi_{n_{mix}})$ is a vector of mixture weights π_i (constrained to be non-negative and with unit sum). The features of the measurement process and environment, the availability of missed or false observations can be captured well by the mixture.

The mixture parameters are estimated by introducing a hierarchical data structure of the mixture model and accounting for the missing data. In particular, Markov chain Monte Carlo (MCMC) techniques are very efficient inference methods. For hierarchical models, Gibbs sampling has proved to be the most effective among various MCMC methods. Gibbs sampling is especially appropriate for the localization problem under unknown measurement parameters. The mixture can be composed of different distributions such as t, Student or Gaussian. We choose Gaussian components since they lead easily to tractable solutions and can model well complex probability density functions even with a small number of components [59].

We estimate the measurement noise parameters with the Gibbs sampler presented in Section 4.5 and next the mobile nodes self-localization is performed by the Multiple Model Auxiliary Particle Filter given in Section 4.3.

4.5 Gibbs Sampling for Noise Parameter Estimation

We suggest to estimate the unknown noise parameters of the measurement model (4.5)-(4.29) with Gibbs sampling (GS), a special form of Bayesian sampling benefitting from the hierarchical structure of the model [60]. Given the observation of a T_m-dimensional vector $\boldsymbol{\eta} = (\eta_1, \dots, \eta_{T_m})' \in \mathbb{R}^{T_m}$ of independent random variables (corresponding to the RSSI measurement error $\boldsymbol{v} = (v_{\ell j,1}, v_{\ell j,2}, \dots, v_{\ell j,T_m})$[1]), the mixture [61] is formed

$$F(\eta_t) = \sum_{i=1}^{n_{mix}} \pi_i F_i(\eta_t), \; t = 1, \dots, T_m, \tag{4.31}$$

where the densities F_i, $i = 1, \dots, n_{mix}$ are known or are known up to a parameter. The weight vector $\boldsymbol{\pi} = (\pi_1, \dots, \pi_{n_{mix}})$ has non-negative components π_i which sum is equal to 1. We are representing the measurement noise by a Gaussian mixture model (GMM): the density $F_i(\eta_t)$ is then Gaussian, $\mathcal{N}(\mu_i, \sigma_i^2)$, where μ_i and σ_i^2 are the mean and variance of the i-th mixture component. The unknown parameters and weights of the GMM, denoted by $\boldsymbol{\theta} = (\theta_1, \dots, \theta_{r_m}) = (\mu_1, \sigma_1^2, \dots, \mu_{n_{mix}}, \sigma_{n_{mix}}^2, \pi_1, \dots \pi_{n_{mix}})$ are iteratively estimated.

According to [61], the mixture model can be represented in terms of missing (or incomplete) data. The model is *hierarchical* with the true parameter vector $\boldsymbol{\theta}$ of the mixture, on the top level and on the bottom are the observed data. The GS relies additionally on the availability of all complete conditional distributions of the elements of $\boldsymbol{\theta}$, breaking down $\boldsymbol{\theta}$ into r_m subsets. It generates $\theta_j, j = 1, \dots, r_m$ conditional on all the other parameters, increasing in this way the number of conditional simulations. The details of the GS algorithm can be found in a number of publications [61, 62]. The n_{mix}-dimensional vectors $\boldsymbol{\delta}_t$, $t = 1, 2, \dots, T_m$ with components $\delta_{t,i} \in \{0, 1\}$, $i = 1, 2, \dots, n_{mix}$ and $\sum_{i=1}^{n_{mix}} \delta_{t,i} = 1$ are defined to indicate that the measurement η_t has density $F_i(\eta_t)$. Next, the missing data distribution depends on $\boldsymbol{\theta}$, $\boldsymbol{\delta} \sim p(\boldsymbol{\delta}|\boldsymbol{\theta})$. The observed data, $\boldsymbol{\eta} \sim p(\boldsymbol{\eta}|\boldsymbol{\theta}, \boldsymbol{\delta})$, are at the bottom level. Bayesian sampling iteratively generates parameter vectors $\boldsymbol{\theta}^{(u)}$ and missing data $\boldsymbol{\delta}^{(u)}$ according to $p(\boldsymbol{\theta}|\boldsymbol{\eta}, \boldsymbol{\delta}^{(u)})$ and $p(\boldsymbol{\delta}|\boldsymbol{\eta}, \boldsymbol{\theta}^{(u+1)})$. Here, u indexes the current iteration.

[1] From the measurement model (4.5), estimated positions (x_ℓ, y_ℓ), the estimated distance $d_{\ell,j}$ and with known transmission constant $\kappa_{\ell,j}$, noise realizations $v_{\ell j,1:T_m}$ are found which serve as measurements to GS.

It is proven in [61], that Bayesian sampling produces an ergodic Markov chain $(\boldsymbol{\theta}^{(u)})$ with stationary distribution $p(\boldsymbol{\theta}|\boldsymbol{\eta})$. After u_0 initial (warming up) steps, a set of U samples $\boldsymbol{\theta}^{(u_0+1)}, \ldots, \boldsymbol{\theta}^{(u_0+U)}$ are approximately distributed as $p(\boldsymbol{\theta}|\boldsymbol{\eta})$. Due to ergodicity, averaging can be made with respect to time and the empirical mean of the last U values can be used as an estimate of $\boldsymbol{\theta}$.

The choice of prior distributions and their hyperparameters is a first, important step of the design of a Gibbs sampler. Conjugate prior distributions [2] are chosen (as in most cases in the literature) since this simplifies the implementation. Since the conjugate prior of $\boldsymbol{\pi}$ is a *Dirichlet* distribution (DD), $\mathscr{D}(\alpha_1,\ldots,\alpha_{n_{mix}})$ [61], $\boldsymbol{\pi}$ is generated according to the DDs with parameters, depending on the missing data. The conjugate priors for σ_i^2 and $\mu_i|\sigma_i^2, i=1,\ldots,n_{mix}$ are the *Inverse Gamma*($\mathscr{IG}$) and *Gaussian* distribution respectively, as recommended in [61]:

$$\boldsymbol{\pi} \sim \mathscr{D}(\alpha_1,\ldots,\alpha_{n_{mix}}), \tag{4.32}$$
$$\sigma_i^2 \sim \mathscr{IG}(\nu_i, s_i^2), \tag{4.33}$$
$$\mu_i|\sigma_i^2 \sim \mathscr{N}(\xi_i, \sigma_i^2/n_i). \tag{4.34}$$

Here, $\alpha_i, \nu_i, s_i^2, \xi_i$, and $n_i, i \in 1,\ldots,n_{mix}$ are the corresponding hyperparameters.

Starting with an initial parameter vector $\boldsymbol{\theta}^{(0)}$, the following *iterative algorithm* is implemented: at the iteration $u, u=0,1,2,\ldots$

a) *generate* $\boldsymbol{\delta}^{(u)} \sim p(\boldsymbol{\delta}|\boldsymbol{\eta},\boldsymbol{\theta}^{(u)})$ from a multinomial distribution with weights proportional to the observation likelihoods, i.e. $p(\delta_{ti}^{(u)}) \propto \pi_i^{(u)} \mathscr{N}\left(\eta_t; \mu_i^{(u)}, \sigma_i^{(u)^2}\right)$;

b) *generate* $\boldsymbol{\pi}^{(u+1)} \sim p(\boldsymbol{\pi}|\boldsymbol{\eta},\boldsymbol{\delta}^{(u)})$:

$$\boldsymbol{\pi}^{(u+1)} \sim \mathscr{D}\left(\alpha_1 + \Delta_1^{(u)}, \ \ldots, \ \alpha_{n_{mix}} + \Delta_{n_{mix}}^{(u)}\right)$$

where $\Delta_i^{(u)} = \sum_{t=1}^{T_m} \delta_{t,i}^{(u)}$ is the number of observations allocated to the mixture component i;

c) *generate* $\mu_i^{(u+1)}|\sigma_i^{(u)^2} \sim \mathscr{N}\left(\xi_i, \sigma_i^{(u)^2}/n_i\right)$:

$$\mu_i^{(u+1)} \sim \mathscr{N}\left(\frac{n_i\xi_i + \sum_{t=1}^{T_m} \delta_{t,i}^{(u)} \eta_t}{n_i + \Delta_i^{(u)}}, \frac{\sigma_i^{(u)^2}}{n_i + \Delta_i^{(u)}}\right)$$

where ξ_i is the average of the observations attributed to the mixture component i;

[2] *Conjugate distributions* are distributions that have the same functional form, e.g., the posterior is the same as the prior distribution and the prior is called a conjugate prior for the likelihood. Conjugate priors play an important role since they lead to elegant Bayesian solutions.

d) *generate*

$$\sigma_i^{(u+1)^2} \sim \mathscr{IG}\left(\nu_i + \upsilon_i, s_i^2 + \widehat{s_i^2} + n_i(\xi_i - \mu_i^{(u+1)})^2\right),$$

where $\upsilon_i = 0.5(\Delta_i^{(u)} + 1)$, $s_i^2 = (T_m(n_{mix} - 1))^{-1}\sum_{t=1}^{T_m}(\eta_i - \overline{\eta})^2$, ($\overline{\eta}$ is the empirical mean of the measurement data), $\widehat{s_i^2} = \sum_{t=1}^{T_m}\delta_{t,i}^{(u)}(\eta_t - \mu_i^{(u+1)})^2$; We consider the additional assumption that $\sigma_i^2 \in [\sigma_{min}^2, \sigma_{max}^2]$. Truncated conjugate priors are still conjugate [61]. Sampling is realized by generating $\sigma_i^{(u)^2}$ from the $\mathscr{IG}$ distribution until $\sigma_i^{(u)^2} \in [\sigma_{min}^2, \sigma_{max}^2]$.

e) *calculate* the output estimate $\hat{\boldsymbol{\theta}} = \frac{1}{U}\sum_{l=1}^{U}\boldsymbol{\theta}^{(u_0+l)}$.

The next section presents results with the Gibbs sampling algorithm.

4.6 Performance Evaluation of the Gibbs Sampling Algorithm for Measurement Noise Parameter Estimation

4.6.1 Measurement Noise Parameter Estimation with Simulated Data

First we consider a scenario where the trajectory of the ℓ-th mobile node is provided by a GPS system, which collects its actual positions during a time interval $t = 1,\ldots,T_m$ with sampling period T. Knowing the distance to the j-th reference node and using the RSSI measurements $z_{\ell j,t}$, the measurement error parameters can be estimated. A sample of measurement errors $v_{\ell j,t}, t = 1,\ldots,T_m$ can be obtained also, if the mobile mode is static for some time interval or if it is moving along a road with known parameters (the route map is available). In the univariate case, where all mobile nodes have the same noise statistics $v_{\ell j,t} = v_t, t = 1,\ldots,T_m$, the noise characteristics can be assessed preliminary, improving in this way the filter performance. GS for estimating mixture parameters is investigated over simulated and real data.

We have selected the following hyperparameters for every $i = 1,\ldots,n_{mix}$: $n_i = 1$, $\nu_i = n_{mix}$, if $\alpha_i = 1$, the Dirichlet distribution reduces to a uniform distribution and the algorithm is initialized with noniformative prior about mixture proportions. The initial values of $\boldsymbol{\theta}^{(0)}$ are chosen based on the prior information about physical restrictions on the parameters: $\sigma_i^{(0)} = 6$ $[dB]$, the bounds of the supplementary assumption $\sigma_i^2 \in [\sigma_{min}^2, \sigma_{max}^2]$ are respectively $[1^2, 20^2]$. Initial mean values $\mu_i^{(0)}$ are calculated based on the observed interval of variation of the data. The initial weights are selected $\pi_i^{(0)} = 1/n_{mix}$. The number of iterations is 250 and the initial "warming up" interval is $u_0 = 100$.

We performed experiments over the scenario shown in Figure 4.1 after evaluating the noise parameters first. A sample of $T_m = 2000$ measurement errors is generated according to the following mixture model with $n_{mix} = 2$ elements: $v_t \sim 0.2\mathcal{N}(-6.5, 2^2) + 0.8\mathcal{N}(8.0, 4^2)$. The histogram of the modeled measurement errors is presented in Figure 4.2 (a). The two modes are well pronounced on this Figure. The estimated mixture parameters are $\widehat{\boldsymbol{\pi}} = (0.19, 0.81)$, $\widehat{\mu}_1 = -6.9$, $\widehat{\sigma}_1^2 = 4$, $\widehat{\mu}_2 = 8.04$, $\widehat{\sigma}_2^2 = 17$. The mixture PDF approximation of the noise is given in Figure 4.2 (b).

It is assumed that the accelerations of the mobile nodes u_x and u_y can change within the range $[-5, 5]$ $[m/s^2]$ and that the command process $\boldsymbol{u}$ takes values among the following acceleration levels $\mathrm{M} = \{(0,0)', (3.5,0)', (0,3.5)', (0,-3.5)', (-3.5,0)'\}$. Thus the number of motion modes is 5. Non-random mobile node trajectories were generated with the dynamic state equation (4.1)-(4.3) without a process noise.

The reference nodes in the scenario from Figure 4.1 are randomly deployed on the observed area. The MM AUX-PF performance with noise statistics estimated by GS is compared with the filter performance with inaccurate noise distributions: in the first case the measurement noise statistics are assumed Gaussian with parameters: $v_t \sim \mathcal{N}(0, \sigma_{mix}^2)$, where $\sigma_{mix}^2 = 6^2$ is the mixture variance, and in the second case $v_t \sim \mathcal{N}(0, 4^2)$, parameters typical for urban environment. The respective position and speed RMSE are given in Figures 4.3 (a) and (b). The experiments show that accurate noise parameters are improving the localization accuracy for deterministically deployed sensor networks. However, they are especially useful for randomly deployed networks. It can be seen from Figure 4.3 (a) and (b) that the estimation errors are larger where the maneuvering phases of mobile nodes coincide with the places of sparsely located reference nodes. The peak dynamic errors increase when the information about noise parameters is insufficient. The position and speed root-mean-square errors (RMSEs) are minimum when the filter operates with estimated noise parameters. If the mean (case 2 in Figure 4.3) or mean and variance (case 3) of the noise are unknown, the errors increase progressively.

The parameters of the state vector initial distribution $\boldsymbol{x}_{1,0} \sim \mathcal{N}(\boldsymbol{m}_{1,0}, \boldsymbol{P}_{1,0})$ are selected as follows: $\boldsymbol{P}_{1,0} = \mathrm{diag}\{\boldsymbol{P}_{\mathrm{x},0}, \boldsymbol{P}_{\mathrm{y},0}\}$, $\boldsymbol{P}_{x,0} = \boldsymbol{P}_{y,0} = \mathrm{diag}\{30\,[\mathrm{m}]^2, 1\,[\mathrm{m/s}]^2, 0.5\,[\mathrm{m/s}^2]^2\}$, and $\boldsymbol{m}_{1,0}$ contains the exact initial node states. Initial mode probabilities are $\tilde{\mu}_{1,0} = 0.8$ and $\tilde{\mu}_{1,0} = 0.05$ for $i = 2, \ldots, 5$. The transition probability matrix $\boldsymbol{\Pi}$ has diagonal elements: $\Pi_{11} = 0.9, \Pi_{ii} = 0.7$, $i = 2, \ldots, 5$. The other parameters of the algorithm are given in Table 1.

The RMSE combined on both position coordinates are used to assess the closeness of the estimated state parameters to the actual dynamic parameters of mobile nodes over $N_{mc} = 50$ Monte Carlo runs.

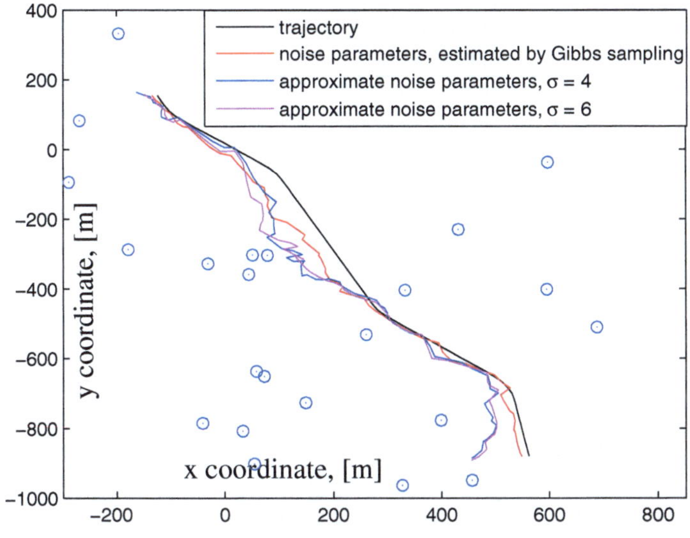

Fig. 4.1. A mobile unit moving in an area covered with randomly deployed wireless sensors with unknown measurement noise characteristics. Estimated trajectories are shown, obtained by the MM AUX-PF: 1. noise parameters estimated with Gibbs Sampling: $v_k \sim \sum_{i=1}^{2} \widehat{\pi}_i \mathcal{N}(\widehat{\mu}_i, \widehat{\sigma}_i^2)$, 2: the measurement noise characteristics are fixed, chosen preliminary as follows: $v_k \sim \mathcal{N}(0, 6^2)$, and respectively, $3: v_k \sim \mathcal{N}(0, 4^2)$

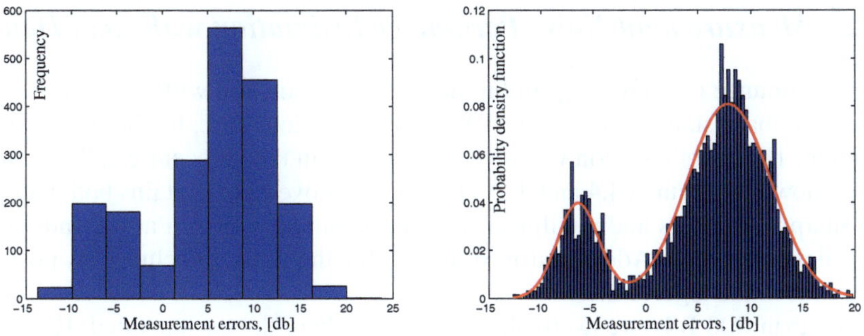

Fig. 4.2. (a) Histogram of the modeled noise, (b) Noise histogram overlayed with mixture density estimate

Table 1. Simulation parameters

Discretization time step T	1.0 [s]
Correlation coefficient a	0.35
Path loss index γ	4
Transmission power κ	30
Variance σ_w^2 of the noise $\boldsymbol{w}_k$ in (4.1)	$0.5^2\ [m/s^2]^2$
Maximum speed $V_{\max}$	$45\ [m/s]$
Number of particles	$N = 500$
Threshold for resampling	$N_{thresh} = N/10$
Number of Monte Carlo runs	$N_{mc} = 50$
Variance σ_v^2 of the noise v_k	$4^2\ [\text{dB}]^2$

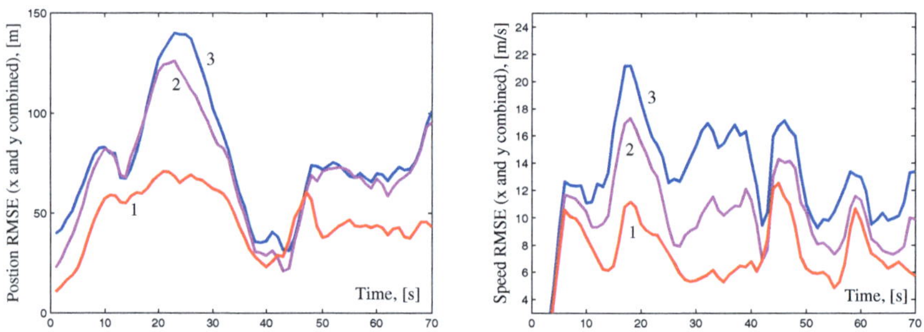

Fig. 4.3. Position RMSE combined (for x and y) and speed RMSEs (for $\dot{x}$ and $\dot{y}$): $1 : v_k \sim \sum_{i=1}^{2} \widehat{\pi}_i \mathscr{N}(\widehat{\mu}_i, \widehat{\sigma}_i^2)$, $2 : v_k \sim \mathscr{N}(0, 6^2)$, $3 : v_k \sim \mathscr{N}(0, 4^2)$

4.6.2 Measurement Noise Parameter Estimation with Real Data

The performance of the GS algorithm has been investigated with real RSSIs, collected from base stations (BSs), by HW Communications Ltd., in Glasgow, United Kingdom. The mobile station was a vehicle driving in the city centre and its trajectory is shown on Figures 4.4 and 4.5. The vehicle movement contains both patterns with sharp maneuvers and rectilinear motion, including a stretch at the end where the vehicle is parked. Additional information for the road is included as position constraints in the algorithms.

More than 400 BSs are available in the area where the car operated. However, only data from the six with the highest RSSIs were provided to the localization algorithm. Also a GPS system collected the actual positions of the moving mobile, for the purposes of validating the performance of the developed algorithms. The data from three BSs are used to evaluate the noise characteristics. The sample size is $T_m = 800$. The noise histogram and density estimate, obtained by fitting the mixture parameters to the data are presented in Figure 4.6 (a), (b) and (c). The number of iterations and the "warming up" bound are selected to be 550 and 450 respectively. The initial standard deviation is chosen as $\sigma^{(0)} = 1.5\ [dB]$, with an additional

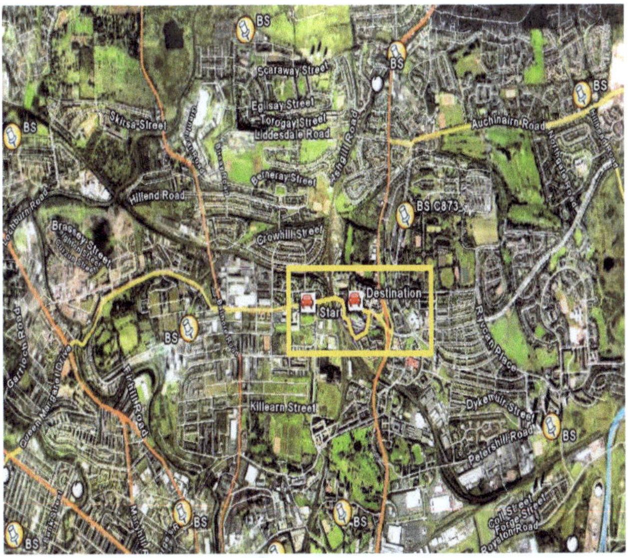

Fig. 4.4. The area in Glasgow, United Kingdom, where the vehicle is moving. The nearest BSs, the start and destination positions are indicated on the map.

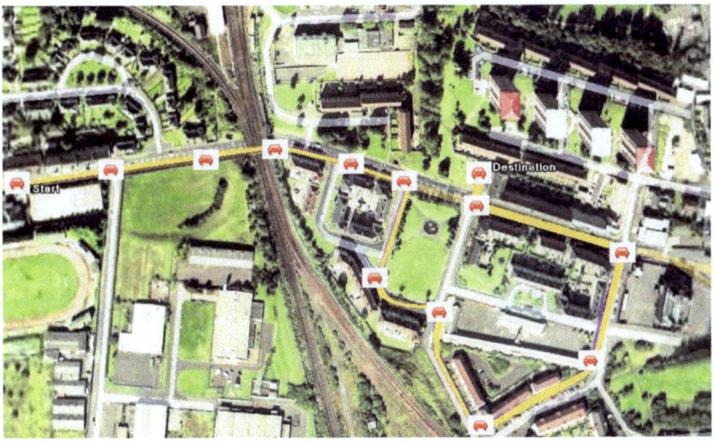

Fig. 4.5. Actual trajectory of the mobile node

sampling restriction of $\sigma^2_{min} = 0.001^2$ and $\sigma^2_{max} = 10^2$. The actual trajectory of the mobile is shown on Figure 4.5 and the estimated trajectories are given on Figure 4.7 both without and with GS for estimating the noise parameters. The position RMSE is presented on Figure 4.8.

For the GS algorithm when the data histograms have clearly differentiated modes (such as in Figure 4.2), the number of mixture components can be easily determined and this facilitates the mixture parameters evaluation. When the modes are difficult to distinguish (such as in Figure 4.6 (c)), a small number of mixture components can

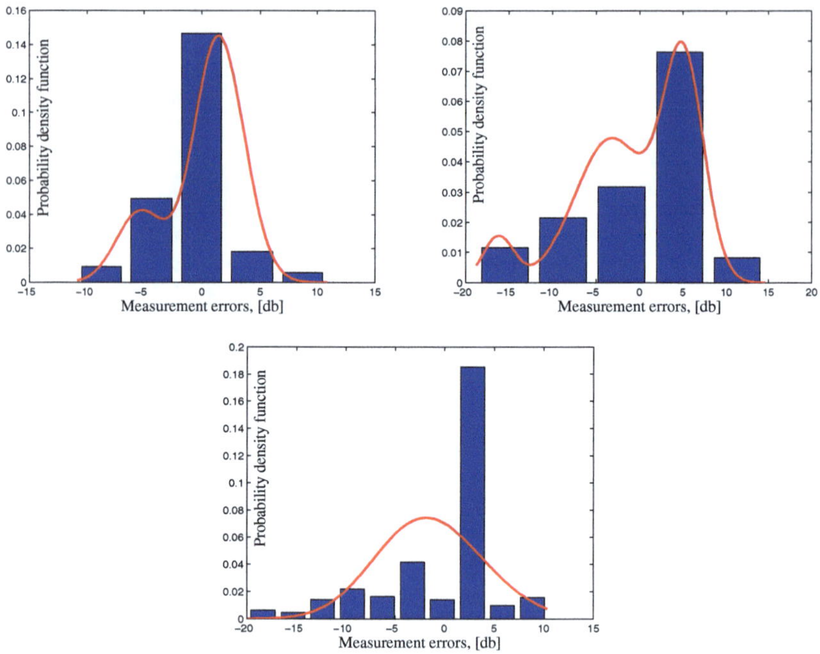

Fig. 4.6. Noise histogram and density estimate with $n_{mix} = 2$ components for: (a) BS1, (b) BS2, (c) BS3

be used. However, this is not an obstacle in the present application. As a result from the GS, the following estimates are obtained for the noise parameters of the RSSI from the three base stations: BS1 ($n_{mix} = 2$) and mixture estimates $\hat{\boldsymbol{\pi}} = (0.2, 0.79)'$, $\hat{\boldsymbol{\mu}} = (-5.4, 1.46)'$, $\sigma_{v,1} = 4.15$, $\sigma_{v,2} = 4.7$; BS2 ($n_{mix} = 3$) with mixture estimates $\hat{\boldsymbol{\pi}} = (0.07, 0.5, 0.4)'$, $\hat{\boldsymbol{\mu}} = (-16.25, -3.26, 5.02)'$, $\sigma_{v,1} = 3.22$, $\sigma_{v,2} = 17.59$, $\sigma_{v,3} = 5.6$; BS3 ($n_{mix} = 2$) with mixture estimates $\hat{\boldsymbol{\pi}} = (0.48, 0.52)'$, $\sigma_{v,1} = 20.27$, $\sigma_{v,1} = 25.6$.

The GS computational time in the case of 550 iterations and sample size $T_m = 800$ is approximately 30 seconds on a conventional PC (AMD Athlon(tm) 64 Processor 1.81 GHz). With a sample size $T_m = 2000$, the computational time is less than 2 minutes.

4.7 Performance Evaluation of the Multiple Model Auxiliary Particle Filter

Two cases have been investigated: for urban and suburban environment. In suburban environment, the correlation coefficient of the shadow process can be regarded as a constant for a wide range of velocities of the mobile [38]. Typical values of the correlation coefficient and shadow process are assumed, as suggested in the literature [37].

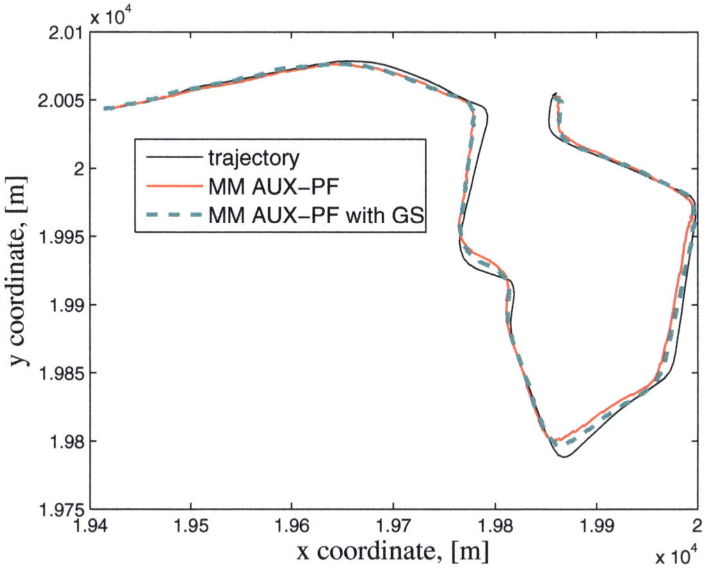

Fig. 4.7. Actual and estimated trajectories obtained by the MM AUX-PF: 1: $v_k \sim \mathcal{N}(0, 6^2)$, 2. noise parameters estimated with the GS, for BS_1, $v_k \sim \sum_{i=1}^{2} \widehat{\pi}_i \mathcal{N}(\widehat{\mu}_i, \widehat{\sigma}_i^2)$, for BS_2 $v_k \sim \sum_{i=1}^{3} \widehat{\pi}_i \mathcal{N}(\widehat{\mu}_i, \widehat{\sigma}_i^2)$ and for BS_3, $v_k \sim \sum_{i=1}^{2} \widehat{\pi}_i \mathcal{N}(\widehat{\mu}_i, \widehat{\sigma}_i^2)$.

4.7.1 Results with Simulated Data

Testing scenario. A sensor deployment architecture is considered, similar to the presented in [35]. Three mobile nodes are moving in an urban area well covered with a wireless sensor network (Figure 4.9). Each mobile node can measure the RSSI to each of the reference nodes, but only the three RSSIs with the highest strength are used for localization.

The MM AUX-PF AS is run for estimating the augmented state vector, consisting of three individual mobile state vectors. Figure 4.9 presents the actual and estimated trajectories of the three mobile nodes. The actual speed of the mobiles is shown in Figure 4.10.

The parameters of the individual state vector initial distribution $\boldsymbol{x}_{i,0} \sim \mathcal{N}(\boldsymbol{m}_{i,0}, \boldsymbol{P}_{i,0})$ are selected as follows: $\boldsymbol{P}_{i,0} = \text{diag}\{\boldsymbol{P}_{\text{x},0}, \boldsymbol{P}_{\text{y},0}\}$, with

$$\boldsymbol{P}_{x,0} = \boldsymbol{P}_{y,0} = \text{diag}\{30\,[\text{m}]^2,\, 1\,[\text{m/s}]^2,\, 0.5\,[\text{m/s}^2]^2\},\, i = 1,2,3$$

and $\boldsymbol{m}_{i,0}$ contains the exact initial node states. Initial mode probabilities are $\tilde{\mu}_{1,0} = 0.8$ and $\tilde{\mu}_{i,0} = 0.05$ for $i = 2,\ldots,5$. The transition probability matrix Π has the following diagonal elements: $\Pi_{11} = 0.9, \Pi_{ii} = 0.7,\ i = 2,\ldots,5$ and the off-diagonal elements (e.g., $\Pi_{i,1} = 0.025,\ i = 2,\ldots,5,\ \Pi_{i,2} = 0.07,\ \ i = 3,\ldots,5$) are chosen equal in each row to guarantee that the some in each row is equal to one. The noise

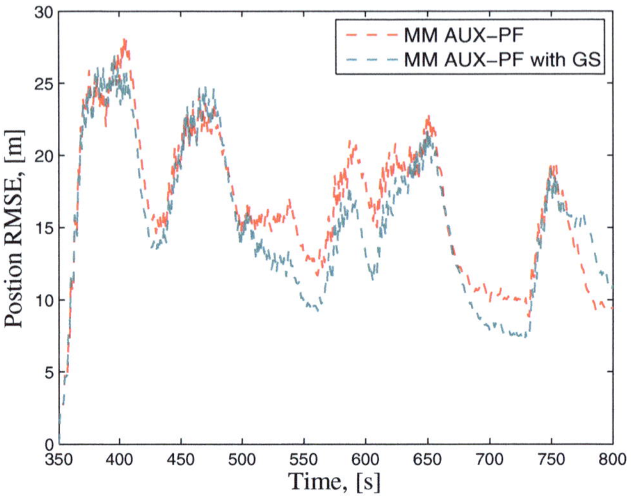

Fig. 4.8. Position RMSE (combined for x and y) of the MM AUX-PF: 1: $v_k \sim \mathcal{N}(0,6^2)$, 2. noise parameters estimated with the GS, for BS_1, $v_k \sim \sum_{i=1}^{2} \widehat{\pi}_i \mathcal{N}(\widehat{\mu}_i, \widehat{\sigma}_i^2)$, for BS_2 $v_k \sim \sum_{i=1}^{3} \widehat{\pi}_i \mathcal{N}(\widehat{\mu}_i, \widehat{\sigma}_i^2)$ and for BS_3, $v_k \sim \sum_{i=1}^{2} \widehat{\pi}_i \mathcal{N}(\widehat{\mu}_i, \widehat{\sigma}_i^2)$.

correlation coefficient was assumed to be equal to a typical value for urban environment, 0.25. The correlated measurement noises are generated by means of Cholesky factorization.

The other parameters of the algorithm are given in Table 2.

Table 2. Simulation parameters for urban environment

Discretization time step T	1.0 [s]
Correlation coefficient a	0.25
Path loss index γ	5
Transmission power κ	30
Variance σ_w^2 of the noise $\boldsymbol{w}_k$ in (4.1)	$0.5^2\ [m/s^2]^2$
Maximum speed $V_{\max}$	$45\ [m/s]$
Number of particles of the MM AUX-PF	$N = 500$
Threshold for resampling	$N_{thresh} = N/10$
Number of Monte Carlo runs	$N_{mc} = 50$
Variance σ_v^2 of the measurement noise v_k	4^2 [dB]^2

It is assumed that the accelerations of the mobile nodes u_x and u_y can change within the range $[-5,5]$ $[m/s^2]$ and that the command process $\boldsymbol{u}$ takes values among the following acceleration levels $\mathrm{M} = \{(0,0)',$ $(3.5,0)', (0,3.5)', (0,-3.5)', (-3.5,0)'\}$. Thus the number of motion modes is

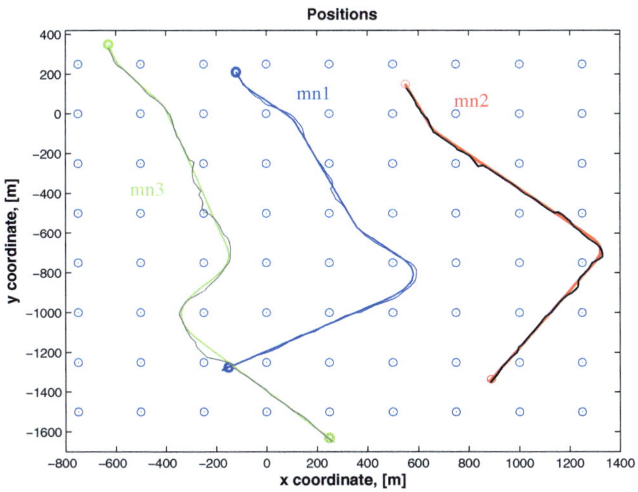

Fig. 4.9. Testing scenario 1: three mobile nodes (mn1, mn2 and mn3) moving in an area covered with a wireless sensor network. The sensors are uniformly deployed and form a rectangular grid.

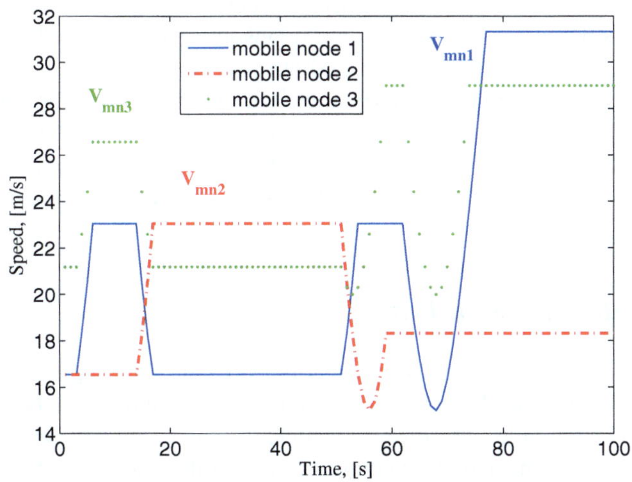

Fig. 4.10. Actual speeds of the three maneuvering mobile nodes

$r = 5$. Non-random mobile node trajectories were generated with the dynamic state equation (4.1)-(4.3) without process noise. The initial state vectors are as follows: $\hat{\boldsymbol{x}}_{1,0} = (-120, 7, 0, 210, -15, 0)'$, $\hat{\boldsymbol{x}}_{2,0} = (550, 7, 0, 150, -15, 0)'$ and $\hat{\boldsymbol{x}}_{3,0} = (-630, 7, 0, 350, -20, 0)'$. The first mobile node performs 3 short-term maneuvers with accelerations from the mode set and a longer maneuver with a control

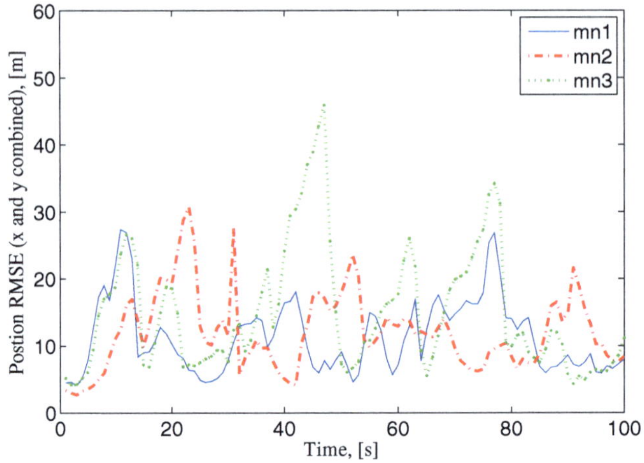

Fig. 4.11. Results for the position RMSE obtained with the MM AUX-PF with an augmented state vector

input $\boldsymbol{u} = (-3.0, 0.0)$, different from the acceleration set. The second mobile node maneuvers are described by the set of accelerations. The third mobile node performs two consecutive maneuvers with opposing accelerations. The root-mean-square errors (RMSE) [42] combined on both position coordinates yield the estimated state parameters to the actual dynamic parameters of each mobile node over $N_{mc} = 50$ Monte Carlo runs. The position and speed RMSEs of the MM AUX-PF with augmented state are shown on Figures 4.11 and 4.12. High position and speed estimates are achieved, with accuracy less than 45m with respect to the mobile nodes position.

4.7.2 Results with Real Data

The performance of the proposed localization MM-AUX-PF algorithms, with AS and AM, respectively, has been investigated over the same real RSSIs, presented in Section 4.6.2. The mobile station was a vehicle driving in the city centre.

Figure 4.13 shows the actual trajectory of the mobile together with the estimated trajectories and Figure 4.14 gives the respective position RMSEs. With the real RSSIs we compared the performance of: *i*) the MM AUX-PF that does not take into account the temporal measurement noise correlation with the MM AUX-PF with AM and MM AUX-PF with AS. From both Figures 4.13 and 4.14 it is evident that the accuracy of the MM AUX-PF with AS and of the MM-AUX PF with AM is higher than the accuracy of the MM AUX-PF neglecting the temporal noise correlation.

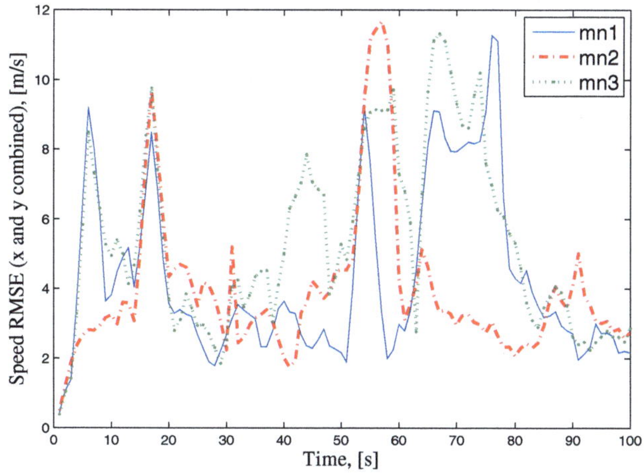

Fig. 4.12. Results for the speed RMSE obtained with the MM AUX-PF with an augmented state

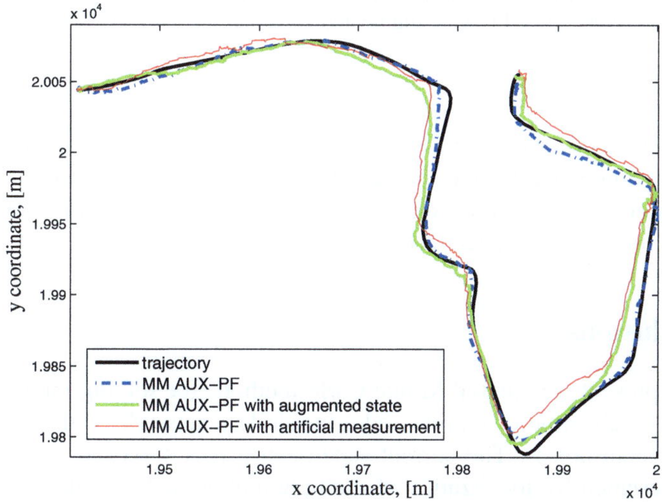

Fig. 4.13. This Figure shows the actual and the estimated trajectory of the vehicle by: *i*) the MM AUX-PF neglecting the measurement noise correlation; *ii*) the MM AUX-PF with augmented state; *iii*) the MM AUX-PF with artificial measurement.

The computational complexity is another important issue that we investigated. The MM AUX-PF execution time increases with the number of maneuvering models used in the implementation. The ratio between the computational time of the MM AUX-PF with 5 models and the computational time of the conventional AUX-PF is approximately 3:1. In the framework of the MATLAB environment, one-step processing time of a mobile node is approximately 2 seconds on a conventional PC

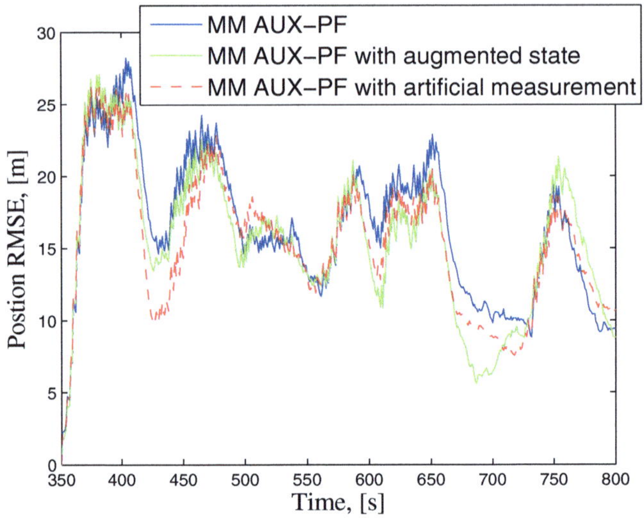

Fig. 4.14. Position RMSE for the: *i*) the MM AUX-PF neglecting the measurement noise correlation; *ii*) the MM AUX-PF with augmented state; *iii*) the MM AUX-PF with artificial measurement.

(AMD Athlon(tm) 64 Processor 1.81 GHz). By using C++ programming tools the computational time is reduced to the sampling interval. In MATLAB environment with a non-optimized code, the execution time for the MM-AUX PF with AS and AM is 2.33 s and 2.38 s, respectively.

4.8 Conclusions

This chapter presents sequential Monte Carlo methods for solving the problem of simultaneous localization of mobile nodes in wireless networks with correlated in time measurement noises. The current state-of-the-art is surveyed and then techniques are presented for localization of multiple mobile nodes, with estimation of the measurement noise parameters based on received signal strengths. Two multiple model auxiliary particle filters (with an augmented state vector and with an artificial measurement, respectively) are proposed for simultaneous localization of a mobile nodes in wireless networks. The algorithms performance has been investigated and validated over different scenarios and it has shown high accuracy for localizing maneuvering nodes.

The developed techniques have the potential to be further extended, with other techniques for noise parameter estimation, and their performance studied from theoretical point of view. The developed techniques can also be used in different applications, such as GPS-free position localization of mobile nodes in wireless networks, for localization of moving vehicles and robots and combined with different data.

The algorithms proposed here can be useful also in scenarios where the location information for the mobile nodes is supporting basic network functions.

Future work will be focused on localization when both fixed and mobile nodes communicate with each other, on techniques for localizing of a large number of nodes, fingerprinting and connectivity issues.

Acknowledgements. We acknowledge the support of the [European Community's] Seventh Framework Programme [FP7/2007-2013] under grant agreement No 238710 (Monte Carlo based Innovative Management and Processing for an Unrivalled Leap in Sensor Exploitation), network activities from the EU COST action TU0702 and the Bulgarian National Science Fund, grant DTK02-28/2009. We are grateful also HW Communications, Ltd., UK for providing us with the real data.

References

1. Juang, P., Oki, H., Wang, Y., Peh, L.S., Rubinstein, D.: Energy-efficient computing for wildlife tracking: Design tradeoffs and early experiences with ZebraNet. In: Proc. Conf. Architectural Support for Programming Languages and Operating Systems, pp. 96–107 (2002)
2. Patwari, N., Ash, J., Kyperountas, S., Hero III, A., Moses, R., Correal, N.: Locating the nodes: Cooperative localization in wireless sensor networks. IEEE Signal Processing Magazine 22(4), 54–69 (2005)
3. Çetin, M., Chen, L., Fisher, J., Ihler III, A., Wainwright, M., Willsky, A.: Distributed fusion in sensor networks. IEEE Signal Proc. Magazine 23(4), 42–55 (2006)
4. Zhao, F., Shin, J., Reich, J.: Information-driven dynamic sensor collaboration for tracking applications. IEEE Signal Processing Magazine 19(2), 61–72 (2002)
5. Xiao, J.J., Ribeiro, A., Luo, Z.-Q.: Distributed compression-estimation using wireless sensor networks. IEEE Signal Proc. Magazine 23(4), 27–41 (2006)
6. Mauve, M., Widmer, J., Hartenstein, H.: A survey on position-based routing in mobile ad hoc networks. IEEE Network Magazine 15(6), 30–39 (2001)
7. Srivastava, M., Muntz, R., Potkonjak, M.: Smart kindergarten: Sensor-based wireless networks for smart developmental problem-solving environments. In: Proc. of the ACM SIGMOBILE 7th Annual International Conf. on Mobile Computing and Networking (2005)
8. Gustafsson, F., Gunnarsson, F.: Mobile positioning using wireless networks: Possibilities and fundamental limitations based on available wireless network measurements. IEEE Signal Processing Magazine 22(4), 41–53 (2005)
9. Moses, R., Krishnamurthy, D., Patterson, R.: A self-localization method for wireless sensor networks. EURASIP Journal on Applied Signal Processing (4), 348–358 (2003)
10. Gustafsson, F.: Particle filter theory and practice with positioning applications. IEEE Transactions on Aerospace and Electronics Systems Magazine Part II: Tutorials 25(7), 53–82 (2010)
11. Djuric, P., Vemula, M., Bugallo, M., Miguez, J.: Non-cooperative localization of binary sensors. In: Proc. of IEEE Statistical Signal Proc. Workshop, France (2005)
12. Sun, G., Chen, J., Guo, W., Liu, K.: Signal processing techniques in network-aided positioning: a survey of state-of-the-art positioning designs. IEEE Signal Processing Magazine 22(4), 12–23 (2005)

13. Chintalapudi, K., Dhariwal, A., Govindan, R., Sukhatme, G.: Ad-hoc localization using ranging and sectoring. In: Proc. of the IEEE Infocomm (2004)
14. Gustafsson, F., Gunnarsson, F.: Localization in sensor networks based on log range observations. In: Proc. of the 10th International Conf. on Information Fusion, Canada (2007)
15. Jirod, J., Estrin, D.: Robust range estimation using acoustic and multimodal sensing. In: Proc. of the IEEE International Conf. on Intelligent Robots and Systems (2001)
16. He, T., Huang, C., Blum, B.M., Stankovic, J.A., Abdelzaher, T.: Range-free localization schemes for large scale sensor networks. In: MobiCom 2003: Proc. of the 9th Annual International Conf. on Mobile Computing and Networking, pp. 81–95. ACM Press, NY (2003)
17. Liu, H., Darabi, H., Banerjee, P., Liu, J.: Survey of wireless indoor positioning techniques and systems. IEEE Transactions on Systems, Man, and Cybernetics - Part C: Applications and Reviews 37(6), 1067–1080 (2007)
18. Vorst, P., Sommer, J., Hoene, C., Schneider, P., Weiss, C., Schairer, T., Rosenstiel, W., Zell, A., Carl, G.: Indoor positioning via three different rf technologies. In: Proceedings of the 4th European Workshop on RFID Systems and Technologies (RFID Sys Tech 2008), June 10-11. vol. 209. ITG-Fachbericht, VDE Verlag, Germany (2008)
19. Pahlavan, K., Akgul, F., Ye, Y., Morgan, T., Alizadeh-Shabodz, F., Heidari, M., Steger, C.: Taking positioning indoors. Wi-Fi localization and GNSS. Inside GNSS (May 2010)
20. Bshara, M., Orguner, U., Gustafsson, F., Van Biesen, L.: Fingerprinting localization in wireless networks based on received-signal-strength measurements: A case study on WiMAX networks. IEEE Transactions on Vehicular Technology 59(1), 283–294 (2010)
21. Takenga, C., Peng, T., Kyamakya, K.: Post-processing of fingerprint localization using Kalman filter and map-matching techniques. In: Proc. of the 9th International Conference on Advanced Communication Technology, vol. 3, pp. 2029–2034 (2007)
22. Engee, P.K.: The global positioning system: Signals, measurements and performance. International Journal of Wireless Information Networks 1(2), 83–105 (1994)
23. Zaidi, Z., Mark, B.: A mobility tracking model for wireless ad hoc networks. In: Proc. of IEEE WCNC 2003, vol. 3, pp. 1790–1795 (March 2003)
24. Zaidi, Z., Mark, B.: Mobility estimation for wireless networks based on an autoregressive model. In: Proc. of the IEEE Globecom, pp. 3405–3409 (2004)
25. Hammes, U., Zoubir, A.M.: Robust mobile terminal tracking in NLOS environments based on data association. IEEE Transactions on Signal Processing 58, 5872–5882 (2010)
26. Morelli, C., Nicoli, M., Rampa, V., Spagnolini, U.: Hidden Markov models for radio localization in mixed LOS/NLOS conditions. IEEE Trans. on Signal Processing 55(4), 1525–1542 (2007)
27. Hu, L., Evans, D.: Localization for mobile sensor networks. In: Proc. of the Tenth Annual Intl. Conf. on Mobile Computing and Networking, USA (2004)
28. Mihaylova, L., Angelova, D., Canagarajah, C.N., Bull, D.R.: Algorithms for Mobile Nodes Self-Localisation in Wireless Ad Hoc Networks. In: Proc. of the 9th International Conf. on Information Fusion, Italy, Florence (2006)
29. Huerta, J.M., Vidal, J., Giremus, A., Tourneret, J.-Y.: Joint particle filter and UKF position tracking in severe non-line-of-sight situations. IEEE Journal of Selected Topics in Signal Processing 3(5), 874–888 (2009)
30. Ihler, A., Fisher, J., Moses, R., Willsky, A.: Nonparametric belief propagation for sensor networks. IEEE Journal on Selected Areas in Communications 23(4), 809–819 (2005)

31. Mihaylova, L., Bull, D., Angelova, D., Canagarajah, N.: Mobility tracking in cellular networks with sequential Monte Carlo filters. In: Proc. of the Eight International Conf. on Information Fusion (2005)
32. Mihaylova, L., Angelova, D., Honary, S., Bull, D.R., Canagarajah, C.N., Ristic, B.: Mobility tracking in cellular networks using particle filtering. IEEE Transactions on Wireless Communications 6(10), 3589–3599 (2007)
33. Karlsson, R., Bergman, N.: Auxiliary particle filters for tracking a manoeuvring target. In: Proceedings of the 39th IEEE Conference on Decision and Control, pp. 3891–3895 (2000)
34. Arulampalam, M.S., Ristic, B., Gordon, N., Mansell, T.: Bearings-only tracking of manoeuvring targets using particle filters. EURASIP Journal on Applied Signal Processing (1), 2351–2365 (2004)
35. Djuric, P.M., Vemula, M., Bugallo, M.F.: Target tracking by particle filtering in binary sensor networks. IEEE Transactions on Signal Processing 56(6), 2229–2238 (2008)
36. Kannan, A., Mao, G., Vucetic, B.: Simulated annealing based wireless sensor network localization with flip ambiguity mitigation. In: IEEE Vehicular Technology Conference Spring (VTC), pp. 1022–1026 (2006)
37. Gudmundson, M.: Correlation model for shadow fading in mobile radio systems. Electronics Letters 27(23), 2145–2146 (1991)
38. Jiang, T., Sidiropoulos, N., Giannakis, G.: Kalman filtering for power estimation in mobile communications. IEEE Transactions on Wireless Communications 2(1), 151–161 (2003)
39. Forkel, I., Schinnenburg, M., Ang, M.: Generation of two-dimensional correlated shadowing for mobile radio network simulation. In: Proceedings of the 7th International Symposium on Wireless Personal Multimedia Communications, WPMC 2004, Abano Terme (Padova), Italy, p. 5 (September 2004)
40. Mihaylova, L., Angelova, D., Bull, D.R., Canagarajah, N.: Localization of mobile nodes in wireless networks with correlated in time measurement noise. IEEE Transactions on Mobile Computing 10, 44–53 (2011)
41. Mihaylova, L., Angelova, D.: Noise parameters estimation with Gibbs sampling for localisation of mobile nodes in wireless networks. In: Proc. of the 13th International Conference on Information Fusion, ISIF, Edinburgh, UK, pp. tu3.5.1–0037 (2010)
42. Bar-Shalom, Y., Rong Li, X., Kirubarajan, T.: Estimation with Applications to Tracking and Navigation. John Wiley and Sons (2001)
43. Camp, T., Boleng, J., Davies, V.: A survey of mobility models for ad hoc network research. Wireless Communications and Mobile Computing 2(5), 483–502 (2002)
44. Mark, B., Zaidi, Z.: Robust mobility tracking for cellular networks. In: Proc. IEEE Intl. Communications Conf., pp. 445–449 (May 2002)
45. Zaidi, Z.R., Mark, B.L.: Real-time mobility tracking algorithms for cellular networks based on Kalman filtering. IEEE Transactions on Mobile Computing 4(2), 195–208 (2005)
46. Moose, R.: An adaptive state estimator solution to the maneuvering target tracking problem. IEEE Transactions on Automatic Control 20(3), 359–362 (1975)
47. Bar-Shalom, Y., Li, X.R.: Estimation and Tracking: Principles, Techniques and Software. Artech House (1993)
48. Li, X.R., Jilkov, V.: A survey of maneuveuvering target tracking. Part I: Dynamic models. IEEE Trans. on Aerosp. and Electr. Systems 39(4), 1333–1364 (2003)

49. Yang, Z., Wang, X.: Joint mobility tracking and hard handoff in cellular networks via sequential Monte Carlo filtering. In: Proc. of the IEEE Conf. on Computer Communications (Infocom), New York, pp. 968–975 (2002)
50. Yang, Z., Wang, X.: Sequential Monte Carlo for mobility management in wireless cellular networks. In: Proc. of the XI European Signal Processing Conf. EUSIPCO (2002)
51. Stüber, G.L.: Principles of Mobile Communication, 2nd edn. Kluwer Academic Publ. (2001)
52. Hissalle, L.P.I., Alahakoon, S.: Estimating signal strengths prior to field trials in wireless local loop networks. In: Proceedings of the International Conference on Industrial and Information Systems, pp. 409–414 (August 2007)
53. Arulampalam, S., Maskell, S., Gordon, N., Clapp, T.: A tutorial on particle filters for online nonlinear/non-Gaussian Bayesian tracking. IEEE Trans. on Signal Proc. 50(2), 174–188 (2002)
54. Liu, J., Chen, R.: Sequential Monte Carlo methods for dynamic systems. Journal of the American Statistical Association 93(443), 1032–1044 (1998)
55. Wan, E., van der Merwe, R.: The Unscented Kalman Filter. In: Haykin, S. (ed.) Ch. 7: Kalman Filtering and Neural Networks, pp. 221–280. Wiley Publishing (September 2001)
56. Pitt, M.K., Shephard, N.: Filtering via simulation: Auxiliary particle filters. Journal of the American Statistical Association 94(446), 590–599 (1999)
57. Saha, S., Ozkan, E., Gustafsson, F., Smidl, V.: Marginalized particle filters for Bayesian estimation of Gaussian noise parameters. In: Proc. of the 13th International Conf. on Information Fusion, ISIF, UK (2010)
58. Ozkan, E., Saha, S., Gustafsson, F., Smidl, V.: Non-parametric bayesian measurement noise density estimation in non-linear filtering. In: Proc. of the 36th International Conference on Acoustics, Speech and Signal Processing (ICASSP), Prague, Czech Republic, IEEE (2011)
59. Kotecha, J.H., Djurić, P.M.: Gaussian sum particle filtering. IEEE Transactions on Signal Processing 51(10), 2602–2612 (2003)
60. German, S., German, D.: Stochastic relaxation, Gibbs distributions, and the Bayesian restoration of images. In: Anderson, J.A., Rosenfeld, E. (eds.) Neurocomputing: foundations of research, pp. 611–634. MIT Press, Cambridge (1988), http://dl.acm.org/citation.cfm?id=65669.104442
61. Diebolt, J., Robert, C.: Estimation of finite mixture distributions through Bayesian sampling. J. Royal Stat. Society B 56(4), 363–375 (1994)
62. Cornebise, J., Maumy, M., Girard, P.: A practical implementation of the Gibbs sampler for mixture of distributions: Application to the determination of specifications in food industry. In: Janssen, J., Lenca, P. (eds.) Proc. of International Symposium on Applied Stochastic Models and Data Analysis (ASMDA), pp. 828–837 (2005)

Chapter 5
A Sequential Monte Carlo Approach for Brain Source Localization

Petia Georgieva, Lyudmila Mihaylova, Filipe Silva, Mariofanna Milanova, Nuno Figueiredo, and Lakhmi C. Jain

Abstract. In this chapter we propose a solution to the Electroencephalography (EEG) inverse problem combining two techniques, which are the Sequential Monte Carlo (SMC) method for estimating the coordinates of the first two non-correlated dominative brain zones (represented by their respective current dipoles) and spatial filtering which is done by beamforming based on EEG data. Beamforming (BF) gives estimates of the respective source moments. In order to validate this novel approach for brain source localization, EEG data from dipoles with known locations and known moments are generated and artificially corrupted with noise. The noise represents the overall influence of other brain sources but they are not brain disturbances. When the power of the EEG signal due to the main brain sources is higher than the summed effect of all other secondary sources, the estimation of the localization of the leading sources is reliable and repetitive over a number of Monte Carlo runs.

Petia Georgieva · Filipe Silva · Mariofanna Milanova · Nuno Figueiredo
Institute of Electrical Engineering and Telematics of Aveiro (IEETA),
Department of Electronics Telecommunications and Informatics,
University of Aveiro, Portugal
e-mail: {fmsilva,petia,nuno.figueiredo}@ua.pt

Lyudmila Mihaylova
School of Computing and Communications, Lancaster University, InfoLab21, South Drive, Lancaster LA1 4WA, United Kingdom
e-mail: mila.mihaylova@lancaster.ac.uk

Mariofanna Milanova
Computer Science Department, University of Arkansas at Little Rock, 2801 S. University Av, Little Rock, AR, USA 72204
e-mail: mgmilanova@ualr.edu

Lakhmi Jain
School of Electrical and Information Engineering, University of South Australia, Adelaide, Mawson Lakes Campus, South Australia SA 5095, Australia
e-mail: Lakhmi.jain@unisa.edu.au

P. Georgieva et al. (Eds.): Adv. in Intelligent Signal Processing and Data Mining, SCI 410, pp. 119–138.
springerlink.com

Keywords: Sequential Monte Carlo methods, Source localization, Brain computer interfaces.

5.1 Introduction

Electroencephalography (EEG) is a widely used technology for brain study because it is noninvasive, relatively cheap, portable and it has an excellent temporal resolution. However, EEG suffers from poor spatial resolution, because only surface sensor measurements can be observed and they are very noisy signals. Therefore, finding the location and time patterns of relevant inner brain sources is difficult. The determination is done by analyzing a set of noisy EEG mixtures in much of research in electromagnetic neuroimaging. Popular approaches include solutions to the ill-posed inverse problem [1]. The construction of spatial filters or beamformers by data-independent [2] or data-driven methods [3] and the blind source separation (BSS) problem is often solved by independent component analysis (ICA) [4, 5, 6].

In this chapter we present a solution to the EEG inverse problem by combining two techniques: a Sequential Monte Carlo method for the source (dipole) localization in the brain and spatial filtering by beamforming for the respective source waves (moments) estimation based on EEG measurements. In brain imaging, the EEG inverse problem can be formulated by using measurements of the electrical potential on the scalp recorded from multiple sensors. The goal is to build a reconstruction system which is able to estimate the location in the brain area. The magnitude and directions of the dominant neural brain sources which have originated the EEG signal. The problem can be divided in two parts:

1. The localization of the principal original source inside the brain, followed by
2. The estimation of the source signal waveforms.

The problem of reconstructing the time pattern of the original source signals obtained from a sensor array without exploiting the a priori knowledge of the transmission channel, can be expressed as follows. A number of related BSS problems have been studied. Choi et al. [7] present a review of various blind source separation and independent component analysis (ICA) algorithms for static and dynamic models in their applications. Beamforming has become a popular procedure for analysis based on non-invasive recording electrophysiological data sets. The idea is to use a set of recordings from sensors and to combine the signals recorded at various sites to increase the signal-to-noise ratio, while being focused on a certain region in space (Region Of Interest (ROI)). In that sense, beamforming uses a different approach. The whole brain is scanned point by point. A spatial filter designed to be fully sensitive to activity from the target location is as insensitive as possible to activity from other brain regions. This is done by constructing a spatial filter in an adaptive way, that is by taking into account the recorded data. More specifically, the beamforming process is carried out by weighting the EEG signals. This is done by adjusting the amplitudes so that when then the components are added together they form the desired source signal.

While phase two is still widely studied, the source localization (phase one) is often assumed as known, particularly when the estimation is based only on EEG recordings. Therefore, this is the primary motivation and in our study we focus on the localization of the principal original source inside the brain.

The second motivation is the potential of application of original brain source localization in brain-computer interfaces (BCI). Despite some encouraging results, e.g. [8, 9, 10, 11], only recently the concept of source-based BCI was adopted in the literature. Therefore, additional research efforts are needed to establish a solid foundation aiming at discovering the driving force behind the growth of source-based BCI as a research area and to expose its implications for the design and implementation of advanced high performance systems.

This chapter is presented as follows: in section 5.2 the Sequential Monte Carlo method is outlined, in section 5.3 the state space model of the source localization problem is formulated. The use of the beamforming technique is considered in section 5.4. Finally, the quality of the reconstruction system is validated by simulations.

5.2 Sequential Monte Carlo Problem Formulation

The methods for solving the EEG source localization problem depend on the location of the sources, fixed or not, on the type of the noise in the measurements and in the sources. We wish to identify active brain zones. In this chapter of the book we focus on the localization of fixed sources. This is a typical assumption for distributed method). We assume a Gaussian measurement noise. The problem of EEG source localization is formulated next as an estimation problem within the Sequential Monte Carlo framework. To define the estimation problem, we consider the evolution of the state sequence $\boldsymbol{x}_k,\ k \in \mathbb{N}$ of a target which is described by

$$\boldsymbol{x}_k = f(\boldsymbol{x}_{k-1}, \boldsymbol{w}_{k-1}). \tag{5.1}$$

Here the state $\boldsymbol{x}_k$ at time k ($\boldsymbol{x}_k \in \mathbb{R}^{n_x}$) is in general a nonlinear function of the state $\boldsymbol{x}_{k-1}$ at the previous time $k-1$ and is also affected by the process noise sequence $\boldsymbol{w}_{k-1}$. $\mathbb{N}$ is the set of natural numbers. The objective is to recursively estimate $\boldsymbol{x}_k$ from measurements

$$\boldsymbol{z}_k = h(\boldsymbol{x}_k, \boldsymbol{v}_k), \tag{5.2}$$

where h is a nonlinear function in general and $\boldsymbol{v}_k$ is the measurement noise sequence. Expressions (5.1)-(5.2) are the state and measurement equations of the general state-space transition model required by the Sequential Monte Carlo estimation [12]. In particular we seek filtered estimates of x_k based on the set of all available measurements up to time k. It is assumed that the observations are taken at discrete time points with a time step T.

Within the Bayesian framework, the estimation problem is to recursively calculate a degree of belief in the state $\boldsymbol{x}_k$ at time k, given the data $\boldsymbol{z}_{1:k}$ up to time k [12]. Hence, it is required to construct the posterior state probability density function (pdf) $p(\boldsymbol{x}_k|\boldsymbol{z}_{1:k})$. It is assumed that the initial pdf of the state vector, $p(\boldsymbol{x}_0|\boldsymbol{z}_0) = p(\boldsymbol{x}_0)$, known also as the prior, is available ($\boldsymbol{z}_0$ is the initial measurement). Then, the posterior conditional pdf $p(\boldsymbol{x}_k|\boldsymbol{z}_{1:k})$ can be obtained recursively, in two stages: *prediction* and *update*. Suppose that the required pdf $p(\boldsymbol{x}_{k-1}|\boldsymbol{z}_{1:k-1})$ is available at time $k-1$. The prediction stage consists in obtaining the prior pdf $p(\boldsymbol{x}_k|\boldsymbol{z}_{1:k-1})$ of the state at time k via the Chapman-Kolmogorov equation

$$p(\boldsymbol{x}_k|\boldsymbol{z}_{1:k-1}) = \int_{\mathbb{R}^{n_x}} p(\boldsymbol{x}_k|\boldsymbol{x}_{k-1})p(\boldsymbol{x}_{k-1}|\boldsymbol{z}_{1:k-1})d\boldsymbol{x}_{k-1}. \tag{5.3}$$

The system model (5.1) describes a first order Markov process with $p(\boldsymbol{x}_k|\boldsymbol{x}_{k-1},\boldsymbol{z}_{1:k-1}) = p(\boldsymbol{x}_k|\boldsymbol{x}_{k-1})$. The probabilistic model of the state evolution $p(\boldsymbol{x}_k|\boldsymbol{x}_{k-1})$ is defined by the system equation (5.1) and the known statistics of $\boldsymbol{w}_{k-1}$. At time step k, a measurement $\boldsymbol{z}_k$ becomes available and this may be used to update the prior (update stage) via the Bayes' rule:

$$p(\boldsymbol{x}_k|\boldsymbol{z}_{1:k}) = \frac{p(\boldsymbol{z}_k|\boldsymbol{x}_k)p(\boldsymbol{x}_k|\boldsymbol{z}_{1:k-1})}{p(\boldsymbol{z}_k|\boldsymbol{z}_{1:k-1})} = \frac{likelihood \times prior}{evidence}, \tag{5.4}$$

where $p(\boldsymbol{z}_k|\boldsymbol{z}_{1:k-1})$ is a normalizing constant defined by the measurement model (5.2) and the known statistics of $\boldsymbol{v}_k$. Hence, the recursive update of $p(\boldsymbol{x}_k|\boldsymbol{z}_k)$ is proportional to

$$p(\boldsymbol{x}_k|\boldsymbol{z}_{1:k}) \propto p(\boldsymbol{z}_k|\boldsymbol{x}_k)p(\boldsymbol{x}_k|\boldsymbol{z}_{1:k-1}). \tag{5.5}$$

In the update stage (5.4) the measurement is used to modify the prior density to find the required posterior density of the current state. The recurrence relations (5.3)-(5.4) form the basis for the optimal Bayesian framework.

The Kalman Filter (KF) [13, 14] gives the optimal solution to this estimation problem assuming linear models (f and h are linear functions) and Gaussian distribution for the process and the observation noises $\boldsymbol{v}_k$ and $\boldsymbol{w}_k$, respectively. The Extended KF (EKF) [14] requires linearization. However, it is difficult to apply it to strongly nonlinear models or models with interconnected components. The unscented KF (also known as central differences filter) and the Gaussian sum filter (assuming a mixtures of Gaussian models) [15] are more recent alternatives to the classical analytical parametric methods as KF. However, in most of the real life problems the recursive propagation of the posterior density cannot be performed analytically (the integral in equation (5.3) is intractable). Usually numerical methods are used and therefore a sample-based construction to represent the state pdf. The family of techniques that solve numerically the estimation problem are known as Sequential Monte Carlo (SMC) methods or Particle filtering or CONDENSATION (CONditional DENsity estimation) algorithms [16, 17]. The bootstrap nonlinear filtering method was introduced in [18] and applied to aerospace navigation.

Various Particle filter (PF) variants such as Sampling Importance Sampling (SIS), Sampling Importance Resampling (SIR), Rao-Blackwellized PFs, Auxiliary PF were introduced to communications, robotics, aerospace navigation, economics and other areas. For more details see the surveys [12, 19, 20]. Hybrid solutions are also considered for models with partial linear structure (a KF is applied), where the nonlinear state vector is estimated with PFs, for example Rao-Blackwellization [21, 19].

In the Sequential Monte Carlo framework, multiple particles (samples) of the state are generated, each one is associated with a weight $W_k^{(\ell)}$ which characterizes the quality of a specific particle ℓ, $\ell = 1,\ldots,N$. Thus, a set of N weighted particles, which has been drawn from the posterior conditional pdf, is used to map integrals to discrete sums. The posterior pdf $p(\boldsymbol{x}_k|\boldsymbol{z}_{1:k})$ is approximated by the weighted a sum of particles

$$\hat{p}(\boldsymbol{x}_k|\boldsymbol{z}_{1:k}) = \sum_{\ell=1}^{N} W_k^{(\ell)} \delta(\boldsymbol{x}_k - \boldsymbol{x}_k^{(\ell)}) \tag{5.6}$$

and the update probability is

$$p(\boldsymbol{x}_k^{(\ell)}|\boldsymbol{z}_{1:k}) = \sum_{\ell=1}^{N} W_k^{(\ell)} \delta(\boldsymbol{x}_k - \boldsymbol{x}_k^{(\ell)}), \tag{5.7}$$

where the new weight $W_k^{(\ell)}$ is calculated based on the likelihood $p(\boldsymbol{z}_k|\boldsymbol{x}_k^{(\ell)})$

$$W_k^{(\ell)} = W_{k-1}^{(\ell)} p(\boldsymbol{z}_k|\boldsymbol{x}_k^{(\ell)}). \tag{5.8}$$

The normalized weights are

$$W_k^{(\ell)} = \frac{W_k^{(\ell)}}{\sum_{\ell=1}^{N} W_k^{(\ell)}}. \tag{5.9}$$

New states are calculated, by giving more weight on particles that are important according to the posterior pdf (5.6). It is often impossible to sample directly from the posterior density function $p(\boldsymbol{x}_k|\boldsymbol{z}_{1:k})$ when it has a complex form or is not precisely known. This difficulty is avoided by making use of importance sampling from a known *proposal distribution* $p(\boldsymbol{x}_k|\boldsymbol{x}_{k-1})$. During the prediction, stage each particle is modified according to the state model (5.1). In the measurement update stage, each particle's weight is re-evaluated based on the new data. Note that the KF requires a low computational cost because it propagates only the first two moments (the mean and covariance of the Gaussian posterior state pdf), while the Sequential Monte Carlo algorithm propagates the state probability density function through the samples (particles). Table 5.1 presents the PF algorithm for brain source localization and links it with the system dynamics and measurement models (See Section 5.3).

Particle Degeneracy Phenomenon

An inherent SMC problem is the particle degeneracy, the case when a small set of particles (or even just one particle) have significant weights. An estimate of the measure of degeneracy, [22], at time k is given as

$$N_{eff} = \frac{1}{\Sigma_{\ell=1}^{N} \left(W_k^{(\ell)}\right)^2}, \tag{5.10}$$

where N_{eff} is the number of effective particles with significant weights. If the value N_{eff} is very low, a resampling procedure is used to avoid degeneracy. A schematic representation of the resampling procedure is depicted in Figure 5.1 [23]. Figure 5.1 shows on the left side $M = 10$ particles before resampling, where the diameters of the circles are proportional to the weights of the particles. The right hand side shows circles of the particles after the resampling step. The large particles are replicated and the small particles are removed. There are various mechanisms to perform the resampling step. Two of them are either resampling at each iteration or resampling when the effective number of particles falls below a user-defined threshold N_{eff}. The resampling step introduces variety in the particles, but increases the variance error of the particle population.

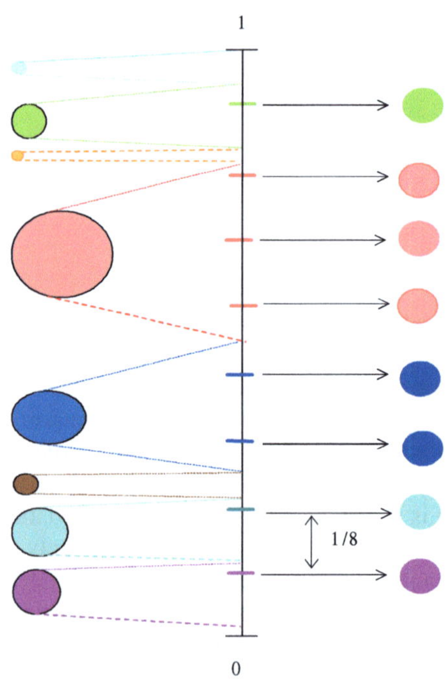

Fig. 5.1. The resampling step of SMC methods, illustrated graphically [23].

In this work, the PF is applied to estimate the localization of the principal brain sources that have generated the available EEG signals. One can consider this problem as the inverse of the EEG problem with respect to the source space localization.

5.3 The State Space EEG Source Localization Model

In order to apply the particle filtering framework, as outlined in the previous section, the state-space transition model of the source localization must to be first defined. The useful frequency spectrum for electrophysiological signals in EEG is typically below 1 kHz, and most studies deal with frequencies between 0.1 and 100 Hz. Consequently the physics of EGG can be described by the quasi-static approximation of Maxwell's equations [24]. The quasi-static total current flow $J(\boldsymbol{x}_s)$ at location $\boldsymbol{x}_s$ can be partitioned into two current flows of distinct physiological significance: a primary (or driving) current flow $J^p(\boldsymbol{x}_s)$ related with the original neural activity and a volume (or passive) current flow $J^v(\boldsymbol{x}_s)$ that results from the effect of the electric field in the volume on extracellular charge carriers:

$$J(\boldsymbol{x}_s) = J^p(\boldsymbol{x}_s) + J^v(\boldsymbol{x}_s) = J^p(\boldsymbol{x}_s) + \sigma(\boldsymbol{x}_s)E(\boldsymbol{x}_s) = J^p(\boldsymbol{x}_s) - \sigma(\boldsymbol{x}_s) \bigtriangledown V(\boldsymbol{x}_s), \quad (5.11)$$

where $\sigma(\boldsymbol{x}_s)$ is the conductivity profile of the head tissues, which is assumed (for simplicity) to be isotropic, and, from the quasi-static assumption, the electric field $E(\boldsymbol{x}_s)$ is the negative gradient of the electric potential, $V(\boldsymbol{x}_s)$. It is also assumed that the head consists of a set of contiguous regions each of constant isotropic conductivity σ_i, $i = 1,2,3$. representing the brain, skull and scalp. Then the electric potential on the surface S_{ij} can be expressed as a sum of the contributions from the primary and volume currents

$$(\sigma_i + \sigma_j)V(\boldsymbol{x}) = 2\sigma_0 V_0(x) - \frac{1}{p\pi}\Sigma_{ij}(\sigma_i + \sigma_j)\int_{S_{ij}} V(\boldsymbol{x}_s)\frac{\boldsymbol{x} - \boldsymbol{x}_s}{\|\boldsymbol{x} - \boldsymbol{x}_s\|^3} dS'_{ij}, \quad (5.12)$$

where $V_0(\boldsymbol{x})$ is the is the potential at location $\boldsymbol{x}$ due to the primary current distribution. The second term is the volume current contribution to the electric potential formed as a sum of surface integrals over the brain skull, skull-scalp and scalp-air boundaries. Equations (5.12) represent the integral solutions to the forward problem. If a primary current distribution $J^p(\boldsymbol{x}_s)$ is specified, the respective primary potential can be calculated

$$V_0(\boldsymbol{x}) = \frac{1}{4\pi\sigma_0}\int_{S_{ij}} J^p(\boldsymbol{x}_s)\frac{\boldsymbol{x} - \boldsymbol{x}_s}{\|\boldsymbol{x} - \boldsymbol{x}_s\|^3} d\boldsymbol{x}_s, \quad (5.13)$$

where σ_0 is the standard deviation for the potential $V_0(\boldsymbol{x})$.

The primary potential $V_0(\boldsymbol{x})$ is then used to solve (5.12) for the potentials on all surfaces, and therefore solves the forward problem for EEG. Unfortunately, the integral in (5.12) has analytic solutions only for special shapes and must otherwise be solved numerically. The primary current distributions depend on the source models.

Source Models: Dipoles and Multipoles

Let us assume that brain activity arises at a small zone of the cortex centered at location $\boldsymbol{x}_s$ and that the observation point with location $\boldsymbol{x}$ is some distance away from this zone. The primary current distribution can be approximated by an equivalent current dipole represented as a point source

$$J^p(\boldsymbol{x}_s) = s\delta(\boldsymbol{x} - \boldsymbol{x}_s), \tag{5.14}$$

where $\delta(\boldsymbol{x})$ is the Dirac delta function, with moment

$$\boldsymbol{s} \equiv \int J^p(\boldsymbol{x}_s) d\boldsymbol{x}_s. \tag{5.15}$$

The current dipole is an extension of the model of the paired-charges dipole in electrostatics. It is important to note that the brain activity does not actually consist of discrete sets of physical current dipoles, but rather than the dipole is a convenient representation for coherent activation of a large number of pyramidal cells, possibly extending over a few square centimeters of gray matter. The current dipole model is the key of EEG processing since a primary current source of arbitrary extent can always be broken down to small regions, each region represented by an equivalent current dipole.

However, an identifiability problem can arise when too many small regions and their dipoles are required to represent a single large region of coherent activation. These sources may be more simply represented by a multipolar model. The multipolar models can be generated by performing a Taylor series expansion of the function $G(\boldsymbol{x}, \boldsymbol{x}_s) = (\boldsymbol{x} - \boldsymbol{x}_s)/\|\boldsymbol{x} - \boldsymbol{x}_s\|^3$ about the centroid of the source. Successive terms in the expansion give rise to the multipolar components: dipole, quadrupole, octupole, and so on. In the present work the current dipole model is adopted.

Spherical Head Model

The computation of the scalp potentials requires solution of the forward equations (5.12), respectively, for a particular source model. When the surface integrals are computed over realistic head shapes, these equations must be solved numerically. Analytic solutions exist, however, for simplified geometries, such as when the head is assumed to consist of a set of nested concentric homogeneous spherical shells representing brain, skull, and scalp. These models are routinely used in most clinical and research applications to EEG source localization. Consider the special case of a current dipole with a moment $\boldsymbol{s}$ located at $\boldsymbol{x}_s$ in a multishell spherical head, the scalp potential $\boldsymbol{z}(\boldsymbol{x})$ measured at location $\boldsymbol{x}$ is

$$\boldsymbol{z}(\boldsymbol{x}) \approx a(\sigma_i) \frac{\boldsymbol{x} \times \boldsymbol{x}_s}{\|\boldsymbol{x} - \boldsymbol{x}_s\|^3} s. \tag{5.16}$$

Assuming that the electrical activity of the brain can be modeled by a number of dipoles, i.e. the measured multichannel EEG signal signals $\boldsymbol{z}_k \in \mathbb{R}^{n_z}$ from n_z sensors at time k are produced by M dipoles, the forward EEG model is given by

$$\boldsymbol{z}_k = \sum_{m=1}^{M} L_m(\boldsymbol{x}_k(m)) \boldsymbol{s}_k(m) + \boldsymbol{v}_k, \tag{5.17}$$

where $\boldsymbol{x}_k(m)$ is a three dimensional localization vector (space directions), $\boldsymbol{L}_m(\boldsymbol{x}_k(m)) \in \mathbb{R}^{n_z \times 3}$ is the lead field matrix for dipole m, $\boldsymbol{s}_k(m)$ is a three dimensional moment vector of the mth dipole of the source signal. By $\boldsymbol{v}_k$ the effect of noise in the measurements is simulated; $\boldsymbol{L}_m(\boldsymbol{x}_k(m))$ is a nonlinear function of the dipole localization, electrodes positions and head geometry, [25]. Its three columns contain the activity that will be measured in the sensors due to a dipole source with unity moment in the x, y, and z directions, respectively, and a zero moment in the other directions. An analytical expression for the forward model exists if the dipole localization, electrodes positions and the head geometry are known. The spherical head model is the simplification that preserves some important electrical characteristics of the head, while reducing the mathematical complexity of the problem. The different electric conductivities of the several layers between the brain and the measuring surface need to be known. The skull is typically assumed to be more resistive than the brain and scalp that, in turn, have similar conductivity properties, [26].

For the dipole source localization problem, the states that have to be estimated are the geometrical positions of M dipoles

$$\boldsymbol{x}_k = \left[\boldsymbol{x}_1^T(1), \ldots, \boldsymbol{x}_k^T(M)\right], \text{ where } \boldsymbol{x}_k(m) = [x(m), y(m), z(m)]^T, \text{ for } \quad m = 1, \ldots, M. \tag{5.18}$$

Then the lead field matrix of M dipoles $\boldsymbol{L}_m(x_k) \in \mathbb{R}^{n_z \times 3M}$ is

$$\boldsymbol{L}_m(x_k) = [\boldsymbol{L}(\boldsymbol{x}_k(1)), \ldots, \boldsymbol{L}(\boldsymbol{x}_k(M))]. \tag{5.19}$$

The vector of moments $\boldsymbol{s}_k \in \mathbb{R}^{3M \times 1}$ is $\boldsymbol{s}_k = [s_k(1), \ldots, s_k(M)]^T$, where each vector $\boldsymbol{s}_k(m)$ consists of the brain source signals in each space direction, $\boldsymbol{s}_k(m) = [s_x(m), s_y(m), s_z(m)]^T$. Equation (5.17) can be reformulated in a matrix form as follows

$$z_k = \boldsymbol{L}(\boldsymbol{x}_k)\boldsymbol{s}_k + \boldsymbol{v}_k. \tag{5.20}$$

Expression (5.20) corresponds to the measurement equation (5.2) of the general transition model. As for the state equation (5.1), since it is unknown how the states (the geometrical positions of M dipoles) evolve over time, a random walk model (first-order Markov chain) is assumed in the source localization space,

$$\boldsymbol{x}_k = \boldsymbol{x}_{k-1} + \boldsymbol{w}_k. \tag{5.21}$$

Equations (5.20)-(5.21) define the dipole source localization model in state space. The PF approach gives the estimates of the 3D location vector $\boldsymbol{x}$ of the M dipoles (assuming number M is known) that originated the EEG measurements z_k. In the above model certain distributions for the process and the measurement noises are assumed and initial values for the states are chosen. The lead field matrix can then be computed. However the moments $\boldsymbol{s}_k(m)$ are unknown. In order to estimate them the beamforming approach is used.

Realistic Head Models

The EEG forward model, equation (5.17), has a closed-form solution for heads with conductivity profiles that can be modeled as a set of nested concentric homogeneous and isotropic spheres. However, in reality, the human heads are anisotropic, inhomogeneous, and not spherical. Though the spherical models work reasonably well, more accurate solutions to the forward problem use anatomical information obtained from high-resolution volumetric brain images obtained with Magnetic Resonance Images (MRI) or X-ray Computed Tomography (CT) imaging.

Many automated and semi-automated methods exist for surface extraction from MRIs. The surfaces can then be included in a boundary element method (BEM) of the forward fields. While this is an improvement on the spherical model, the BEM calculations are very time consuming and the use of a realistic head model may appear impractical when incorporated as part of an iterative inverse solution. In the present work, the spherical head model approach is chosen. The described models are embedded into the PF for EEG localization and jointly with the beamforming step (detailed in Section 5.4) are presented in Table 5.1.

5.4 Beamforming as a Spatial Filter

Beamforming deals with the estimation of the time patterns in the three space directions of the mth current dipole $\boldsymbol{s}_k(m) = [s_x(m), s_y(m), s_z(m)]^T$. This is located at $\boldsymbol{x}_k(m) = [x(m), y(m), z(m)]^T$. The measurements of electrical potentials on the scalp are recorded from N sensors. These are located at the head surface. The beamformer filter consists of weight coefficients, forming the weighting matrix $\boldsymbol{B} \in \mathbb{R}^{n_z \times 3M}$, that when multiplied by the electrode measurements gives an estimate of the dipole moment at time k:

$$\boldsymbol{s}_k = \boldsymbol{B}^T \boldsymbol{z}_k. \tag{5.22}$$

The choice of the beamformer weights is based on the statistics of the signal vector $\boldsymbol{z}_k$ received at the electrodes. The objective is to optimize the beamformer response with respect to a prescribed criterion, so that the output $\boldsymbol{s}_k$ contains minimum noises and interferences. There is a number of criteria for choosing the optimum weights. The method described below represents a linear transformation where the transformation matrix is designed according to the solution of a constrained optimization problem (the early work is attributed to [27].

The basic approach consists in the following. Assuming that the desired signal and its direction are both unknown, accurate signal estimation can be provided by minimizing the output signal variance. To ensure that the desired signal is passed with a specific (unit) gain, a constraint may be used so that the response of the beamformer to the desired signal is:

$$\boldsymbol{B}^T \boldsymbol{L}(\boldsymbol{x}_k) = \boldsymbol{I}, \tag{5.23}$$

Table 5.1. Sequential Monte Carlo Algorithm for EEG source localization

Initialization

I. $k = 0$ for $\ell = 1, 2, \ldots, N$

* Generate N samples according to a chosen distribution $\boldsymbol{x}_0^{(\ell)} \sim p(\boldsymbol{x}_0)$ around the initial vector $\boldsymbol{x}_0$ generated in the range $[min(\boldsymbol{D}), max(\boldsymbol{D})]$, where $\boldsymbol{D}$ is the vector containing the dipole coordinates

* Set initial weights $W_0^{(\ell)=1/N}$ (equal importance to all samples)

II. For $k = 1, 2, \ldots$

Prediction step

* For $\ell = 1, 2, \ldots, N$ compute the state prediction according to the random walk state equation (5.21), with system dynamics noise $\boldsymbol{w}_k \sim \mathcal{N}(0, \boldsymbol{Q})$ assumed to be a Gaussian process, with $E[\boldsymbol{w}_k \boldsymbol{w}_{k+j}^T] = 0$ for $j \neq 0$.

* The covariance matrix of $\boldsymbol{w}_k$ is $\boldsymbol{Q} = \sigma_w^2 I$. The value of σ_w is chosen as a percentage (0-50%) from the previously estimated state vector $\boldsymbol{x}_{k-1}$.

Beamforming step

i) compute the transfer function $\boldsymbol{L}(\boldsymbol{x}_k)$

ii) apply the BF technique to define the spatial filter using equation (5.25)

iii) compute the amplitudes at time k of the source signal propagated in 3 directions, for all estimated sources.

Measurement update: evaluate the importance weights

* For $\ell = 1, 2, \ldots, N$, on the receipt of a new measurement, compute the output according to the measurement equation (5.20) and compute the weights $W_k^{(\ell)} = W_{k-1}^{(\ell)} p(z_k | \boldsymbol{x}_k^\ell)$.

* The likelihood is calculated as $p(z_k | \boldsymbol{x}_k^\ell) \sim \mathcal{N}(h(\boldsymbol{x}_k^{(\ell)}), \sigma_v \boldsymbol{I})$.

$p(z_k | \boldsymbol{x}_k^\ell) = exp[-0.5 * (z_k - \hat{z}_k) \boldsymbol{R}^{-1} (z_k - \hat{z}_k)^T]$, where $\hat{z}_k$ is the predicted measurement and $\boldsymbol{R}_z = cov(z_k) = E[z_k z_k^T]$ is the measurement error covariance matrix.

* Normalize the weights, for $\ell = 1, \ldots, N$, and obtain $W_k^{(\ell)} = W_k^{(\ell)} / \Sigma_{\ell=1}^N W_k^{(\ell)}$.

Output

* Calculate the posterior mean $E[\boldsymbol{x}_k | z_{1:k}]$ as $\hat{\boldsymbol{x}}_k = E[\boldsymbol{x}_k | z_{1:k}] = \Sigma_{\ell=1}^N W_k^{(\ell)} \boldsymbol{x}_k^{(\ell)}$

* **Compute the effective sample size**

$N_{eff} = \frac{1}{\Sigma_{\ell=1}^N}$

* **Selection step (resampling)** if $N_{eff} < N_{thresh}$

* Multiply/ suppress samples $\boldsymbol{x}_k^{(\ell)}$ with high/low importance weights $W_k^{(\ell)}$, in order to obtain N new random samples approximately distributed according to the posterior state distribution. The residual resampling algorithm [22] is applied.

* Finally weights are reset.

For $\ell = 1, 2, \ldots, N$ set $W_k^{(\ell)} = 1/N$.

where $\boldsymbol{I}$ denotes the identity matrix. Minimization of contributions to the output by interference is accomplished by choosing weights to minimize the variance of the filter output:

$$Var\{\boldsymbol{y}_k\} = tr\{\boldsymbol{B}^T \boldsymbol{R}_z \boldsymbol{B}\}, \tag{5.24}$$

where, $tr\{\}$ is the trace of the matrix in brackets and $\boldsymbol{R}_z$ is the covariance matrix of the EEG signals. In practice, $\boldsymbol{R}_z$ is estimated from the EEG signals during a given time window. Therefore, the filter is derived by minimizing the output variance subject to the constraint defined in (5.24). This constraint ensures that the desired signal is passed with unit gain. Finally, the optimal solution can be derived by constrained minimization using Lagrange multipliers [28] and it can be expressed as:

$$\boldsymbol{B}^{opt} = \boldsymbol{R}_z^{-1} \boldsymbol{L}^T(\boldsymbol{x}_k) \left(\boldsymbol{L}^T(\boldsymbol{x}_k) \boldsymbol{R}_{z_k}^{-1} \boldsymbol{L}^T(\boldsymbol{x}_k)\right)^{-1}. \tag{5.25}$$

Such beamformer is often called the Linearly Constrained Minimum Variance (LCMV) beamformer. The LCMV provides not only an estimate of the source activity, but also its orientation and the reason why it is classified as vector beamforming. The differences and similarities among beamformers based on this criterion for the choo ice of the optimum weights are discussed in [28].

5.5 Experimental Results

EEG Simulation

Simulated EEG signals were generated by using equation (5.17) at 30 scalp locations (Fp1, AF3, F7, F3, FC1, FC5, C3, CP1, CP5, P7, P3, Pz, PO3, O1, Oz, O2, PO4, P4, P8, CP6, CP2, C4, FC6, FC2, F4, F8, AF4, Fp2, Fz, Cz) covering the entire hemisphere according to the standard 10/20 International system (Figure 5.2). In this study, a 3-sphere model describes the head geometry which includes three concentric layers for the brain, skull and scalp. The radii of the three concentric spheres are 8.7, 9.2 and 10 cm respectively and the corresponding conductivity values are 0.33, 0.0165 and 0.33 S/m. The origin of the reference coordinate system is located in the centre of the spheres with the x-axis pointing from right to left, the y-axis pointing back to front and the z-axis pointing bottom to top.

The solution of the EEG source localization problem requires a significant number of forward model evaluations. The proposed algorithm can require its evaluation at thousands of different source locations. In spite of their simplicity and ease of computation, the multilayer spherical model is computed off-line by applying a grid based method to generate dipoles, assuming that the state space is discrete and consists of a finite number of states (dipoles). Two approaches were implemented for generating the grid of dipoles (Table 5.2):

- The dipoles' positions ($\boldsymbol{D}$) are randomly generated with a uniform distribution inside a spherical head model, (with radius R= 10 cm)
  $\boldsymbol{D} = \{d_{ij} : \lfloor x_{ij} \in [-9,9]cm, y_{ij} \in [-9,9]cm, z_{ij} \in [9,9]cm \rfloor\}$, where $\lfloor\ \rfloor$ denotes the rounding operation;

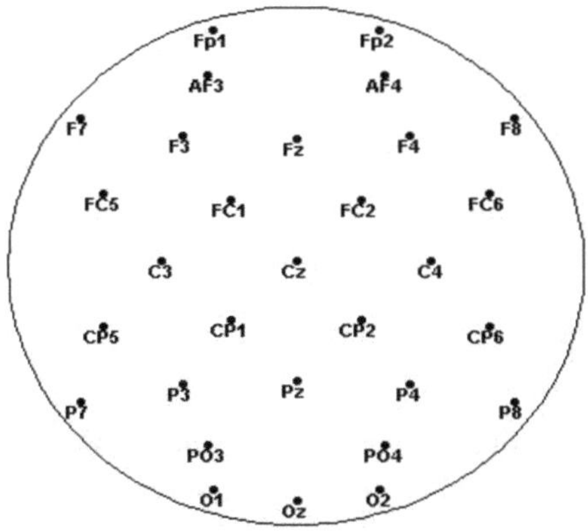

Fig. 5.2. EEG electrode location according to the 10/20 International system

Table 5.2. Grid of dipoles

Number of grid dipoles
512 (8^3): linear grid
125: random uniform distribution
25: random uniform distribution

- The dipoles are linearly distributed inside a cube with limits minimum -9cm and maximum 9cm.

The results show that the random uniform grid distribution is more suitable for the SMC estimation framework than the linear grid. Based on the information for the localization of the grid of brain dipoles and the localization of the surface scalp electrodes the Leadfield matrix $\boldsymbol{L}(\boldsymbol{x})$ is obtained. The ground true for the SMC algorithm is generated by randomly choosing coordinates of two uncorrelated dipoles $\boldsymbol{d}_{true} = [x_1, y_1, z_1, x_2, y_2, z_2]^T$ inside the spherical head model (R = 10 cm). Most probably they will not coincide with any of the grid dipoles and therefore they are substituted by the nearest grid dipole location. Sinusoidal wave-forms are simulated as the brain signals originated at these dipoles in three space directions,

$$\boldsymbol{Y}_{true} \in \mathbb{R}^{(2\times3)\times t} : \; [source_i(x,y,z) = a_i(x,y,z) sin(2\pi f_i(x,y,z)t)], \quad i = 1,2, \quad (5.26)$$

where t is the time over which the EEG signal is recorded.

Correlation Sources

The main assumption in the Beamformer analysis is that the activity of any neural source is not linearly correlated to any other source. In order to guarantee that the

EEG data is generated only by uncorrelated brain sources, the following correlation test is performed. The covariance matrices of the dipole signals in each direction (x, y, z) are obtained as for example

$$\boldsymbol{P}_x = \boldsymbol{source}(\boldsymbol{x})\boldsymbol{source}(\boldsymbol{x})^T \in \mathbb{R}^{2\times 2}, \tag{5.27}$$

where $\boldsymbol{source}(\boldsymbol{x}) \in \mathbb{R}^{2\times t}$ is the matrix with the signal propagations of all dipoles in x direction. The correlation factor is a matrix $\boldsymbol{C}_x \in \mathbb{R}^{2\times 2}$ with elements

$$\boldsymbol{C}_x(i,j) = \frac{P_x(i,j)}{\sqrt{P_x(i,i)}\sqrt{P_x(j,j)}}, \quad i \neq j. \tag{5.28}$$

If one element of $\boldsymbol{C}_x$ has a small value, for example less than 0.05, then the correlation between the two signals in direction x is significant. In order to facilitate the analysis of the SMC method in the experiments provided, propagation only in one direction is assumed for each dipole signal

$$\boldsymbol{Y}_{true} = [0, 0, a\sin(2\pi \mathrm{f}_1 \mathrm{t}, 0, 0, \mathrm{a}\sin(2\pi \mathrm{f}_2 \mathrm{t}))], \tag{5.29}$$

where an amplitude $a = 0.1$ and frequencies $f_1 = 100Hz$, $f_2 = 15Hz$ were chosen. It is assumed that the EEG recording system operates with 1 kHz sampling rate. Then over a duration of 1 s are recorded 1000 samples. Hence, the EEG signals are computed assuming that the measured brain electrical activity is mainly due to the activity initiated at these two dipoles. The forward model (5.16) takes the following form:

$$\boldsymbol{z} = \boldsymbol{L}(\boldsymbol{d}_{true})\boldsymbol{Y}_{true} + \boldsymbol{v}, \tag{5.30}$$

where the noise $\boldsymbol{v}$ is assumed to be Gaussian, with the following parameters:

$$\boldsymbol{v} \sim \mathcal{N}(0, \sigma_v^2 \boldsymbol{I}), \sigma_v = \frac{Var\{\boldsymbol{L}(\boldsymbol{d}_{true})\boldsymbol{Y}_{true}\}}{SNR}, \quad \boldsymbol{L}(\boldsymbol{d}_{true}) \in \mathbb{R}^{30\times 3\times 2}, \tag{5.31}$$

where σ_v is the noise standard deviation.

The white noise v represents the influence of other brain sources and the effect of possible external sources not generated by brain activity (for example artifacts). The signal-to-noise ratio (SNR) is a parameter and experiments with various values were tested (-5, 0.75, 5, 10). The noise power was defined in such a way that the maximum SNR never exceeds 10.

SMC for EEG Source Estimation

The beamforming-based SMC algorithm for recursive source estimation is summarized below and its accuracy and robustness are discussed. The algorithm was inspired by related previous works [29, 30, 31].

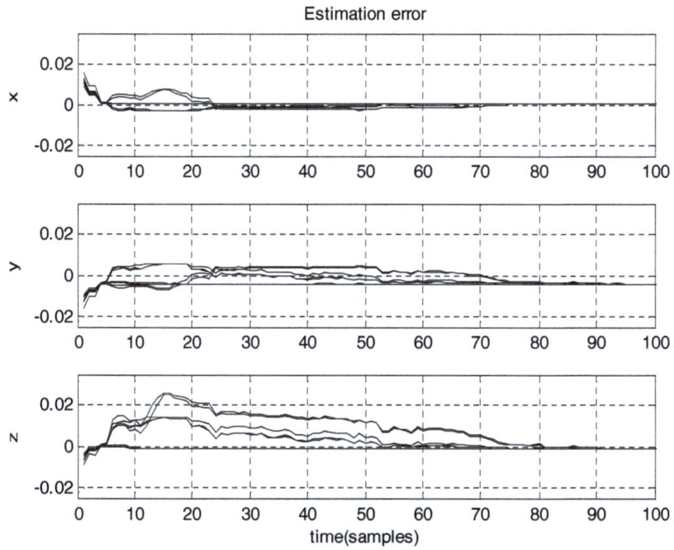

Fig. 5.3. The graphs show results with SNR=5dB, a random grid consisting of 125 dipoles and based on MC=5 runs.

Since the EEG signal has been originated by activity in two principal dipoles, the dimension of the state vector at each time k is $\boldsymbol{x}_k \in \mathbb{R}^{6\times 1}$ (three coordinates per dipole). The initial state vector $\boldsymbol{x}_0 = [x_{10}, y_{10}, z_{10}, x_{20}, y_{20}, z_{20}]^T$ is chosen randomly

$$\{d_{i0} : x_{i0} \in [-9,9], y_{i0} \in [-9,9], z_{i0} \in [9,9]cm, i = 1,2\}. \tag{5.32}$$

In all the tests included in this work the number of initially generated particles is $N = 500$. The complete algorithm is evaluated over 5 Monte Carlo runs (MC=5).

Figures 5.3 and 5.4 present results from five sequential MC runs and randomly generated grid of 125 dipoles for one of the estimated dipoles. When the SNR=5 dB (Figure 5.3), the estimation error $\boldsymbol{d}_{true} - \boldsymbol{d}_{estimated}$ of all dipole coordinates (x,y,z) over the five MC runs nicely converge shortly after the first 80 samples. However, when SNR= -5 dB (Figure 5.4) for most of the MC runs the errors do not converge. The interpretation is that when the power of the EEG signal due to the main brain sources is higher than the sum effect of all other secondary sources (that may have influenced the scalp recording), it is possible to estimate correctly the location of the main dipoles.

Figure 5.5 shows the average prediction error over five sequential MC runs and linearly build grid of 512 dipoles for the coordinates of the two estimated dipoles. The convergence speed of the error is acceptable. However, the steady state error in most of the experiments is higher than in the experiment with the randomly generated grid (Figure 5.3). This result is possibly related with the stochastic nature of the SMC approach.

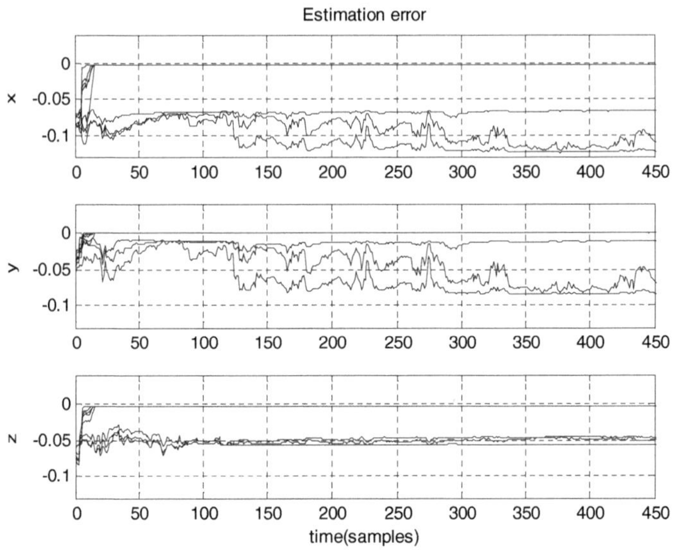

Fig. 5.4. The graphs show results with SNR=-5dB, a random grid consisting of 125 dipoles and based on MC=5 runs.

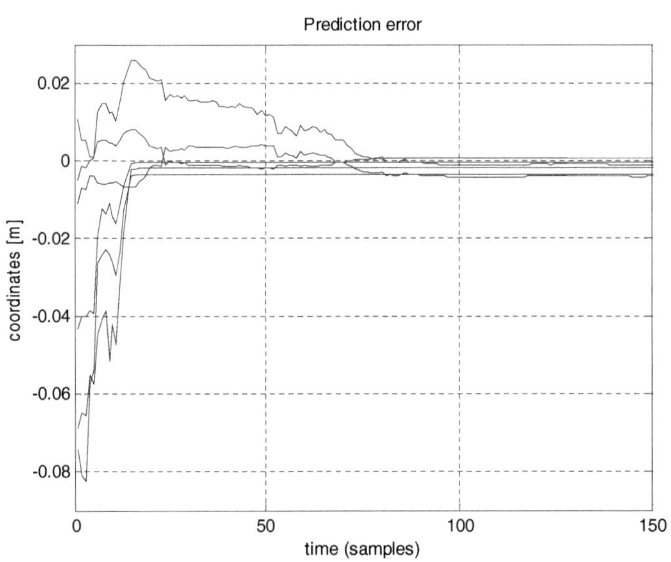

Fig. 5.5. The graphs show the average prediction error, for SNR=5dB, a linear grid consisting of 512 dipoles, based on MC=5 runs.

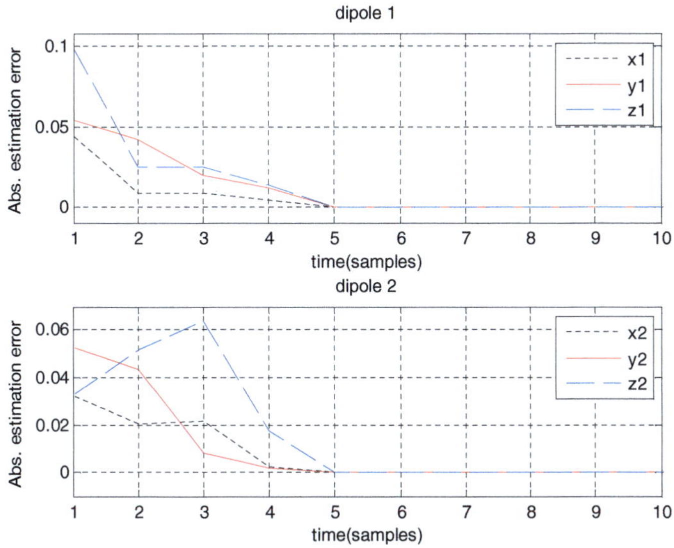

Fig. 5.6. The graphs show results for the absolute estimation error for the coordinates of two dipoles, with SNR=5 dB, a random grid consisting of 25 dipoles, (5 % perturbation of the random walk model) and based on MC = 5 runs.

Finally in Figures 5.6-5.7 the absolute estimation errors are shown with respect to the co-ordinates of two dipoles when SNR =5 dB and a random grid of 25 dipoles is generated. Different levels of the Gaussian noise $\boldsymbol{w}_k \sim \mathcal{N}(0, \sigma_w^2 \boldsymbol{I})$ are assumed in the random walk model. The standard deviation for the results shown in Figure 5.6 corresponds to 5% perturbation of the components of the state vector, (5.21), while for the results presented in Figure 5.7 the standard deviation of the system noise corresponds to 25% perturbation of the components of the state vector. Note that, for lower level of noise the convergence is fast (over five iterations) and there is practically no steady state error (as obvious from Figure 5.6), while for stronger perturbations the estimation errors fluctuate (Figure 5.7). In order to improve the visibility on Figure 5.7 a) and b) and also on Figure 5.7 c) and d) are depicted the same curves for dipole 1 and dipole 2, respectively over different iterations.

The reconstruction of the two leading brain activity sources in Figure 5.6 seems perfect. However, this is not a very realistic situation because it restricts the brain space to only 25 inner sources. For low dimensional grid the distance between the "allowed" zones is higher (a coarse space grid) and in the lack of perturbations the estimation will converge quickly. For higher dimensional brain space grid the probability to associate the major brain activity with wrong but still closely located sources is higher. Therefore, for higher dimensional grid small steady state estimation errors can be expected.

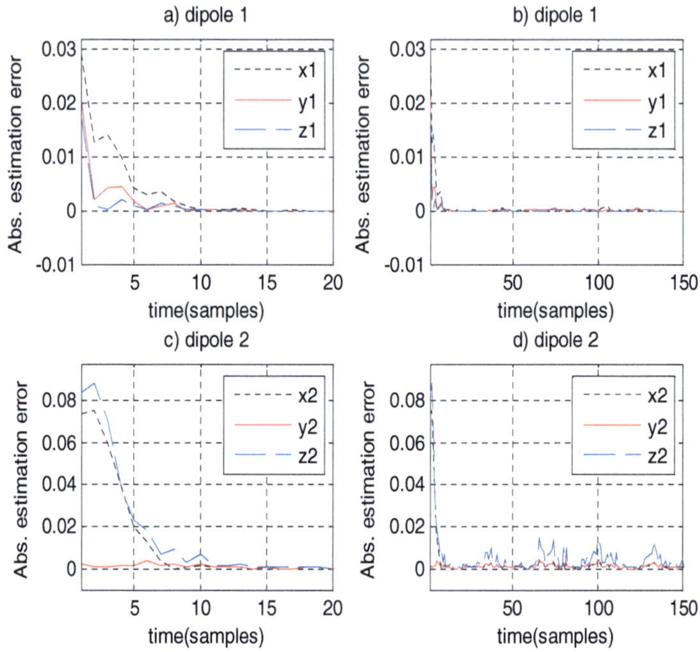

Fig. 5.7. The graphs show results for the absolute estimation error for the coordinates of two dipoles, with SNR=5 dB, a random grid consisting of 25 dipoles, (25 % perturbation of the random walk model) and based on MC=5 runs.

5.6 Conclusions

In this chapter the problem of EEG source estimation is formulated in a Bayesian framework by applying the SMC approach. The localization of two principle zones is performed by uncorrelated dipoles. The location coordinates are calculated by starting with many randomly chosen initial guesses. These correspond to the particles in the designed particle filter. Even for very wrong initial values of the location coordinates, the estimation process converges to the correct values.

Current research is focused on increasing the number of the estimated dipoles (more than two), however due to the restriction of the beamformer to reconstruct only uncorrelated sources, the results so far are not conclusive. It may appear necessary to use a different BSS technique to recover the source waveform. A major drawback of the SMC is that it can become prohibitively expensive when a large number of particles is used or the state vector has a very high dimension. For example, for localization of multiple (hundreds) of brain sources, the computational complexity can be reduced by a Rao-Blackwellisation procedure.

Our next research objective is to implement the SMC with real EEG data, in particular, for Visually Evoked Potentials (VEPs). In order to validate the success of any brain source estimation method we need to know the ground truth and it is not always possible. However, the VEPs are well studied, it is known that the visual cortex zone is the most active zone during visual stimulus and that the EEG channels around the visual cortex (e.g. the occipital channels) exhibits the largest peaks. We expect the proposed technique to converge to dipole coordinates corresponding to the visual cortex area. The combination of the SMC method with other brain imaging methods as MRI or functional MRI can increase the estimation accuracy.

Acknowledgments. This work was supported by the Portuguese Foundation for Science and Technology under the grant SFRH/BSAB/1092/2010 and the Institute of Electronic Engineering and Telematics of Aveiro (IEETA), Portugal.

References

1. Michel, C.M., Murray, M.M., Lantz, G., Gonzalez, S., Spinelli, L., Grave De Peralta, R.: EEG source imaging. Clinical Neurophysiology 115(10), 2195–2222 (2004)
2. Gross, J., Ioannides, A.A.: Linear transformation of data space in MEG. Physics in Medicine and Biology 44(1), 87–103 (1999)
3. Van Veen, B.D., Van Drongelen, W., Yuchtman, M., Suzuki, A.: Localization of brain electrical activity via linearly constrained minimum variance spatial filtering. IEEE Transactions on Biomedical Engineering 44(9), 867–880 (1997)
4. Cichocki, A., Amari, S.-I.: Adaptive Blind Signal and Image Processing: Learning Algorithms and Applications. John Wiley & Sons, Inc., New York (2002)
5. Hyvärinen, A., Karhunen, J., Oja, E.: Independent Component Analysis. Wiley-Interscience (2001)
6. Sanei, S., Chambers, J.A.: EEG Signal Processing. Wiley-Interscience (September 2007)
7. Choi, S., Cichocki, A., Park, H.-M., Lee, S.-Y.: Blind Source Separation and Independent Component Analysis: A Review. Neural Information Processing - Letters and Reviews 6(1), 1–57 (2005)
8. Grosse-Wentrup, M., Liefhold, C., Gramann, K., Buss, M.: Beamforming in noninvasive brain-computer interfaces. IEEE Transactions on Biomedical Engineering 56(4), 1209–1219 (2009)
9. Kamousi, B., Liu, Z., He, B.: An EEG inverse solution based brain-computer interface. The International Journal of Bioelectromagnetism 7(2), 292–294 (2005)
10. Noirhomme, Q., Kitney, R.I., Macq, B.: Single-trial EEG source reconstruction for brain-computer interface. IEEE Transactions on Biomedical Engineering 55(5), 1592–1601 (2008)
11. Qin, L., Ding, L., He, B.: Motor imagery classification by means of source analysis for brain computer interface applications. Journal of Neural Engineering 1(3), 133–141 (2004)
12. Arulampalam, M., Maskell, S., Gordon, N., Clapp, T.: A tutorial on particle filters for online nonlinear/non-Gaussian Bayesian tracking. IEEE Trans. on Signal Proc. 50(2), 174–188 (2002)
13. Kalman, R.: A new approach to linear filtering and prediction problem. Trans. ASME, Ser. D, J. Basic Eng. 82, 34–45 (1960)
14. Jaswinski, A.: Stochastic Processes and Filtering Theory. Academic Press, New York (1970)
15. Alspach, D.L.: Gaussian sum approximations in nonlinear filtering and control. Information Sciences 7, 271–290 (1974)

16. Isard, M., Blake, A.: Contour tracking by stochastic propagation of conditional density. In: European Conf. on Comp. Vision, Cambridge, UK, pp. 343–356 (1996)
17. Isard, M., Blake, A.: Condensation – conditional density propagation for visual tracking. International Journal of Computer Vision 28(1), 5–28 (1998)
18. Gordon, N., Salmond, D., Smith, A.: A novel approach to nonlinear / non-Gaussian Bayesian state estimation. IEE Proceedings-F 140, 107–113 (1993)
19. Gustafsson, F.: Particle filter theory and practice with positioning applications. IEEE Transactions on Aerospace and Electronics Systems Magazine Part II: Tutorials 25(7), 53–82 (2010)
20. Cappé, O., Godsill, S., Mouline, E.: An overview of existing methods and recent advances in sequential Monte Carlo. IEEE Proceedings 95(5), 899–924 (2007)
21. Mihaylova, L., Angelova, D., Honary, S., Bull, D.R., Canagarajah, C.N., Ristic, B.: Mobility tracking in cellular networks using particle filtering. IEEE Transactions on Wireless Communications 6(10), 3589–3599 (2007)
22. Liu, J., Chen, R.: Sequential Monte Carlo methods for dynamic systems. Journal of the American Statistical Association 93(443), 1032–1044 (1998)
23. Djurić, P.M., Kotecha, J.H., Zhang, J., Huang, Y., Ghirmai, T., Bugallo, M., Míguez, J.: Applications of particle filtering to selected problems in communications. IEEE Signal Processing Magazine 20(5), 19–38 (2003)
24. Baillet, S., Mosher, J.C., Leahy, R.M.: Electromagnetic brain mapping. IEEE Signal Processing Magazine 18(6), 14–30 (2001)
25. Salu, Y., Cohen, L.G., Rose, D., Sxato, S., Kufta, C., Hallett, M.: An improved method for localizing electric brain dipoles. IEEE Transactions on Biomedical Engineering 37(7), 699–705 (1990)
26. Lai, Y., van Drongelen, W., Ding, L., Hecox, K.E., Towle, V.L., Frim, D.M., He, B.: Estimation of in vivo human brain-to-skull conductivity ratio from simultaneous extra- and intra-cranial electrical potential recordings. Clinical Neurophysiology 116(2), 456–465 (2005)
27. Capon, J.: High-resolution frequency-wavenumber spectrum analysis. Proceedings of the IEEE 57(8), 1408–1418 (1969)
28. Huang, M.-X., Shih, J.J., Lee, R.R., Harrington, D.L., Thoma, R.J., Weisend, M.P., Hanlon, F., Paulson, K.M., Li, T., Martin, K., Miller, G.A., Canive, J.M.: Commonalities and differences among vectorized beamformers in electromagnetic source imaging. Brain Topography 16, 139–158 (2004)
29. Mohseni, H.R., Nazarpour, K., Wilding, E.L., Sanei, S.: The application of particle filters in single trial event-related potential estimation. Physiological Measurement 30(10), 1101–1116 (2009)
30. Anteli, J.M., Mingue, J.: Eeg source localization based on dynamic Bayesian estimation techniques. International Journal of Bioelectromagnetism 11(4), 179–184 (2009)
31. Mohseni, H.-R., Ghaderi, F., Wilding, E.E., Sanei, S.: A beamforming particle filter for EEG dipole source localization (4), 337–340 (2009)

Chapter 6
Computational Intelligence in Automotive Applications

Yifei Wang, Naim Dahnoun, and Alin Achim

Abstract. In this chapter, we discuss one of the most popular machine vision applications in the automotive industry: lane detection and tracking. Model-based lane detection algorithms can be separated into lane modeling, feature extraction and model parameter estimation. Each of these steps is discussed in detail with examples and results. A recently proposed lane feature extraction approach, which is called the Global Lane Feature Refinement Algorithm (GLFRA), is also introduced. It provides a generalized framework to significantly improve various types of gradient-based lane feature maps by utilizing the global shape information and subsequently improves the parameter estimation and the tracking performance. Another important aspect of this application lies in the tracking stage. We compare the performances of three different types of particle filters (the sampling importance resampling particle filter, the Gaussian particle filter and the Gaussian sum particles filter) quantitatively and provide insightful result analysis and suggestions. Furthermore, the influence of feature maps on the tracking performance is also investigated.

6.1 Introduction

Throughout the last two decades, a significant amount of research has been conducted in the area of road/lane analysis. This chapter's main focus is on the image-based boundary detection and tracking of structured lanes.

6.1.1 Lane Detection

For the image-based lane detection algorithms, most of the existing solutions can be divided into two categories: feature-based and model-based methods. The

Yifei Wang · Naim Dahnoun · Alin Achim
Visual Information Lab, Department of Electrical & Electronic Engineering,
University of Bristol, Merchant Venturers Building,
Woodland Road, Bristol, BS8 1UB, UK
e-mail: `{Yifei.Wang,Naim.Dahnoun,Alin.Achim}@bristol.ac.uk`

P. Georgieva et al. (Eds.): Adv. in Intelligent Signal Processing and Data Mining, SCI 410, pp. 139–174.
springerlink.com

feature-based methods locate the road areas using segmentation whereas the model-based methods represent the lane boundaries by mathematical models. The use of the two different methods is mainly determined by the road condition and the applications. Feature-based methods extract and analyse local lane features in order to separate the lane from the background pixel by pixel. Many feature-based algorithms are able to detect unmarked or unstructured roads based on the segmentation-based approaches [34]. Unlike the feature-based methods, model-based methods normally incorporate various constraints (such as a flat road or parallel road boundaries) during the parameter estimation stage to minimize the error and provide a simple description of the lane with mathematical models such as parabolas [14, 35, 18] or splines [30, 32]. Existing parameter estimation algorithms include, among others, the Hough transform [18], active contour [32, 12] and maximum a posterior estimation [14, 35]. This chapter focuses on the model-based approach since we are only concerned with the detection and tracking of structured road boundaries.

A typical model-based lane detection algorithm can be separated into lane modelling, feature extraction and the model parameter estimation. Lane modelling is performed off-line and determines the mathematical model to represent the lane boundaries. Models vary from restrictive straight lines to flexible splines. It is important to know that flexible models require more parameters. Adding extra parameters means creating extra dimensions while estimating their values. Therefore, in order to pick a suitable model, we have to be very clear about the available system resources, the objectives and the applications of our systems. If the system is designed to issue lane departure warnings, the straight line model is normally sufficient since the goal is to find only the lateral offset of the vehicle. However, if the aim of the system is path planning for an autonomous vehicle, a more sophisticated model is required to enclose information in the far distance.

After the model is decided, we need to find the values of the model parameters that provide the best description of the road. This can be achieved with various model (line/curve) fitting algorithms such as the Hough transform and the optimization algorithms. However, before any parameter estimation, feature points that correspond to the lane boundaries need to be extracted. These points will be used as source data for the parameter estimation algorithms. A sophisticated estimation algorithm provides a high probability of finding the best matches to the extracted feature points. If the feature points are correctly matched to lanes, the estimation will likely be accurate. This indicates that the performance of the estimation algorithms is heavily dependant on the feature extraction procedure. On a system level, the output depends not only on the parameter estimation but also on the feature extraction step.

A feature map should include the most important information that defines the lane boundaries. This information includes intensity variations (gradient magnitudes), gradient orientations, directivity (adjacent pixels sharing similar gradient orientation), colours, textures, lane marking patterns and so on. Among all this information, the gradient is the most widely used. Gradient calculation is efficient. It provides not only the magnitudes and orientation information but most importantly, the lane boundaries are almost always defined with some degree of intensity change. Even if the lane is unmarked, the on-road and off-road materials have usually different reflectivity which results in different intensities under ambient light. However, edges are not merely produced by the lane boundaries on the images, but rather any object in the scene will introduce edges. In complicated scenarios, the degree of over detection could be intolerable. Later developments of feature extraction approaches are mainly focused on extracting the lane features based on more lane properties.

One problem with most existing feature extraction algorithms is that the global shape information of the lane geometry is not included [27]. In complicated situations where the road is cluttered with heavy shadows or other objects, image content in various locations share similar properties with the lane boundaries. If the feature extraction algorithm is only based on a local image area, unwanted feature points will be detected. These features will then be included during the parameter estimation and tracking stages that cause inaccuracies and errors.

In this chapter, we introduce a Global Lane Feature Refinement Algorithm (GLFRA) which readily solves the above problem. It significantly improves the feature extraction performance based on traditional methods and subsequently improves lane detection and tracking performance. This approach utilizes the global lane shape information and is able to extract feature points that are most likely to belong to lanes. Another important factor is that this algorithm is designed to iteratively refine the feature map obtained by most of the traditional extractors such as the edge operator, intensity-bump filter and steerable filter. We call all of the local feature extractor the local feature extraction algorithm. The only requirement for applying the GLFRA is that the lane features are able to be represented on the image space. The computational complexity of the algorithm can vary according to the chosen local feature extraction algorithm. If the local extractor is likely to include a large amount of noize, more iterations are required.

After feature extraction, parameter estimation algorithms are applied for optimization. In this chapter, a classical model-based parameter estimation algorithm, proposed in [14], is described. The performance of the algorithm is evaluated by comparing the results with the manually selected ground truth. The performance improvements based on GLFRA over the local feature extractors are also illustrated.

6.1.2 Lane Tracking

In general, after the lane boundaries are detected, lane tracking is the final building block of a complete system. This step is often omitted in many existing lane analysis systems and only few tracking algorithms have been tested for this application [28, 35, 19]. However, the tracking step is very important since it gathers valuable information from previous results and utilizes this information to find the lane parameters quickly and accurately.

The most popular tracking algorithms employ various types of Kalman filters and particle filters. These algorithms normally have lower computational complexity compared with the parameter estimation algorithms from the detection stage. The particle filter algorithm and its derivatives normally demand much higher computing power than Kalman filters. However, due to the rapidly growing parallel computing systems, the particle filters can also be implemented very efficiently. The particle filters are capable of operating under severe non-linear and non-Gaussian environments. They also have a higher ability to keep tracking accurately when encountering discontinuities. This chapter focuses on different types of particle filters since the functions included during lane tracking are highly non-linear and non-Gaussian.

Although the traditional Sampling Importance Resampling (SIR) particle filter has been applied for lane tracking purposes, the performance of derivatives of particle filters has not yet been exploited. Possibly, the later developments of particle filters are more suitable for the lane tracking applications. Also, one of the problems of applying the SIR particle filter lies in the parameter selection stage. Even though the posterior probability density of the parameters is available, it is still difficult to pick a single set of parameters to represent the actual lanes. In complicated environments, if the mean is chosen, the final results could be influenced by non-zero mean noize and become inaccurate. If the values corresponding to the highest probability are chosen, the results could be unstable or drift away completely. Furthermore, the resampling step of the SIR particle filter cannot be implemented efficiently in parallel, which limits the speed of the tracking step significantly in a hardware implementation.

Motivated by the above reasons, three types of particle filters are tested and compared in this chapter: SIR [9], Gaussian Particle Filter (GPF) [15], and Gaussian Sum Particle Filter (GSPF) [16]. Other types of particle filters, such as the recently developed alpha-stable particle filter [22], are beyond the scope of this chapter. Furthermore, the tracking results based on the GLFRA and the local extractor are compared.

Better tracking results are not only achieved by applying a suitable tracking algorithm, but also by incorporating a better feature map. A good feature map allows a higher detection rate and more accurate detection results. Superior detection results then lead to higher tracking quality. On the contrary, even with identical parameter initializations, the particle filter performance can be significantly enhanced if a more sophisticated feature map is chosen.

6.1.3 Chapter Structure

The rest of this chapter is organized as follows. Section 6.2 introduces a popular lane model based on parabolas. Section 6.3 presents the GLFRA followed by Section 6.4 which describes the lane parameter estimation. Section 6.5 describes the implementation of the three different particle filters for lane tracking. Section 6.6 illustrates the lane detection and tracking results and Section 6.7 concludes this chapter.

6.2 Lane Modelling

As mentioned in Section 6.1, the first step of a lane detection system is lane modelling. During the past years, various lane models have already been suggested, ranging from straight line segments [12] to flexible splines [32]. Simple models

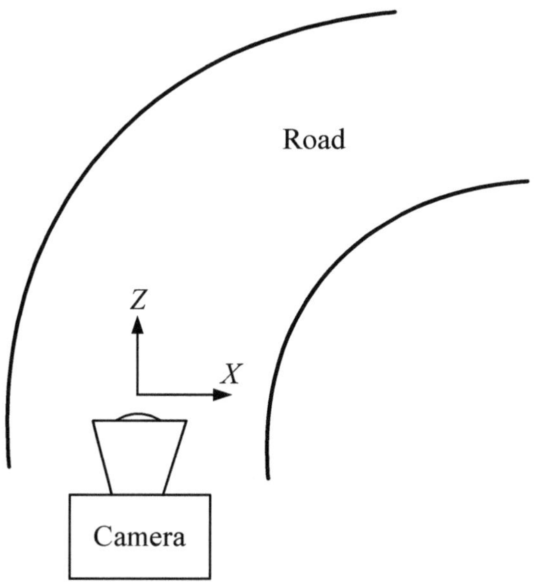

Fig. 6.1 Bird's eye view of the curved road and camera.

cannot represent the lane shapes accurately, especially in the far-field. On the other hand, sophisticated lane models result in much heavier computational costs and may increase the amount of detection error.

In this section, we introduce the classic parabola lane model. This 2D model is very widely used with certain degrees of flexibility. It is sufficient to represent the lane boundaries correctly in most highway scenarios. One of the most important attributes of the parabola model is simplicity. Only four parameters are needed to represent the left and right curves. The spline lane model is much more flexible but includes more parameters and is more suitable for urban environments. In this chapter, we only concentrate on the 2D models which assume a flat road plane. Detailed studies of 3D models can be found in [2, 8, 23].

In most highway scenarios, the lane curvatures are kept constant for very long distances [8]. It is reasonable to assume each of the lane boundaries is a circular arc on the road plane. In order to easily represent mathematically the arc shape, the parabola is used as an approximation as shown in Eq. (6.1).

$$X = AZ^2 + BZ + C \tag{6.1}$$

where X represents the horizontal position of the curve and Z denotes the distance from the image plane in pixel unit as shown in Fig. 6.1.

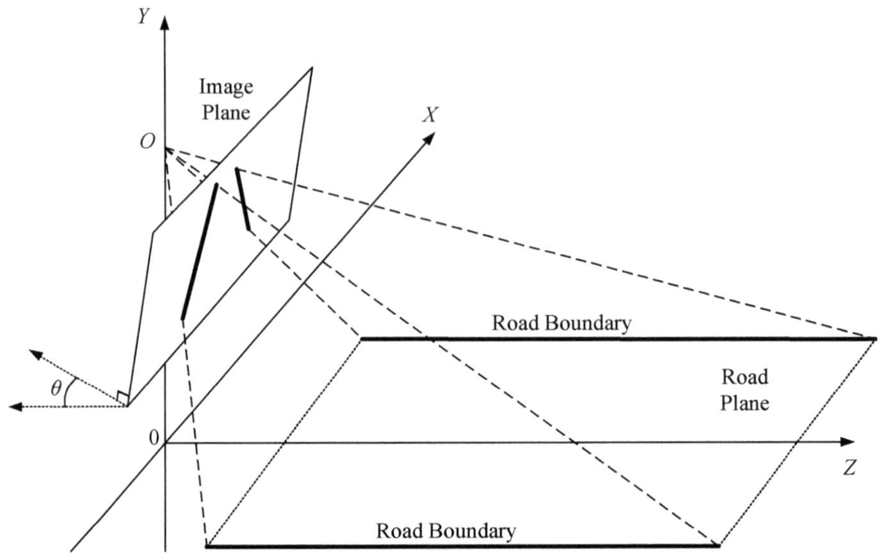

Fig. 6.2 Perspective projection of straight lane boundaries on to the image plane.

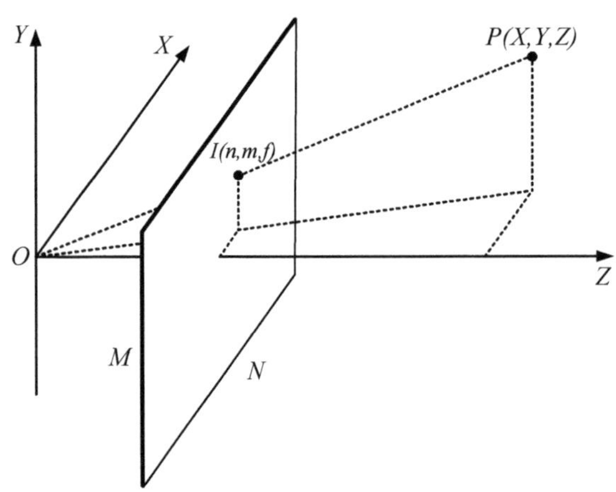

Fig. 6.3 Illustrates perspective projection of a point on to the image plane.

The immediate problem we face is that the parabola lane model does not represent the lane boundaries on the image plane due to the perspective projection effect as shown in Fig. 6.2 where O represents the optical centre and θ denotes the camera pitch angle. There are two ways of solving this problem. We can either transform the image to the bird's eye view using Inverse Perspective Mapping (IPM) with fixed camera parameters, or transform the parabola model to the image plane using perspective projection. The second method is certainly better since it saves a large

Fig. 6.4 Shows a forward pointing camera with no pitch angle above a flat slope surface. The origin is at the optical centre. This is equivalent to a tilted camera above a surface with no tilting angle.

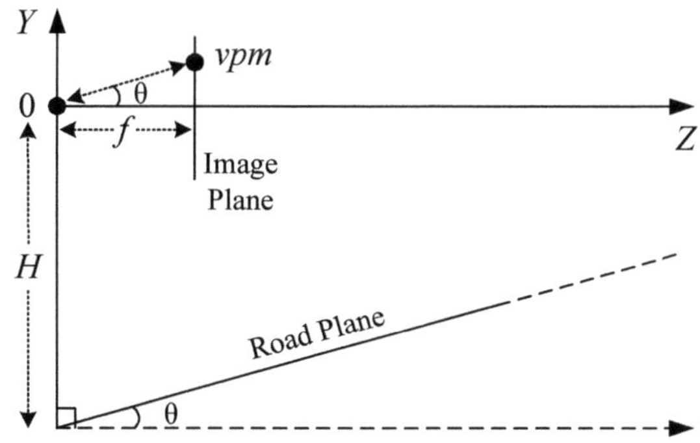

amount of calculation. Instead of transforming each frame in the video sequences, we only need to modify off-line the lane model to integrate the perspective projection effect. As illustrated in Fig. 6.3, a simple relationship of perspective projection can be derived based on trigonometry (shown in Eq. (6.2) and the origin is at the optical centre).

$$\frac{Y}{m} = \frac{X}{n} = \frac{Z}{f} \tag{6.2}$$

where f is the focal length. With this relationship, the parabola model on the road plane can be transformed to the image plane. In order to simplify the derivation, we set the origin at the optical centre and represent the camera pitch angle with a sloped road surface so that the optical-axis overlaps with the Z-axis (shown in Fig. 6.4). By doing so, the parabola model has been changed into:

$$X = A\left(\frac{Z}{\cos\theta}\right)^2 + B\left(\frac{Z}{\cos\theta}\right) + C \tag{6.3}$$

Since we want to represent the model on the image plane using n and m, X and Z need to be replaced. A direct relationship, which can be found in Eq. (6.2), is

$$n = \frac{Xf}{Z} \tag{6.4}$$

Substitute Eq. (6.3) into Eq. (6.4):

$$n = A\left(\frac{Zf}{\cos^2\theta}\right) + B\left(\frac{f}{\cos\theta}\right) + C\left(\frac{f}{Z}\right) \tag{6.5}$$

Now that X is replaced, we need to substitute Z as well. This depends on another relationship including the vanishing line position vpm and the camera height H. As shown in Fig. 6.4, under the flat road assumption, the line connecting O and vpm is parallel to the ground plane and subsequently we can see that

$$\frac{Z}{f} = \frac{H - Y}{vpm} \tag{6.6}$$

Using Eq. (6.6) and Eq. (6.2) to solve Y and substituting Y into Eq. (6.2) yields

$$Z = \frac{Hf}{m + vpm} \tag{6.7}$$

The transformed parabola model can be then represented as

$$n = \frac{AHf^2}{\cos^2\theta}\frac{1}{(m+vpm)} + \frac{Bf}{\cos\theta} + \frac{C}{H}(m+vpm) \tag{6.8}$$

Changing the origin from the optical centre to the image origin and combining the camera parameters with the parabola parameters yields

$$n = \frac{a}{m - vpm} + b(m - vpm) + c \tag{6.9}$$

This final image plane lane model for each lane boundary requires three parameters (vpm is not included since it can be calculated based on known camera parameters) and can be written as

$$LM_1(\mathbf{x}, m) = \frac{a_1}{m - vpm} + b_1(m - vpm) + c_1 \tag{6.10}$$

$$LM_2(\mathbf{x}, m) = \frac{a_2}{m - vpm} + b_2(m - vpm) + c_2 \tag{6.11}$$

where $\mathbf{x} = [a_1, a_2, b_1, b_2, c_1, c_2]$ is the array storing all the unknown parameters. $LM_1(\mathbf{x}, m)$ and $LM_2(\mathbf{x}, m)$ are defined as the models for the left and right lane boundaries. They also represent the n-axis positions of the sample points on image row m for the left and right lane model respectively.

At this point, the lane model perspective mapping is finished and we want to focus on minimising the number of parameters. The most significant correlation between the left and the right lane boundaries is the parallelism. If we want to restrict the two parabolas on the ground plane to be parallel, the parameter A (curvature) and B (gradient) for both boundaries need to be identical. In order to reflect this on the projected lane model, both parameters a_1 and a_2 are set to be a and both parameters c_1 and c_2 are set to be c. Therefore, the above equations can be represented as

$$LM_1(\mathbf{x}, m) = \frac{a}{m - vpm} + b_1(m - vpm) + c \tag{6.12}$$

$$LM_2(\mathbf{x}, m) = \frac{a}{m - vpm} + b_2(m - vpm) + c \tag{6.13}$$

As a result, the six parameters in the model are reduced to four. This allows for more prior information to be included in the model while minimising the computational complexity during the parameter estimation stage. However, the downside of this restriction is obviously that the model lacks the ability to represent non-parallel lane boundaries accurately. Nevertheless, this is rarely the case. The dynamic range of the parameter a which controls the lane curvature normally varies from -1500 to 1500. b_1 and b_2 correspond to the horizontal positions (this controls the positions of the intersections between the curves and the lowest row of the image) of the left and right curves. Their values normally vary between -2 and 2. The vanishing point position parameter c represents the intersection between the vanishing line and the two tangent lines to the curves at the lowest row of the image. Its dynamic range is typically from 0 to the image width. The suggested dynamic ranges of the parameters are based on the most common situations, in some extreme cases the values of some parameters can exceed the above ranges.

6.3 Lane Feature Extraction

In this section, we start by introducing some of the existing feature extractors and then, devote most of our attention to presenting the GLFRA. Some of the most important lane features are the edges [24, 14, 35, 18, 4, 28, 30, 32, 21] since the lane boundaries are normally assumed to be clearly marked with white or yellow paint in high contrast. It requires small computational power and defines sharp transitions in the image intensity. Well-painted lane markings produce strong edges, which benefits the detection of lanes. However, as the environment changes, the lane edges may not be as strong and may be heavily affected by shadows or adverse weather conditions. The choice of the edge threshold has always been a difficult task and some existing systems choose a very small value or use the image gradient directly without thresholding [14, 13]. This means that many unwanted features, such as edges corresponding to trees, cars, buildings, shadows and so on, are included. As a result, if the driving environment is noisy or varying rapidly, the over detected edges can mislead the parameter optimization algorithms. There are also systems that use the intensity-bump (dark-bright-dark pattern) filter to extract the lane markings [3, 11]. This feature extractor introduces less noize in the feature map but it only works with marked roads and the threshold is still difficult to decide. Colour information has not been extensively used as a primary feature until recently, due to the fact that the lane boundary colour is not unique and it can be easily influenced by illumination changes. In order to counter these difficulties, a lane detection system based on the colour sampled from the road surface is introduced [10]. The algorithm first locates the rough positions of the lane boundaries using the algorithm proposed in [25]. Pixel values in an area between the lane boundaries are used as samples to construct the Gaussian distributed colour profile of the road. By doing so, the road colour profile is updated frame by frame to adapt to different lighting conditions. However, if large areas between the lanes are occupied by other objects, the Gaussian colour model may be inaccurate and cause the inclusion of other objects. In [5], the authors addressed this problem by not only by using the colour information but also by evaluating the shapes and movements of the highlighted areas to create an accurate colour model. Another important lane feature can be extracted since its line structure contains strong orientation information. For the gradient-based methods, since the edge magnitudes are measured with intensity changes, they cannot

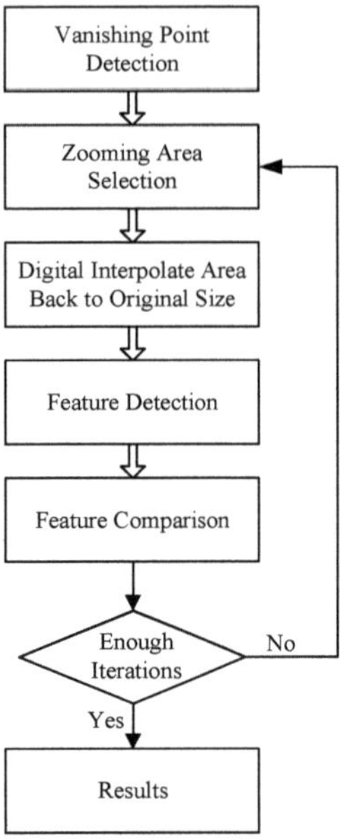

Fig. 6.5 Block diagram of the GLFRA.

be used to represent the strength of the directivity. In [17], the DCT coefficients are used to locate the diagonally orientated high-frequency locations. However, this type of feature is not general enough to accurately represent the lane boundaries in all situations. In [34], the authors used texture anisotropy fields as features, based on the covariance matrix of the grey value changes in an image, to find the image areas with distinct orientations. Another well-known algorithm using the orientation information is introduced in [20, 21] where the authors propose a feature extraction algorithm based on the steerable filter. The second derivative of Gaussian is used as the basis function and this extractor is able to find the orientations corresponding to the maximum and minimum filter responses. This difference in the filter response reveals the strength of the directivity at the current pixel position. Since it is still based on local image areas, distractive features could also be included. There are also algorithms that combine various information to achieve better results. In [33], the authors combine the line intensity, position and the gradient direction to be used as lane features.

The GLFRA considers the characteristics of the lanes and its global shape information. It is able to significantly improve the feature maps extracted using various local extractors. In this chapter, only the Sobel gradient and the steerable filter extractors are considered as examples. The idea is to gradually zoom towards the vanishing point of the lanes on a single frame in order to overlap the lane boundaries. The feature points found on the zoomed images are compared with the original image feature points and the previously zoomed feature maps. Most of the irrelevant features can be removed after this process. The algorithm block diagram is shown in Fig. 6.5.

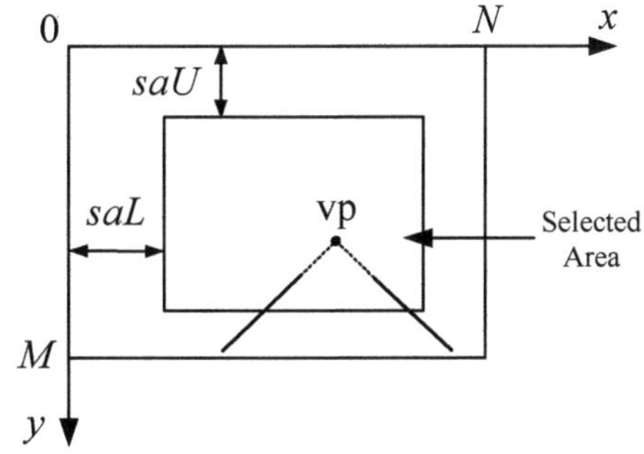

Fig. 6.6 Selected zooming area of an image.

6.3.1 Theoretical Preliminaries

The purpose of the GLFRA is to find, in the image, those features corresponding solely to the lanes. Suppose a vehicle is driven on a clear straight road with continuous lane marking and maintaining a constant lateral offset from the lanes. It is easy to notice that the positions of lane markings, from the driver's point of view, do not change over short periods of time. Of course, from frame to frame, the lane markings are actually moving backwards as the vehicle moves forward, but since the colour and the width of the markings are similar within each local image area, the driver does not feel this effect and perceives the lane markings as static objects. This algorithm takes advantage of the above phenomenon and tries to find the non-moving features from the scene. Additionally, the GLFRA does not require multiple consecutive frames. Its operation is based on a single image to avoid any lags.

Similar to the driver's view, if digital zooming is applied towards the vanishing point of the lanes, the above property of the lane boundaries still holds. By carefully selecting a region of the image according to the vanishing point position and interpolating this region back to the original image size, most of the objects except the lane boundaries will be misaligned. The first task is to select an appropriate area on the image. As the goal is to zoom towards the vanishing point, the position of the vanishing point must be determined. Also, after interpolation, the vanishing point should stay at the same position on the image space. As illustrated in Fig. 6.6, defining the position of the vanishing point, vp, as (vpm, vpn), the total number of rows as M, and the number of columns as N, the width and height of the selected area could be calculated as

$$saN = z \times N \tag{6.14}$$

$$saM = z \times M \tag{6.15}$$

where z is the zooming ratio and z^2 is the ratio between the area of the selected region and the original image area. The selection of the zooming area must follow the following rule:

$$\frac{vpn}{N - vpn} = \frac{vpn - saL}{saL + saN - vpn} \tag{6.16}$$

$$\frac{vpm}{M - vpm} = \frac{vpm - saU}{saU + saM - vpm} \tag{6.17}$$

where saL and saU are the position of the left and upper border of the selected area respectively.

Subsequently, the selected area is interpolated back to the original size. This operation moves all points except the vanishing point to new positions, which are calculated as

$$n_I(t+1) = vpn + (n_I(t) - vpn) \cdot \frac{1}{z} \tag{6.18}$$

$$m_I(t+1) = vpm + (m_I(t) - vpm) \cdot \frac{1}{z} \tag{6.19}$$

where $n_I(t)$, $m_I(t)$ and $n_I(t+1)$, $m_I(t+1)$ represent the column and row coordinates of point I before and after the interpolation respectively. A point on a straight lane boundary before interpolation needs to stay on the same line after interpolation. To prove this, we assume a straight line

$$m_I = a \cdot n_I + b \tag{6.20}$$

which passes through the vanishing point and consider I a point on the line. Substituting Eq. (6.20) into Eq. (6.19):

$$m_I(t+1) = a \cdot vpn + b + (a \cdot n_I(t) + b - a \cdot vpn - b) \cdot \frac{1}{z} \tag{6.21}$$

which could be rearranged to give

$$m_I(t+1) = a \cdot \left[vpn + (n_I(t) - vpn) \cdot \frac{1}{z} \right] + b \tag{6.22}$$

Substitute Eq. (6.18) into Eq. (6.22), we get

$$m_I(t+1) = a \cdot n_I(t+1) + b \tag{6.23}$$

Thus, Eq. (6.23) proves that the points on the lane will stay on the same line after interpolation.

So far we have assumed straight lanes and continuous lane markings. However, a multiple vanishing points detection algorithm, along with the iterative zooming process readily solves the problem for the cases of curved and discontinuous lanes. This will be discussed in detail in Section 6.3.2 and 6.3.3.

6.3.2 *Vanishing Point Detection*

Vanishing point detection constitutes the first step of the algorithm. Its location is vitally important for the rest of the task. Although a few pixels variation in the vanishing point position does not significantly influence the system performance, the detected vanishing point has to correspond to the lanes. Most of the vanishing point detection algorithms are based on the Hough transform [32, 26, 31]. However, these methods require the choice of hard thresholds for both edge detection and the Hough space accumulator. It is very difficult to find a suitable set of thresholds for various tasks and environments. Since we assume the road is flat, the vanishing line position can be calculated using the known camera parameters. Therefore, the two dimensional detection of the vanishing point is reduced to a one-dimensional search.

First, the gradient map is generated by means of a Sobel edge detector. A very small threshold is applied to reduce the computation. The threshold in our case is between 20 and 40 for the non-normalized gradient, which is small enough to locate lane features under various conditions. Assuming an edge point belongs to a lane, it is likely that the orientation of this edge is perpendicular to the gradient direction of the lanes. In this case, a line passing through this edge with the direction normal to its gradient orientation is generated to estimate the lanes. The intersection between this line and the vanishing line contributes to the estimation of the vanishing point position.

A one-dimensional accumulator with a length equal to $2N$ ($-0.5N \sim 1.5N$) is created to account for the possibility of the vanishing point being outside the image. Each edge produces a line and each time the line intersects the vanishing line, the corresponding element in the accumulator increments by $(gm+1)$ where gm is the normalized gradient magnitude. The accumulator is then smoothed with a Gaussian low pass filter to compensate for the inaccuracy of the edge orientation. The element with the most votes corresponds to the vanishing point position. The problem is that if the lane is a curve, the vanishing point position of the lane changes gradually with distance. If the zooming process is applied based on a static vanishing point, the lane boundaries could be misaligned in some areas of the image. In order to solve this problem, each frame is partitioned into a few horizontal sections as shown in Fig. 6.7.

The lane boundaries within each image section are treated as piecewize straight lines. This allows us to find the correct vanishing point positions for different image regions so that the zooming process for different image sections can also be separated. The area of each image section decreases linearly from the lowest image section to the highest section to accommodate the perspective effect. The vanishing points for different image sections are detected only using the edge points in the current section. In the far-field, the number of lane edges is lower than that of the near field. This indicates the lane boundaries' edge points contribute less in the accumulation process. In this case, the search region of the upper section is based on the vanishing point position in the lower section and the vanishing point position difference between the two lower image sections. Finally, during lane tracking, the search region of the lowest section is constrained by the corresponding vanishing point position in the previous frame to reduce error and computation. An example of the multiple vanishing point detection results is given in Fig. 6.7. The vanishing

Fig. 6.7 Vanishing point detection result. The image is horizontally partitioned and each partition is labelled. The white horizontal line is the horizon or the vanishing line. The vanishing point of each region is labelled respectively.

point corresponding to each image band is labelled seperately. In most of our experiments the image contents below the vanishing line are separated into three sections to avoid the existence of very narrow image regions.

6.3.3 Feature Extraction

The task is now to extract the lane features. In this chapter, we illustrate the effect of the GLFRA operating on two different feature maps produced by the Sobel operator and the steerable filter proposed in [21]. The steerable filter is a well-known method for lane feature detection. It is able to extract features with strong directivity. For the implementation of the steerable filter, only the line structure detector is built for comparison (the original work includes the extraction of circular reflectors).

Since the points on the lane will still stay on the same line after zooming and interpolation, the simplest idea is to apply the logical 'and' operator to the original image feature map and to the interpolated image feature map pixel by pixel. This means that if the interpolated features overlap with the original image features, these features are likely to belong to the lanes.

For the gradient map, the orientation of the overlapping features should be similar. The allowed direction difference is set to be between $0 \sim \pi/2$ *rads* in order to tolerate curves, edge detection errors and the orientation change caused by severe shadows. For the steerable filter feature map, the orientation information is already included during the filtering and thresholding. The comparison between the feature orientation is not included while applying our algorithm.

However, unwanted features still have the potential to overlay and have similar orientations due to the large amount of unwanted feature points that will be detected if a low threshold is used. In this case, an iterative zooming process is suggested. Based on experiments, 15 iterations of a gradually zooming process are normally sufficient to remove most of the noize even under very noisy conditions for the gradient map. The steerable filtered feature map contains less noize and thus fewer iterations are required. Our experiments show that 7 iterations of zooming are sufficient in most cases.

For the Sobel gradient map, since the edge orientation is needed at each iteration, we need to either interpolate the x and y directed edge maps separately or interpolate the whole image and apply edge detection at every iteration. The edge map is then compared with the original image edge map. Only the positions occupied by similar edges throughout the whole process are preserved. Specifically, if the orientation of $I(m,n)_{original}, I(m,n)_1, \ldots, I(m,n)_{iter}$ are similar, then

$$\begin{aligned} I(m,n)_{final} = I(m,n)_{original} \& I(m,n)_1 \& \\ I(m,n)_2 \& \ldots \& I(m,n)_{iter} \end{aligned} \tag{6.24}$$

Another possibility is to accumulate the overlapping edges and set a threshold to ensure the final feature points have been overlapped with most of the edge points. Since no orientation comparison is needed for the steerable filtered results, the original feature map is interpolated at each iteration and then compared. The steerable filter requires a little more computational power compared with the edge detection. However, during the zooming stage the computation is saved. This also shows the adaptability of the GLFRA. If the feature map contains a large amount of noize, more iterations of the zooming are required. If the feature map contains less noize, less iterations are needed.

During the interpolation stage, bilinear interpolation is chosen for its low complexity and satisfactory performance. The one-dimensional bilinear interpolation between two points (m_0, n_0) and (m_1, n_1) is given by

$$m = m_0 + \frac{(n - n_0)(m_1 - m_0)}{n_1 - n_0} \tag{6.25}$$

In the 2D cases, interpolation is first applied in the n-direction then in the m-direction.

The degree of zooming or the zooming ratio z is also an important parameter. It is unreasonable to select a very small zooming area. This introduces large image distortion and also, causes the system to remove a large number of lane features if the lanes are discontinued. This will not be a problem if an adequately large zooming area is selected. It has been found by experiment that the minimum zooming ratio z should be above 90%. Consequently, a large zooming ratio is applied at each iteration which decrements linearly from 100% to 90%. In this case, only a very small percentage of each discontinuous lane marking will be erased during the entire process and will not introduce any problem during the parameter estimation and tracking stage. This allows the algorithm to deal with segmented or dashed lines.

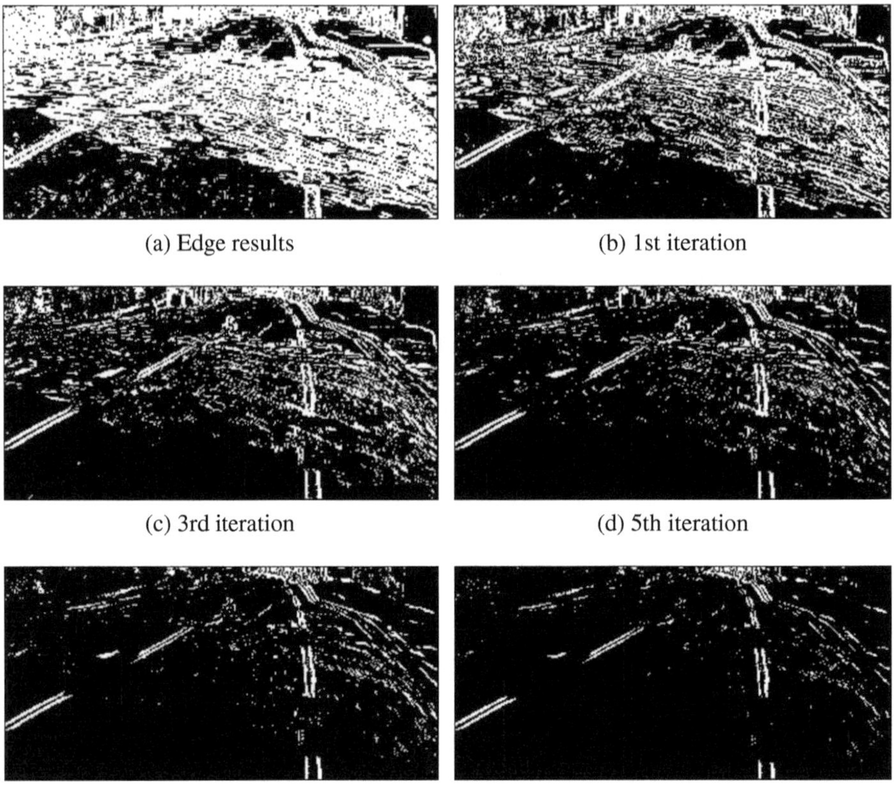

(a) Edge results (b) 1st iteration

(c) 3rd iteration (d) 5th iteration

(e) 7th iteration (f) 10th iteration

Fig. 6.8 Iterative results for noize removal with the GLFRA.

Fig. 6.8 shows the intermediate stage results of the iterative zooming process. In order to illustrate the significant difference of applying or not applying the GLFRA, the feature map is plotted as a binary image. All feature points above the low edge threshold are shown as white.

To extend this algorithm to curved lanes, a lane is separated into piecewize straight lines as described in Section 6.3.2. The zooming process is applied to each section of the image according to the corresponding vanishing point. By doing so, the lane boundaries' edges will move in the correct direction in different image areas.

Finally, most of the unwanted features are removed and the remaining features are marked as '1's. In order to give each pixel a different weighting, the '1's are replaced by the corresponding feature magnitudes provided by the local extractors. Furthermore, a weighted sum of this result with the original feature map produces a new feature map with magnified correct lane features.

The GLFRA can be extended easily to work with non-flat roads if different vanishing line positions are allowed for each image section. However, this is not included in this chapter since vanishing point voting with varying vanishing line provides less reliable results.

Experimental results on testing the GLFRA and the comparison between the gradient based and the steerable filter based feature maps can be found in Section 6.6.1.

6.4 Lane Model Parameter Estimation

After feature extraction, the lane model parameters need to be optimized to produce an accurate representation of the lane boundaries.

In this section, we introduce the parameter optimization algorithm, proposed in the LOIS system [14], based on the maximum a posterior (MAP) estimation. The performance of the algorithm is evaluated with the feature maps extracted using the GLFRA and the Sobel operator.

This algorithm can be separated into two steps. Firstly, we need to define the posterior density function (combination of the prior and the likelihood function) so that the curves similar to the ground truth produce high probability values and the ones far way produce low values. Then we have to optimize the lane model parameters to create a pair of curves that produces high probability values.

In order to construct a likelihood function that captures all the properties of the gradient map, the characteristics of the gradient map need to be analysed. Three types of information can be extracted from the gradient map: pixel positions, gradient magnitudes and the gradient orientations. Each type of information needs to be included so that the probability values are accurate descriptions of goodness of fit.

The parameters to be estimated are denoted by an array $\mathbf{x}$ where $\mathbf{x} = [a, b_1, b_2, c]$. In order to simplify the likelihood function, we use $LM_{l=1}(\mathbf{x}, m)$ and $LM_{l=2}(\mathbf{x}, m)$ to represent the left and right lane curves respectively. The likelihood function is defined as:

$$L(\mathbf{x}) \propto \sum_{l=1}^{2} \sum_{m,n} F(m,n) f(\alpha_{prox}, n - LM_l(\mathbf{x}, m)) \\ f\left(\alpha_{orient}, \cos\left(\vec{F}(m,n) - \vec{LM}_l(\mathbf{x}, m)\right)\right) \tag{6.26}$$

where $F(m,n)$ represents the gradient magnitude at pixel (m,n) on the feature map. $LM_l(\mathbf{x}, m)$ denotes the column number of the sample point on the estimated curve on image row m. $\vec{F}$ and $\vec{LM}$ denote the orientations of the image gradient and the lane curve gradient. Function f is defined as:

$$f(\alpha, x) = \frac{1}{1 + \alpha \cdot x^2} \tag{6.27}$$

The parameters α_{prox} and α_{orient} control the decaying speed of function f. In our experiments, $\alpha_{prox} = 0.01$ and $\alpha_{orient} = 1.13$. Although Eq. (6.27) does not decay as fast as exponential functions, it creates less distinct local maxima.

Eq. (6.26) incorporates all three types of gradient map information by multiplying them together. In order to maximize the likelihood function, the estimated curves need to cross or stay close to a large number of edge points with high magnitudes. Also, the orientations of the close pixels' gradients need to be approximately perpendicular to the orientations of the estimated curve gradients on the same image row. This is due to the 90 degree difference between the edge points and the line orientation.

The prior probability density function $p_{prior}(\mathbf{x})$, shown in Eq. (6.28), adds tunable constraints to the estimation. With this prior constraint, the width of the lane cannot be too narrow and LM_1 must be on the left side of LM_2.

$$p_{prior}(\mathbf{x}) \propto \frac{2}{\pi} \times \arctan(\alpha_{width} \times (b_r - b_l)) \tag{6.28}$$

The final posterior probability density $p_{posterior}$ is found by combining the normalized Eq. (6.26) and Eq. (6.28).

$$p_{posterior} = p_{prior}(\mathbf{x}) \times L(\mathbf{x}) \tag{6.29}$$

Based on the above posterior density function, the parameters are then optimized using the Metropolis algorithm [1] (details of the Metropolis algorithm are not discussed since they are beyond the scope of this chapter). It is an iterative optimization algorithm with the certain ability of escaping from the local maxima. However, since the global maximum is not guaranteed, the estimation could be inaccurate or even completely fail. The algorithms that find the global maximum such as simulated annealing require unacceptably high computational power which would introduce difficulties for real-time implementation.

The reason why the Metropolis algorithm is preferred in this system is that GLFRA removes most of the unwanted features. This directly reduces the number of local maxima in the posterior probability distribution and consequently the Metropolis algorithm is more likely to find the global maximum.

The initialization of the model is based on the detected vanishing point positions ($c = vpn$) and the most probable lane positions. The parameters are then updated iteratively with the Metropolis algorithm. In order to reduce the number of iterations and increase the accuracy of the estimation, the variances of the parameters are allowed to vary with time. If the variance of the parameters are normalized according to their dynamic range, choosing different variances shifts the priority of the estimation to the parameters with larger normalized variances (parameters with larger normalized variances converge faster at the expense of accuracy). Due to the near field lane boundaries being normally approximately straight, b_1 and b_2 are updated first with high priority. Then, the priority shifts to the vanishing point position c.

Finally, the curvature parameter a is estimated. Separating the priority of estimation in time also allows parameters to be fine-tuned while other parameters are the primary estimation target. This indicates time varying variances are able to produce accurate results more efficiently. Defining i as the current number of iterations, I as the total number of iterations and $Var(\mathbf{x})$ as an array containing the variances of the parameters: $[a, b_1, b_2, c]$, the complete process proceeds as shown in Algorithm 1.

Algorithm 1. Parameter optimization.

for $i = 0$ to $0.35I$ **do**
  Set $Var(\mathbf{x}) = [0, 0.1, 0.1, 1]$
  Update b_1 and b_2 with high priority.
end for
for $i = 0.35I + 1$ to $0.6I$ **do**
  Set $Var(\mathbf{x}) = [0, 0.01, 0.01, 20]$
  Update c with high priority.
end for
for $i = 0.6I + 1$ to I **do**
  Set $Var(\mathbf{x}) = [300, 0.01, 0.01, 1]$
  Update a with high priority.
end for

The complete process normally requires approximately 400 iterations to converge, depending on the road scene and the parameter initialization. The GLFRA includes less unwanted feature points and boosts the speed of the optimization process significantly. Some of the parameter estimation results based on different feature maps and the analysis of those results can be found in Section 6.6.2.

6.5 Lane Tracking

After the lane boundaries are detected, tracking is initiated to follow the changes of the model parameters. The main motivation of applying the tracking algorithm is to minimize the computational complexity while producing accurate results. By using the prior information generated during the detection stage, these goals can be achieved.

As discussed in Section 6.2, the lane is modelled as a circular arc. It is a reasonable assumption since the curves on the highways are constructed with constant curvature. However, when the vehicle is heading into or coming out of a curve, this assumption is violated. In order to address this problem during the tracking stage, in which more prior information is available, we modify the lane model to include another parameter which reflects the linear curvature changes. The lane model is defined as $LM_l(\mathbf{x}, m_f)$ where $\mathbf{x} = [a, b_1, b_2, c, d]$:

$$LM_1(\mathbf{x},m) = \frac{a}{m-vpm} + \frac{d}{(m-vpm)^2} + b_1(m-vpm) + c$$
$$LM_2(\mathbf{x},m) = \frac{a}{m-vpm} + \frac{d}{(m-vpm)^2} + b_2(m-vpm) + c \tag{6.30}$$

where parameters a, b_1, b_2 and c are identical to the lane model parameters in the detection stage. Parameter d is included to represent the spatial derivative of the lane curvature. Without this parameter, the description would be inaccurate when the vehicle is entering or getting out of an accentuated curve. This may severely influence the tracking performance or even result in losing track of the lane. The derivation follows a similar approach as described in Section 6.2

In the lane tracking applications, the density function of the parameters is unknown and the likelihood functions are non-linear. In order to present a non-linear filtering problem, we define x_{k-1} as the vector containing the values of the lane model parameters at time $k-1$ and want to estimate x_k at time k. The change of parameter values in time is given as

$$x_k = g(x_{k-1}, w_{k-1}) \tag{6.31}$$

The (possibly known) non-linear time update function g is applied to update the parameter array from time k to $k+1$. w_k is the process noize which may be non-Gaussian.

The measurement function h, shown in Eq. (6.32) defines the relationship between the parameter vector and the measurements.

$$z_k = h(x_k, v_k) \tag{6.32}$$

where z_k denotes the measurement matrix at time k and v_k is the measurement noize with possibly non-Gaussian distribution. This equation can also be addressed as the likelihood function or the posterior density function which will be discussed in more detail in Section 6.5.2.

If the time update and measurement update equations are linear, and x_k is described with a Gaussian density function, it is viable to propagate the density function through these linear equations. With a non-Gaussian environment, it is extremely difficult to represent the probability density with simple equations and it will be even harder to propagate the complicated equation through non-linear equations. The particle filter, which is also referred to as the sequential Monte Carlo (SMC) approach, is applied to solve the problem above. It is a sophisticated estimation algorithm which is able to work in non-linear and non-Gaussian environments. The objective is to estimate the a posteriori density of the hidden state or unknown parameter. The unknown probability density function is represented with samples/particles instead of equations. The propagation of the equation is then reduced to the propagation of samples which can be easily calculated. Due to the existence of various forms of particle filters, the performance of three different types of particle filters were evaluated: the SIR, GPF and GSPF.

The GPF is based on a Gaussian pdf assumption. The mean and covariance of the weighted particles are calculated at each time instance to describe the Gaussian posterior density. By doing so, the time consuming resampling step is avoided since the particles can be easily drawn from a Gaussian density. The entire GPF process can thus be completed with highly paralleled implementations.

In the lane tracking application, although the posterior density of parameters is normally non-Gaussian, it does not mean using a Gaussian density as an approximation is inappropriate. If the feature map contains too many unwanted feature points, the Gaussian density is normally not a good assumption. However, after applying the feature maps, most of the unwanted features are already removed. The effect of outliers is minimized and the GPF can be a suitable choice. The complete GPF algorithm and mathematical derivation of GPF can be found in [15].

There is still another possibility to avoid resampling without the Gaussian assumption. The GSPF models the noize and the posterior density with a mixture of Gaussian densities. One of the most important properties of the Gaussian mixture model is that the posterior density are separated into a group of individual Gaussian densities. This provides more flexibility during the parameter selection process. By utilising this property to fit the requirement of this particular application, the accuracy of the final estimations of the parameter values surpasses the one provided by the traditional SIR and GPF.

6.5.1 Time Update

In a typical computer vision based lane tracking system, the actual time update equation can be estimated by using the vehicle motions. However, we assume that this information is not available and incorporate a random Gaussian noize to follow the parameter variations. During the time update stage, particles are drawn from the previous posterior density and propagated through the time update equation as shown in Eq. (6.33).

$$\mathbf{x}_k^i = \mathbf{x}_{k-1}^i + \mathbf{w}_{k-1}^i \tag{6.33}$$

where

$$\mathbf{w}_{k-1}^i \sim N(\mu, Q_{k-1})$$

$$Q_{k-1} = \begin{pmatrix} \sigma_a^2 & 0 & 0 & 0 & 0 \\ 0 & \sigma_{b_1}^2 & 0 & 0 & 0 \\ 0 & 0 & \sigma_{b_2}^2 & 0 & 0 \\ 0 & 0 & 0 & \sigma_c^2 & 0 \\ 0 & 0 & 0 & 0 & \sigma_d^2 \end{pmatrix}$$

where $w_{k-1}^i \sim N(\mu, Q_{k-1})$ is the process noize added in the time update equation to follow changes of the lanes and prevent sample impoverishment. For the SIR and PF, $\mu = 0$. For the GSPF, each Gaussian component represents a change in lane shape or position. Five Gaussian components are used, corresponding to a small change of lane shape ($\mu = 0$), fast positive or negative change of lane curvature ($\mu_a = \pm 100$) and the spatial change of curvature ($\mu_d = \pm 100$) respectively. Q_{k-1} denotes the covariance matrix of the Gaussian noize. In order to cope with rapid changes in lane parameters and avoid sample impoverishment, we have chosen $\sigma_a = 100$, $\sigma_{b_1} = 0.17$, $\sigma_{b_2} = 0.17$, $\sigma_c = 4$ and $\sigma_d = 100$ for both the PF and GPF and $\sigma_a = 60$, $\sigma_{b_1} = 0.17$, $\sigma_{b_2} = 0.17$, $\sigma_c = 4$ and $\sigma_d = 60$ for the GSPF. The variance of a and d for the GSPF is smaller due to the fact that the additive noize has a non-zero mean.

6.5.2 Measurement Update

The measurement update stage evaluates the weighting of the samples and finally produces an estimation of the posterior probability density of the lane model parameters $\mathbf{x} = [a, b_1, b_2, c, d]$. In this chapter, the likelihood function chosen for tracking is different from the one for detection but is similar to the one used in [28] and is also improved specifically for the lane tracking application.

As mentioned in Section 6.4, the gradient map contains three important types of information: pixel positions, gradient magnitudes and gradient orientations. The likelihood function needs to utilize this information and produces a high likelihood when the curve is fitted to the correct feature points. During the parameter estimation stage, all pixels are included in the calculation of likelihood since little prior information is available and the initialization of the algorithm could be inaccurate. During the tracking stage, previous detection or tracking results are available so that most of the feature points that are far away from the estimated curve can be excluded for likelihood calculation. Furthermore, a fast decaying function is needed so that the parameters which correspond to the mean value of the posterior density function can be used as the final estimation. For the above reasons, the likelihood function is constructed as an exponential function, and for each lane boundary, only one feature point from each image row is included in the calculation. The selection of this point is based on evaluating the distance and the orientation differences between the curve sample point and the feature points along with the gradient magnitudes. The feature point which produces the minimum distance on an image row is selected.

Similar to the detection stage, the normalized gradient magnitudes of the edge points are represented as $F(m,n)$. Parameter R defines the width of the search region centred at $LM_l(\mathbf{x},m)$. The n-axis coordinate of any feature point found inside the search region is denoted as $n(i)$ where $i \in (1, T_m)$, and T_m denotes the total number of feature points on the image line m. The likelihood function is shown in Eq. (6.34).

$$p_{image}(z|x) \propto exp\Bigg(-\sum_{l=1}^{2}\sum_{m=1}^{M}\min_{i=1}^{T_m}\Bigg[\frac{LM_l(\mathbf{x},m)-n(i)}{R} + \alpha\left|\cos(\vec{LM_l}(\mathbf{x},m)-\vec{F}(m,n(i)))\right| + \beta(1-F(m,n(i)))\Bigg]/M\cdot S\Bigg) \quad (6.34)$$

where M denotes the number of image rows under the vanishing line. S is a coefficient controlling the weight difference of particles. In all our experiments, S was set to 10. $\overrightarrow{LM}(\mathbf{x},m)$ and $\overrightarrow{F}(m,n(i))$ denote the orientation of the lane curve gradient and the orientation of the feature point $F(m,n(i))$, respectively. The cos function is used here since the image gradient is perpendicular to the edge lines. α and β are weighting factors controlling the importance of distance, orientation difference, and the gradient magnitude to the overall likelihood. In our experiments, α is set to 1. β is set to 0.25 which indicates the gradient magnitude contributes much less to the overall likelihood.

If there is no feature point on an image line m, the contribution of that image line to the likelihood is suppressed to the minimum value by setting:

$$min_{i=1}^{T_m}\left[\frac{LM_l(\mathbf{x},m)-n(i)}{R}+\right.$$
$$\left.\alpha\left|\cos(\overrightarrow{LM}_l(\mathbf{x},m)-\overrightarrow{F}(m,n(i)))\right|+\beta(1-F(m,n(i)))\right]$$
$$=\frac{R}{R}+\alpha+\beta=1+\alpha+\beta; \tag{6.35}$$

The result of Eq. (6.34) is then combined with the prior probability distribution, shown in Eq. (6.28), to constrain the lane width and results in the posterior probability distribution as shown in Eq. (6.36).

$$p(z|x) \propto p_{image}(z|x) \times p_{prior} \tag{6.36}$$

where p_{prior} is defined in Eq. (6.28) and $p(z|x)$ is normalized so that it integrates to 1.

In Eq. (6.36) the feature point chosen for each image line is selected based on the combination of edges to lane model sample points distances, edge and curve gradient orientation differences and also the gradient magnitudes. By including all the information during the edge point selection process, the possibility of ignoring useful feature points is minimized. Furthermore, the exponential factor is averaged with the number of image lines and then multiplied by a scaling factor S instead of summation. Summation results in a very large exponential factor which, in turn, makes the weighting difference unacceptably large. If only very few particles have dominating weights, the 'spiky' posterior pdf lacks the ability to accurately represent the probability of each particle. Finally, the likelihood value should be independent of the image size. Therefore, averaging the values of all image lines is the most appropriate choice.

The main reason for choosing a local likelihood function is to make sure the estimated curves stay reasonably close to the previous results to minimize the effect of the outliers and reduce the computational complexity.

In implementing the SIR, the effective sample size is used to decide if resampling of the particles is required. For the GSPF, after the posterior probability density becomes available for the first frame, the Expectation Maximization (EM) [7] algorithm is used to initialize the Gaussian components for the GSPF. For the following frames, the Gaussian components weights are updated using the sum of weights of the samples drawn from the corresponding components to multiply with the previous Gaussian weights. The minimum weight of a Gaussian component is limited to avoid the situations where the weights of few components are considerably higher than the rest.

It is important to note that the output of the particle filters is a probability density of the parameters represented by samples, not the values of the final estimation. In order to generate the final estimation, this probability density needs to be carefully analysed according to the characteristics of different applications. However, since it is almost impossible in the lane tracking case to obtain a reasonable characterization of the surrounding noize, the final multivariate probability density is complex to analyse. That is why most of the existing lane tracking systems choose the mean as the final parameter estimation, which is not always suitable especially in complicated environments.

6.5.3 Parameter Selection

In this section, we describe an intuitive parameter selection process based on the posterior density function produced by the GSPF. The mean values of the Gaussian component with the highest weight is considered to be the final estimation result. Based on experiments, the variance of each Gaussian component is much lower than the variance of the posterior density obtained with the SIR or the GPF. This indicates the GSPF divides the influence of the feature points into different groups. Each group of feature points contribute to the likelihood values of some closely related particles. For example, if a group of feature points corresponds to the lane boundaries, they will contribute to the parameter values described with a single Gaussian component. Similarly, if a group of feature points correspond to some outliers, they will contribute to the parameter values described with another Gaussian component. As a result, if the Gaussian with the highest weight corresponds to the lane boundaries, then the mean of this Gaussian constitutes an accurate estimator. The final output of the PF and GPF is chosen to be the mean of the posterior probability density.

6.6 Experimental Results

6.6.1 Lane Feature Extraction Results

The feature extraction performance of the GLFRA is compared with the original Sobel gradient map and the steerable filter approach [21]. The experiments show that the GLFRA is able to remove most of the unwanted features based on various gradient-based feature maps if the features can be represented in the image space. There are other types of feature extraction algorithms and not all methods are presented here.

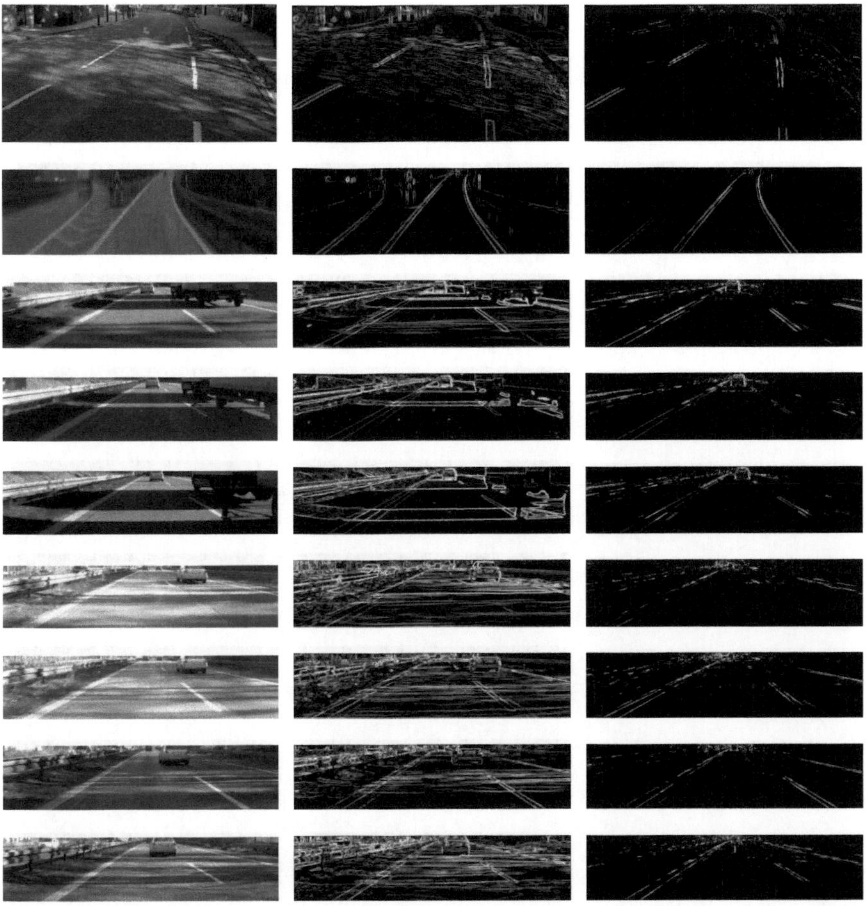

Fig. 6.9 Lane feature extraction results. The first column shows the original images. The second column shows the image gradient maps extracted using the Sobel operator. The third column shows the GLFRA results based on the gradient maps.

While choosing the minimum zooming ratio z in our experiments, it has been found experimentally that its value should be above 90%. Consequently, a large zooming ratio is applied at each iteration (decremented linearly from 100% to 90%). By doing so, only a very small portion of each discontinuous lane marking will be erased during the process and this will not introduce any problem during the parameter estimation and the tracking stage. In all our experiments, z is set to be 0.92. Modifying this value from 0.9 to 0.93 does not induce a large difference in the feature extraction results. This value is chosen based on a large number of subjective tests. Values above 0.93 start to influence the results since the noize removal ability of the process is reduced, and in complicated environments, more remaining noize will be included. Some example gradient-based feature extraction results

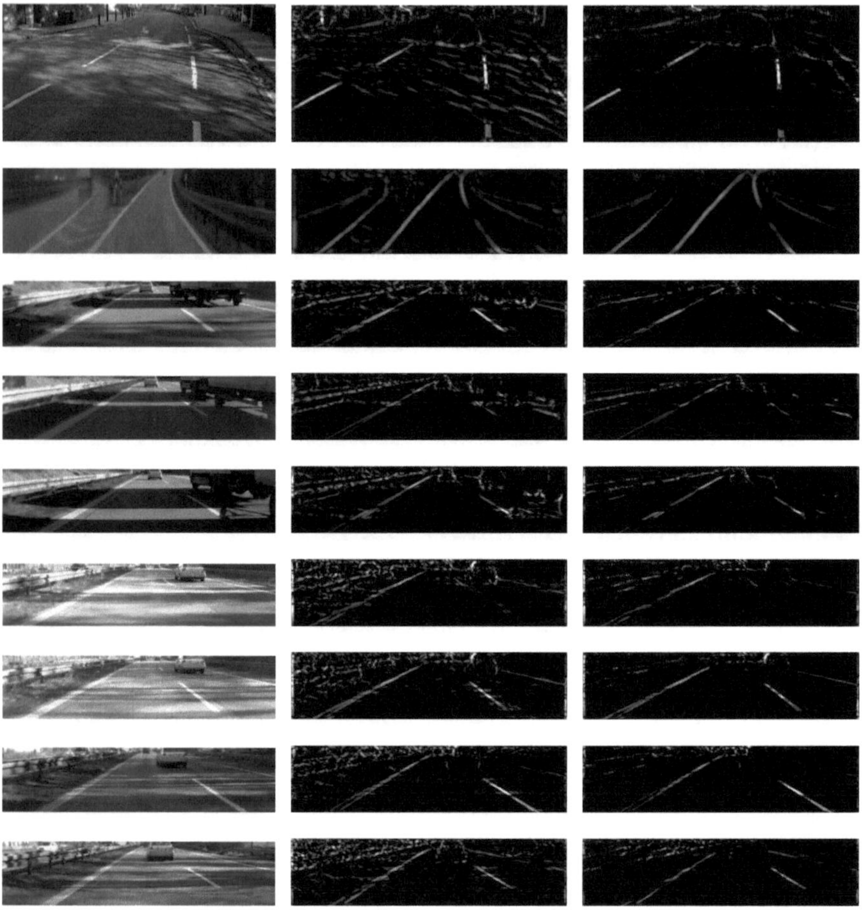

Fig. 6.10 Lane feature extraction results. The first column shows the original images. The second column shows the feature maps extracted using the steerable filter. The third column shows the GLFRA results based on feature maps obtained using the steerable filter.

are shown in Fig. 6.9 and steerable filter based results are shown in Fig. 6.10. As these figures illustrate, the GLFRA improves feature extraction results significantly. In complicated situations (as shown in the first column in Fig. 6.9 and 6.10.) the traditional feature maps include numerous unwanted feature points (as shown in the second column in Fig. 6.9 and 6.10.). Column three of Fig. 6.9 and 6.10 show results achieved using the GLFRA based on the two different feature maps. Most of the unwanted feature points are removed after fifteen iterations. With the improved feature map, the performance of the parameter estimation algorithm and the tracking algorithm will be subsequently improved. Fifteen iterations are chosen because very little difference is observed with a higher number of iterations.

Comparing these steerable filter results with the gradient based feature map (Column two in Fig. 6.9 and 6.10), significant improvement can be observed since less unwanted features are included and the magnitudes of some of the unwanted feature points are generally smaller. However, the steerable filter results can still be improved by removing the outliers. Once the results are refined with the GLFRA, the feature maps are much cleaner after only seven iterations (chosen based on experiments) and the final feature points are more adequate representions of the road boundaries (as shown in column three in Fig. 6.10.) As mentioned in Section 6.3.3, a fewer number of iterations of refinement are needed to achieve satisfactory results for better local extractors. In practical situations, the number of iterations must also be determined according to the driving environment. The chosen system parameters are applied to various road images and video sequences. During our subjective tests, the GLFRA achieved high performance.

Finally, it is interesting to see that the refined feature maps based on both edge detection and steerable filter become similar. Originally, significant performance differences can be observed between the two algorithms (as shown in column 2 in Fig. 6.9 and Fig. 6.10.). After refinement, the results based on the steerable filter are only marginally better than those based on edge detection (as shown in column 3 in Fig. 6.9 and Fig. 6.10.). This demonstrates the capability of the GLFRA to improve various feature maps and produce high quality output.

It is worth noting at this point that GLFRA was also implemented on the TMS320DM6437 DSP platform from Texas Instruments to achieve real-time performance. The system is able to achieve above 23 frames per second (fps) with a 240×360 video input. The frame rate can be further increased by code optimization [6]. More details about the implementation of the algorithm can be found in [29].

6.6.2 *Lane Model Parameter Estimation Results*

This section presents the experimental results of the model parameter estimation process. The main purpose of this experiment is to show that after applying the GLFRA, the curve fitting results can be improved. It is worth noting that only the image gradient map is chosen here for comparison since the original likelihood function of the parameter estimation algorithm proposed in [14] is based on the gradient map. Also, the feature orientations extracted with the steerable filter are not accurate and cannot be used as data source for the curve fitting.

The test images included in this section are chosen from the most difficult scenes in several video sequences. In Fig. 6.11 (a) and (d), both scenes are affected heavily by shadows and a diverging lane scene is included in Fig. 6.11 (g). The corresponding gradient maps are shown in Fig. 6.11 (b), (e) and (h). All of these gradient maps contain a large number of unwanted feature points. Fig. 6.11 (c), (f) and (i) show the feature map obtained using the GLFRA. Most of the unwanted features are removed. Compared with gradient maps, the GLFRA feature maps are less noisy and the lane features are well preserved.

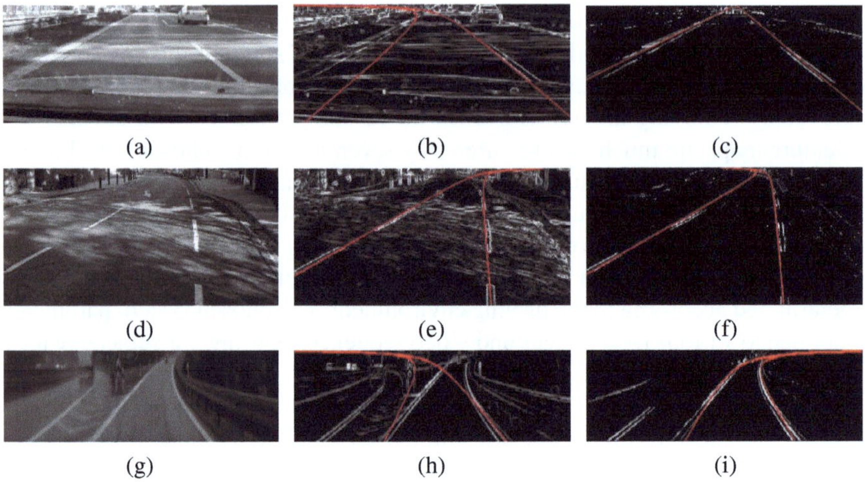

Fig. 6.11 (a), (d) and (g): input images excluding the parts above the vanishing lines. (b), (e) and (h): detection results of (a), (d) and (g) respectively based on image gradient. (c), (f) and (i): detection results of (a), (d) and (g) respectively based on feature maps generated by the GLFRA.

The detection of the lanes is based on the Metropolis algorithm, which does not guarantee to find the global maximum since the update of the parameters is based on a random selection process. In this case, the detection result varies even if the same feature map is processed multiple times. Therefore, the feature map of the input images shown in Fig. 6.11 have been processed repeatedly (200 times in our experiments), with six hundred iterations each time. The parameter values are then averaged for comparison purposes. The parameter settings during the detection stage are identical for both feature maps.

The averaged parameters: a, b_1, b_2 and c from Eqs. (6.12) and (6.13), are compared with the manually selected true parameters. The average absolute error for each of the parameters is calculated. As the required accuracies and dynamic ranges of a, b_1, b_2 and c are different (suggested dynamic ranges of the parameters can be found in Section 6.2), a comparison between the detection results based on different feature maps would be more appropriate. Defining the parameter estimation error based on the GLFRA as $EP(s)$ and the parameter estimation error based on the gradient map as $EG(s)$, the relationship between $EP(s)$ and $EG(s)$ could be represented as:

$$ER(s) = \frac{EP(s)}{EG(s)} \tag{6.37}$$

The pre-set variances of all the parameters are shown in Section 6.4. Table 6.1 shows the ER value corresponding to different parameters calculated from Fig. 6.11 (a), (d) and (g) as well as the detection time ratio T_P/T_G.

Table 6.1 *ER* values corresponding to a, b_1, b_2 and c and the Time ratio T_P/T_G calculated from Fig. 6.11 (a), (d) and (g).

	Fig. 6.11 (a)	Fig. 6.11 (d)	Fig. 6.11 (g)
$ER(a)$	0.11	0.36	1.06
$ER(b_1)$	0.27	0.99	0.41
$ER(b_2)$	0.38	0.27	0.19
$ER(c)$	0.71	0.80	1.01
T_P/T_G	0.19	0.18	0.56

It is important to note that large curve fitting errors can be observed by subjective testing because of errors in a few parameter values. If the far-field of the scene is noisy, then parameter estimation results may only contain large errors in a, since the change of a (curvature) has little influence on the near-field curve shape. Similarly, if only the estimations of b_1 and b_2 are not accurate, the largest difference between the estimated curve and the true road boundaries should lie in the near-field on the image plane. The accuracy of parameter c influences the accuracy of all other parameters. As illustrated in the second and the third column in Table 6.1, remarkable curve fitting performance improvement is achieved by applying the GLFRA. However, for Fig. 6.11 (g) (fourth column in Table 6.1), improvements are mainly in the estimations of b_1 and b_2 where estimation errors of b_1 and b_2 are only 41% and 19% of the ones based on the traditional gradient map. This improvement is very significant in terms of near-field lane position estimation. The values of $ER(a)$ and $ER(c)$ for Fig. 6.11 (g) are very close to 1. This indicates that the estimations of a and c based on both feature maps are quite accurate and there is little room for improvement.

In terms of performance, Table 6.1 shows that the detection results based on the GLFRA have been significantly improved over the results based on the traditional gradient map. Finally, the parameter estimation time based on the GLFRA is also massively reduced because fewer feature points are included during curve fitting. This also compensates for the extra computational complexity that the GLFRA introduces. Since the feature extraction processes normally need only a fraction of the computational power used by the detection algorithms, on a system level, the overall processing time is reduced by applying the GLFRA.

6.6.3 Lane Tracking Results

In our experiments, the tracking results were obtained based on several video sequences. These videos contain different road scenes, including severe shadows, heavy curvatures, sudden curvature changes and the presence of other vehicles. These videos are mainly recorded on motorways and include some degree of camera shaking. All the experimental results included in this section are based on two selected video sequences, both containing discontinuous lane markings. Video a

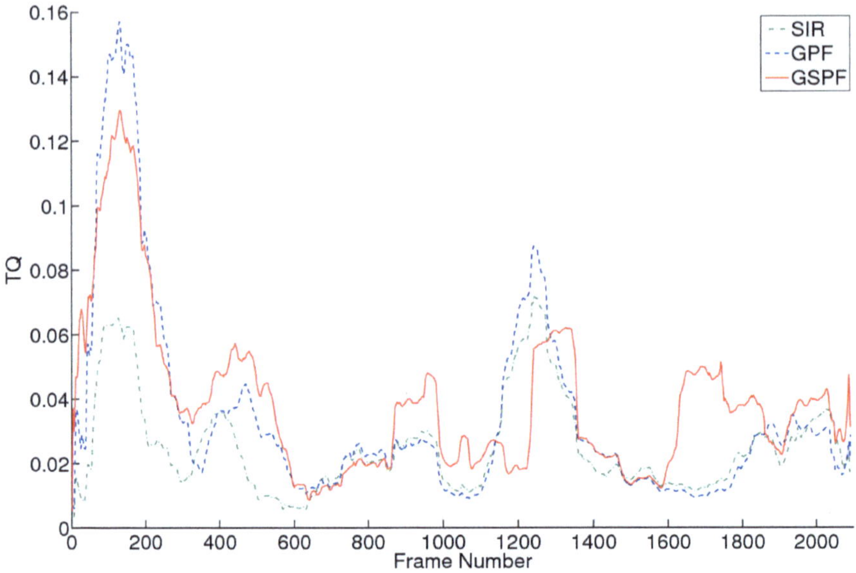

(a) TQ measurements of three particle filters based on the GLFRA.

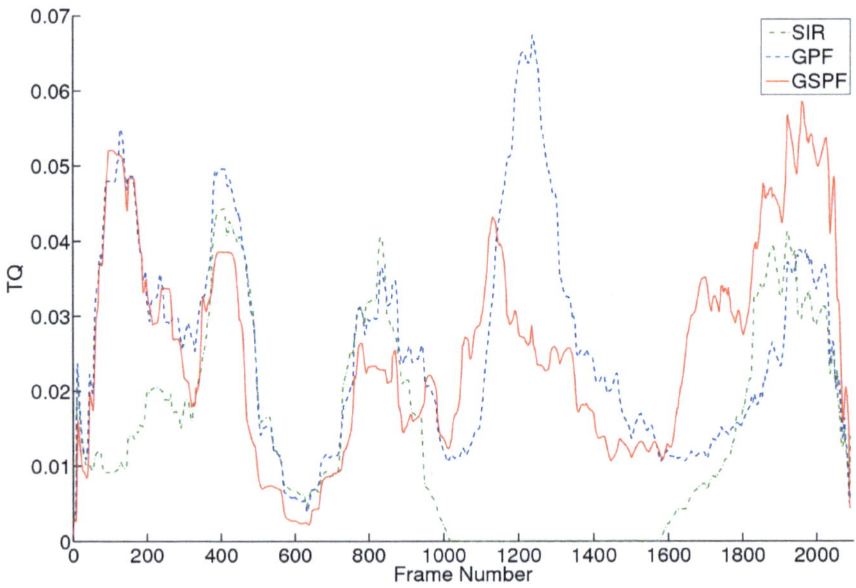

(b) TQ measurements of three particle filters based on the Sobel gradient map.

Fig. 6.12 Smoothed tracking quality (TQ) measurements for video *a* based on different feature maps.

(selected frames are shown in Fig. 6.13) includes scenes with heavy shadows, other vehicles and off-road objects. Video *b* (selected frames are shown in Fig. 6.14) is not as noisy as video *a* but shows rapid curvature changes when the vehicle exits a section of a heavily curved road.

The Tracking Quality (TQ) and the variance of the unknown parameters are measured in order to compare and evaluate the performance of the three types of particle filters, which are the PF, GPF and GSPF. TQ is defined as the inverse of the Mean Squared Error (MSE) between the sample points' positions on the estimated curves and the manually chosen ground truth (as shown in Eq. (6.38)). The ground truth is generated by marking a few key points ($(\mathbf{K}_n, \mathbf{K}_m)$ where $\mathbf{K}_n$ and $\mathbf{K}_m$ are arrays storing the x- and y-coordinates of these points respectively) on the left and right lane boundaries. Furthermore, the comparisons between the tracking performance based on the traditional gradient map and the GLFRA are included.

Throughout the experiments, 500 particles are used since no significant performance enhancement was observed once the particle size reaches 500. A correct detection result is used for initialising the particle filters and the covariance of the added noize is the same for all particle filters. Furthermore, reinitialization is switched off during our TQ measurement so that large errors are allowed to ensure a thorough evaluation of the algorithms.

The TQ for each frame is defined as:

$$TQ = 1/\mathrm{E}[(LM(\mathbf{x}, \mathbf{K}_m) - \mathbf{K}_n)^2] \tag{6.38}$$

Eq. (6.38) shows that the TQ is closely related to the distance between the samples on the predicted lane and the ground truth.

The TQ measurement of video *a* is smoothed and illustrated in Fig. 6.12. As Fig. 6.12 (a) shows, all three types of filters are able to correctly track the lanes in complicated environments, based on the GLFRA. For the complete video sequence, the averaged TQ is calculated and shown in Table 6.2. On average, the GSPF achieved the highest performance which means the final model parameters produce lines closest to the ground truth. This significant observation proves that the GSPF, combined with the new parameter selection approach, is the most suitable algorithm for lane tracking applications. Not only does it achieve better tracking results, but it can also be easily implemented in parallel.

The variances of the lane model parameters are also recorded after each frame as another indicator of the tracking performance. In order to illustrate the variance

Table 6.2 Comparison of average TQ based on the Sobel gradient map (S) and the GLFRA (P)).

	TQ		
	SIR	GPF	GSPF
P	0.0256	0.0351	0.0392
S	0.0141	0.0264	0.0258

difference of different parameters, the parameter variances are averaged over all frames as shown in Table 6.3. Since the final estimation of the GSPF is the mean of the highest weighted Gaussian, the listed averaged variances are the variances of the chosen Gaussian components. For the PF and GPF, the variances of the parameter posterior probability densities are given. In the P row (tracking based on the GLFRA) for both videos, the GSPF achieved the minimum variance for most of the parameters among the three. The variance of the GPF estimation is also smaller than that of the tradition PF. This observation is very important since it proves that estimating the posterior probability density with a Gaussian or a Gaussian mixture is accurate. This also means that if the GLFRA is used for measurements, the GSPF and the GPF are two viable alternatives to the SIR particle filter in lane tracking applications and offer better performance. The variance of the parameter d when using GSPF is surprisingly high, based on both the GLFRA and the Sobel feature map. This is caused mainly by our time update strategy for the GSPF which adds non-zero mean noize to accommodate sudden curvature changes (details in Section 6.5.1).

Table 6.3 Comparison of lane model parameters averaged variances between the three particle filters and based on the different feature maps. P represents the GLFRA and S denotes the Sobel gradient map.

		Video a		
		PF	GPF	GSPF
P	a	19649	11334	8973
	b_1	0.004	0.003	0.001
	b_2	0.010	0.008	0.002
	c	31.18	20.81	13.39
	d	72385	48194	62937
S	a	27331	17709	25470
	b_1	0.005	0.004	0.002
	b_2	0.015	0.008	0.003
	c	36.34	27.71	24.20
	d	81681	98187	297862

		Video b		
		PF	GPF	GSPF
P	a	20960	14640	12727
	b_1	0.0058	0.0044	0.004
	b_2	0.0052	0.01	0.004
	c	33.47	22.49	29.14
	d	201820	116060	98094
S	a	22700	15870	31374
	b_1	0.0056	0.0039	0.0097
	b_2	0.0051	0.0035	0.0133
	c	35.37	20.77	45.93
	d	174420	109820	127100

Next, the tracking results based on the GLFRA are compared with those based on the traditional edge map. The edge threshold set for the traditional edge map and the GLFRA are identical. As illustrated in Fig. 6.12 (a) and (b), the tracking performance improvements for all three types of tracking algorithms are significant, due to most of the outliers being removed. The traditional SIR lost track of the right lane boundary from approximately frame 1000-1600 which results in very low TQ values. Note that the SIR is able to successfully track the lanes based on the GLFRA. The tracking performance improvement achieved using the feature map generated

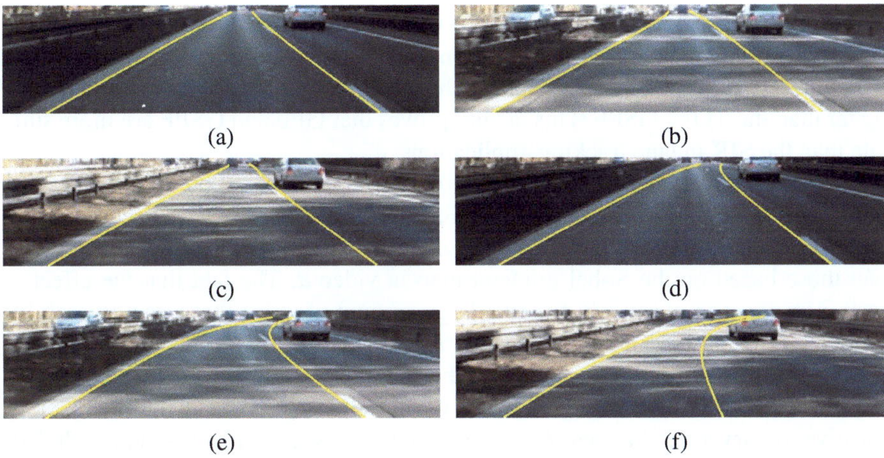

Fig. 6.13 Lane tracking results based on video *a*. Three frames are chosen for illustration. The first row shows the tracking results based on the GLFRA feature maps. The second row shows the results based on traditional gradient maps.

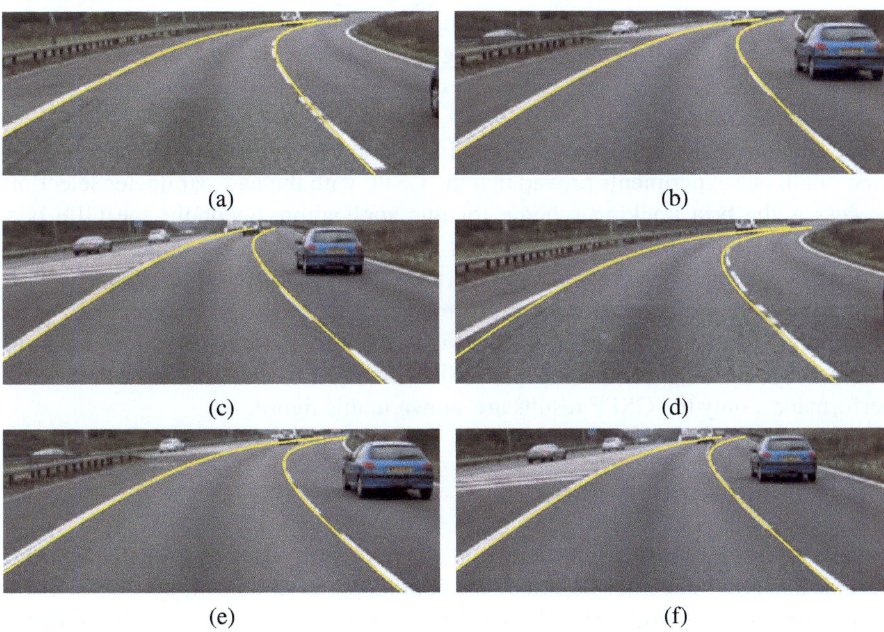

Fig. 6.14 Lane tracking results based on video *b*. Three frames are chosen for illustration. The first row shows the tracking results based on the GLFRA feature maps. The second row shows the results based on traditional gradient maps.

by the GLFRA is more significant than that achieved using a better tracking algorithm. The TQ achieved using the GSPF based on the GLFRA is 2.7 times the TQ achieved using SIR based on the traditional feature map. Another important point is that the TQ of the GPF is the highest among the three, although only marginally higher than the TQ of GSPF. This, again, proves that GPF and GSPF are more suitable than the SIR in lane tracking applications.

The averaged variance of each parameter is compared based on different feature maps. The results are shown in Table 6.3. From this table, it can be seen that the variances of the estimated parameters based on the GLFRA are significantly lower than those based on the Sobel gradient map in video *a*. The fact that the effect is more significant in video *a* is due to a large amount of noize that was successfully removed by using the GLFRA. In video *b*, the noize level is much lower than in video *a* so the variance differences of the parameters are not as distinct. In our subjective tests, when tracking is based on the GLFRA, particle filters are more sensitive to curvature changes. Also, due to less noisy edge pixels being included, the effect caused by outliers is minimized. This leads to more accurate tracking results. Finally, if the previous estimation is inaccurate, the particle filters recover much faster based on the GLFRA.

In our experiments, a thorough comparison between three types of particles filters is carried out by evaluating the tracking quality (TQ) and the variance of the posterior density. Furthermore, the tracking performance based on the feature maps generated using the GLFRA and the original feature maps are compared. During the experiments, all the system parameters are chosen so as to achieve the optimum tracking performance.

To summarize the analysis of the tracking results, three main points can be made. First of all, our experiments proved that the GSPF with the new parameter selection process is the best tracking scheme for this application. Secondly, the GPF is a viable alternative to the SIR particle filter which minimizes the computational cost and achieves higher performance. Finally, the GLFRA generates feature maps with minimum noize which allows the performance of all three tracking algorithms to be improved dramatically. Fig. 6.13 and Fig. 6.14 show some selected lane tracking results based on different feature maps. Since the GSPF achieves the best tracking performance, only the GSPF results are shown in this figure.

6.7 Conclusions

In this chapter, a complete lane detection and tracking system is presented. During the lane detection stage, we described the popular parabola lane model, the GLFRA for feature extraction and a MAP based parameter estimation algorithm. By utilising the global lane shape information, the GLFRA is able to improve most of the existing gradient-based feature maps by removing the irrelevant feature points produced by unwanted objects in the scene. The detection rate, accuracy, and processing speed are significantly improved after the unwanted feature points are removed. When investigating the original PF, the GPF, and the GPSF for lane tracking, the GPF and the

GSPF are found to be more efficient alternatives to the SIR particle filter. Both GSPF and the GPF achieve more accurate tracking results and do not require the resampling stage, which cannot be implemented in parallel. The GSPF achieves the best performance by selecting the mean of the highest weighted Gaussian component as the final output. Finally, the tracking performance is also significantly improved by incorporating the feature map extracted with the GLFRA instead of the traditional local feature extractors.

References

1. Beichl, I., Sullivan, F.: The Metropolis algorithm. Computing in Science and Engineering 2, 65–69 (2000)
2. Benmansour, N., Labayrade, R., Aubert, D., Glaser, S.: Stereovision-based 3D lane detection system: a model driven approach. In: Proceedings of the 11th International IEEE Conference on Intelligent Transportation Systems (2008)
3. Bertozzi, M., Broggi, A.: GOLD: a parallel real-time stereo vision system for generic obstacle and lane detection. IEEE Transactions on Image Processing 7, 62–81 (1998)
4. Chapuis, R., Aufrere, R., Chausse, F.: Accurate road following and reconstruction by computer vision. IEEE Transactions on Intelligent Transportation Systems 3(4), 261–270 (2002), doi:10.1109/TITS.2002.804751
5. Cheng, H.Y., Jeng, B.S., Tseng, P.T., Fan, K.C.: Lane detection with moving vehicles in the traffic scenes. IEEE Transactions on Intelligent Transportation Systems 7(4), 571–582 (2006), doi:10.1109/TITS.2006.883940
6. Dahnoun, N.: Digital Signal Processing Implementation: Using the TMS320C6000 Processors. Prentice-Hall PTR (2000)
7. Dempster, A.P., Laird, N.M., Rubin, D.B.: Maximum likelihood from in-complete data via the EM algorithm. Journal of the Royal Statistical Society: Series B 39, 1–38 (1977)
8. Dickmanns, E.D., Mysliwetz, B.D.: Recursive 3-d road and relative ego-state recognition. IEEE Trans. Pattern Anal. Mach. Intell. 14, 199–213 (1992), `http://portal.acm.org/citation.cfm?id=132013.132020`, doi:10.1109/34.121789
9. Doucet, A., de Freitas, N., Gordon, N.: Sequential Monte Carlo Methods in Practice. Springer (2001)
10. He, Y., Wang, H., Zhang, B.: Color-based road detection in urban traffic scenes. IEEE Transactions on Intelligent Transportation Systems 5, 309–317 (2004)
11. Ieng, S.S., Tarel, J.P., Labayrade, R.: On the design of a single lane-markings detectors regardless the on-board camera's position. In: Proceedings of IEEE Intelligent Vehicles Symposium 2003, pp. 564–569. IEEE (2003)
12. Kang, D.J., Choi, J.W., Kweon, L.S.: Finding and tracking road lanes using line-snakes. In: Proceedings of Conference on Intelligent Vehicle (1996)
13. Kang, D.-J., Jung, M.-H.: Road lane segmentation using dynamic programming for active safety vehicles. Pattern Recognition Letters 24, 3177–3185 (2003)
14. Kluge, K., Lakshmanan, S.: A deformable-template approach to lane detection. In: Proceedings of the Intelligent Vehicles 1995 Symposium, pp. 54–59 (1995)
15. Kotecha, J.H., Djuric, P.M.: Gaussian particle filtering. IEEE Transactions on Signal Processing 51, 2592–2601 (2003)
16. Kotecha, J.H., Djuric, P.M.: Gaussian sum particle filtering. IEEE Transactions on Signal Processing 51, 2602–2612 (2003)

17. Kreucher, C., Lakshmanan, S.: LANA: a lane extraction algorithm that uses frequency domain features. IEEE Transactions on Robotics and Automation 15, 343–350 (1995)
18. Li, Q., Zheng, N., Cheng, H.: Springrobot: A prototype autonomous vehicle and its algorithms for lane detection. IEEE Transactions on Intelligent Transportation Systems 5, 300–308 (2004)
19. Lim, K.H., Seng, K.P., Ang, L.-M., Chin, S.W.: Lane detection and Kalman-based linear-parabolic lane tracking. In: International Conference on Intelligent Human-Machine Systems and Cybernetics, IHMSC 2009, vol. 2, pp. 351–354 (2009), doi:10.1109/IHMSC.2009.211
20. McCall, J.C., Trivedi, M.M.: An integrated, robust approach to lane marking detection and lane tracking. In: Proceedings of IEEE Intelligent Vehicles Symposium, pp. 533–537 (2004)
21. McCall, J.C., Trivedi, M.M.: Video-based lane estimation and tracking for driver assistance: Survey, system, and evaluation. IEEE Transactions on Intelligent Transportation Systems 7, 20–37 (2006)
22. Mihaylova, L., Brasnett, P., Achim, A., Bull, D., Canagarajah, N.: Particle filtering with alpha-stable distributions. In: Proceedings of the 13th IEEE Statistical Signal Processing Workshop (SSP), pp. 381–386 (2005)
23. Nedevschi, S., Schmidt, R., Graf, T., Danescu, R.: 3D lane detection system based on stereo vision. In: IEEE Intelligent Transportation Systems Conference (2004)
24. Park, J.W., Lee, J.W., Jhang, K.Y.: A lane-curve detection based on an lcf. Pattern Recognition Letters 24, 2301–2313 (2003)
25. Pomerleau, D.: RALPH: rapidly adapting lateral position handler. In: Proceedings of the IEEE Symposium on Intelligent Vehicles, pp. 506–511 (1995)
26. Shufelt, J.A.: Performance evaluation and analysis of vanishing point detection techniques. IEEE Pattern Analysis and Machine Intelligence 21, 282–288 (1999)
27. Veit, T., Tarel, J.P., Nicolle, P., Charbonnier, P.: Evaluation of road marking feature extraction. In: 11th International IEEE Conference on Intelligent Transportation Systems, ITSC 2008, pp. 174–181 (2008)
28. Wang, Y., Bai, L., Michael, F.: Robust road modeling and tracking using condenzation. IEEE Transactions on Intelligent Transportation Systems 9, 570–579 (2008)
29. Wang, Y., Dahnoun, N., Achim, A.: A novel lane feature extraction algorithm implemented on the TMS320DM6437 DSP platform. In: Proceedings of the 16th International Conference on Digital Signal Processing, pp. 733–738 (2009)
30. Wang, Y., Shen, D., Teoh, E.K.: Lane detection using spline model. Pattern Recognition Letters 21(8), 677–689 (2000), doi:10.1016/S0167-8655(00)00021-0
31. Wang, Y., Shen, D., Teoh, E.K.: Lane detection using spline model. Pattern Recognition Letters 21, 677–689 (2000)
32. Wang, Y., Teoh, E.K., Shen, D.: Lane detection and tracking using B-snake. Image and Vision Computing 22, 269–280 (2004)
33. Yim, Y., Oh, S.-Y.: Three-feature based automatic lane detection algorithm (TFALDA) for autonomous driving. In: Proceedings of International Conference on Intelligent Transportation Systems, pp. 929–932 (1999)
34. Zhang, J., Nagel, H.-H.: Texture-based segmentation of road images. In: Proceedings of the Intelligent Vehicles 1994 Symposium, pp. 260–265 (1994)
35. Zhou, Y., Xu, R., Hu, X., Ye, Q.: A robust lane detection and tracking method based on computer vision. Measurement Science and Technology 17, 736–745 (2006), doi:10.1088/0957-0233/17/4/020

Chapter 7
Detecting Anomalies in Sensor Signals Using Database Technology

Gereon Schüller, Andreas Behrend, and Wolfgang Koch

Abstract. Signals are usually post-processed in order to enhance their accuracy and reliability. For instance, sensor data from moving objects data are often processed by tracking systems which allows for enhancing the provided kinematic information. In high-level fusion systems, this kinematic information can be combined with additional domain-specific data which allows for detecting object behavior and threat patterns. These systems contribute to situation awareness by employing patterns which characterize situations of interest. The used patterns may vary over time and depend on the specific questions to be investigated. Database systems provide a flexible way of combining data, and continuous queries allowing an ongoing automatic evaluation of search patterns. In this chapter, we present a way of using database systems as the central component in a higher-level fusion system. We discuss how patterns for the detection of anomalies in tracking scenarios can be expressed in relational algebra. Finally, we present an application of such a system for monitoring and analyzing air traffic using a commercial database management system.

7.1 Introduction

Anomaly detection is the processing of information from several sources to alert human decision makers that some state is "irregular" and requires special attention or action. Context information is an important instrument for anomaly detection because it provides insight into otherwise unobservable events or improves the accuracy, reliability or costs of the result that could be produced from a single information source and thus enhances situation awareness. These pieces of information can

Gereon Schüller · Wolfgang Koch
Department of Sensor Data and Information Fusion, Fraunhofer FKIE, Wachtberg, Germany
e-mail: `{gereon.schueller,wolfgang.koch}@fkie.fraunhofer.de`

Andreas Behrend
Department of Computer Science, University of Bonn, Germany
e-mail: `behrend@cs.uni-bonn.de`

P. Georgieva et al. (Eds.): Adv. in Intelligent Signal Processing and Data Mining, SCI 410, pp. 175–196.
springerlink.com

be of different origin, but we can generally distinguish between dynamically changing data, e. g. kinematic or attribute data gained by sensors, and rather static, slowly changing data derived from context information. What is assessed to be a "regular" or "irregular" situation depends on the data. It is related with the nature of the events themselves or, on the quality of the (ir-)regularity of the behavior of the observed objects. Both may vary from time to time, some criteria may be abandoned and new criteria may be introduced. Thus, a flexible anomaly detection system would be desirable, that could be adapted to provide new insight without the need for a complete redevelopment of the fusion system.

A database system (DBS) is a powerful way of fusing different data sources and exploit them by means of queries posed to the DBS. DBSs are often regarded as systems of rather slowly changing data, organized by a centralized database management system (DBMS) that answers pre-defined questions. However DBSs are more powerful as they are able to fuse data and can perform complicated computations in an efficient way. In a recent paper [6], we have shown how a database can be constructed that acquires sensor data, performs a tracking algorithm and stores the intermediate data and results for faster performance. In this chapter, we will present a way of augmenting such a system, that can be seen as a level-1-fusion-system, to obtain a higher-JDL-level-system [19] and provide anomaly awareness.

This chapter is organized as follows: Section 7.2 provides a few examples of applications for a tracking and awareness system. In Section 7.3, possible criteria for anomalies are presented. Section 7.4 gives a motivation for the usage of a database system for processing sensor signals and we briefly introduce the basic concepts of relational databases and relational algebra. Examples for expressing various anomaly criteria using relational algebra are presented in Section 7.5. Afterwards, a practical realization of a high-level fusion system is presented which is concerned with the monitoring of air traffic. The chapter closes with an outlook to future work and a conclusion.

7.2 Driving Factors for a Tracking and Awareness System

The rapid development of information, communication and sensor technology has led to the development of ubiquitous sensor networks and to a real-time availability of large amounts of data. Together with these technical abilities, a growing demand in the civilian and military area for exploitation of these data can be recognized. In the moment, we have arrived in a situation where the acquisition of data is a minor problem, but drowning in data, without drawing the best possible benefit from it becomes a more and more critical issue. Let us consider for example the amount of data in the Internet. According to experts, the amount of Internet data doubles every 2 months and has reached 492 trillion bytes in August of 2009 [17]. It is rather clear that no human being could ever explore the entire Internet; and while it may be possible to read the 32 volumes of the *Encyclopædia Britannica*, it is absolutely impossible to read the 3 million articles of *Wikipedia*.

The same holds for sensor data. In 2009, 150 active military satellites, including 54 pure sensor satellites acquire data, potentially around the clock and around the world, not counting all the sensors that reside in airplanes, ships, land vehicles etc. Each measurement can be inaccurate, incomplete, ambiguous, or irresolvable. Measurements may be false or corrupted by hostile measures. These problems are common in practice. So the extraction of senor information is not trivial and has led to sophisticated tracking methods [4, 7, 19, 18]. In this chapter, we assume that target tracking and data fusion techniques have become a key technology, but they bear potential for substantial contribution to a situation awareness system.

Awareness systems based on tracking have a vast amount of applications. For instance, a military awareness systems could be:

- a system that identifies the intrusion of adversarial airplanes into the own airspace
- a system that distinguishes civilian ground vehicles from military or terrorist vehicles
- a system that identifies pirate attacks on maritime ships.

In the civilian area, an awareness system could be used for

- the identification of traffic jams, accidents or cars driving against the direction on motorways
- detecting swarms of birds endangering air traffic
- identifying persons carrying hazardous material (in combination with chemical sensors).

In all these cases, an automatic system would free human operators from the task of watching the tracked results themselves in order to detect patterns, a fatiguing and thus error prone task due to the rareness of these events. Depending on the granularity of the data fusion process, different levels are usually distinguished. The common JDL-model [19], for example, distinguishes the following fusion levels:

Level 1: Object refinement. Association of data (including products of prior fusion) to estimate an object or entity's position, kinematics, or attributes (including identity).

Level 2: Situation refinement. Aggregation of objects/events to perform relational analysis and estimation of their relationships in the context of the operational environment (e.g., force structure, network participation and dependencies).

Level 3: Impact Assessment. Projection of the current situation to perform event prediction, threat intent estimation, own force vulnerability, and consequence analysis. Routinely used as the basis for actionable information.

Level 4: Process Refinement. Evaluation of the ongoing fusion process to provide user advisories and adaptive fusion control or to the request additional sensor/source data.

A tracking and awareness system could be seen between level 1 and level 2. In the following, such a system will be called “higher-level-fusion system” in order to indicate that the system exceeds level 1.

7.3 Criteria for Anomaly Detection

One can imagine that monitoring real life situations in order to detect threats depends on knowledge about “abnormal” events defined either directly or recognized as a deviation from a pattern of normality. We will now classify the parameters that can be expected to be indicators of such abnormal events.

7.3.1 Pattern Based Filtering for Improved Classification and Threat Detection

The detection of a certain type of anomalies can be based on object classification combined with context information. Mutual object interrelations relevant to anomaly detection can be inferred from multiple target tracking techniques. For example, if objects move in a group from the same source and/or to the same destination, it can be concluded that these objects belong together and can be classified as “belonging to group X”. If one object of this group can be classified as belonging to a certain object class, it can serve as a classification of the entire group. Also split-off events can be interesting, e.g., they can be classified as missile launches or as debarkation of a speed boat from a mother ship.

Standard target tracking applications gain information related to ‘Level-1-Fusion’, according to the JDL-model. Fusing the time-series of data allows for predicting future object movements in case new measurements are missed. On the other hand, new knowledge about the most current track data can be used for enhancing historical track data, too. This process is called “retrodiction”, a made-up word derived from “retrospective” and “prediction”.

As an example, Figure 7.1 shows the effect of retrodiction for analyzing a dog fight between two aircrafts. The dots represent single radar measurements while the line fragments refer to indicated tracks. In the upper image, various hypothetical tracks are shown because the tracking algorithm did not recognize the actual number of aircrafts. At a certain point, the system identifies that two planes are in the field of view (FoV). In the lower image, this information has been used for enhancing historical track data and loose track fragments could be connected. After improving

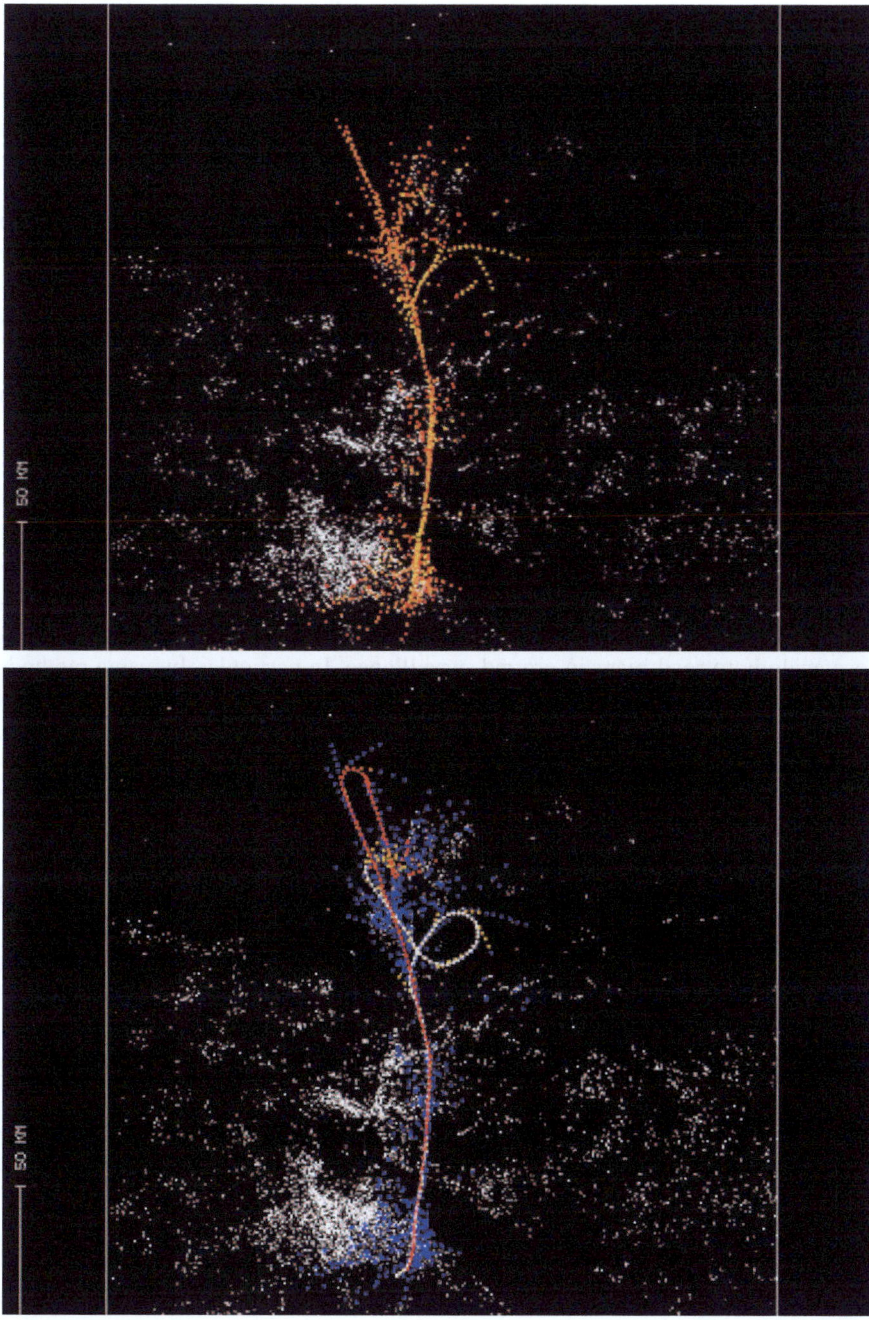

Fig. 7.1. Air scenario with two aircrafts and a high residual clutter background. The upper image shows the results of a normal filtering process, leading to multiple track hypotheses. The lower image shows the tracks after retrodiction, yielding one track for each target and allowing for the identification of a dog fight.

the track quality by retrodiction, one can classify the objects represented by the tracks with respect to the following information aspects:

1. *Velocity History*: Objects moving at high speed can be distinguished from objects moving at low speed, e. g., boats vs. vessels, or helicopters vs. jet aircrafts. Helicopters could also be identified if an object "stops" in the air.
2. *Acceleration History*: A fighter aircraft, for example, can perform higher accelerations than civil airplanes. Certain weapon systems will limit acceleration of fighters, so a high acceleration means that this aircraft does not carry these weapons (anymore).
3. *Heading, Aspect Angle:* These can be combined with the Doppler resolution spectra to classify the object.

The results of these filter processes based on the tracking products after retrodiction and genuine context knowledge may reveal either "standard" object behavior or deviation from this behavior.

7.3.2 Violation of Space-Time Regularity Patterns

Certain conclusions from JDL-level-1 tracking information can be deduced using available context information. In maritime surveillance, sea and air lanes exist in close analogy to road-maps in ground surveillance. In particular, by using methods developed for road-map assisted vehicle tracking, it is possible to test the hypothesis 'Off-lane vessel/aircraft' versus 'On-lane vessel/aircraft' and vice versa. Evidently, a result such as 'A target moving on a lane now moves off-lane' provides higher-JDL-level information with many implications on the target type and its possible intentions. As a side remark, precise and up-to-date sea/air lane maps, i.e., higher level context information, can be constructed by analyzing tracks of "lane-moving" targets. A closer analysis of JDL-level-1 tracks can thus be the key to detecting anomalous events of this type:

1. *Rare events:* The combination of kinematic tracks with context information like road maps can for example identify a truck on a dirt road in the forest at night.
2. *Common history:* Multiple target tracking can identify moving groups. If a group identified as 'hostile' splits off, it can be concluded that each object is a hostile one.
3. *Object Sources and Destinations:* The analysis of many tracks allows to identify common start and end points, that can then be classified as sinks and sources of objects, like an airport or a submarine pen.

7.3.3 Exploiting Poor-Resolution Sensor Attributes

Emerging applications in harbor protection or ship access control require non-standard information sources such as chemical sensors for detecting 'anomalous' materials (i.e. Chemical, Biological, Radiological, Nuclear and Explosive materials, also known as CBRNE). Chemical sensors, for example, are fundamentally limited with respect to their poor space-time resolution, since the detection of a chemical signature always occurs with a certain time delay. Moreover, as the non-directional spread of chemical substances is hard to calculate [30], we are usually unable to localize its source, to associate it with a certain individual, or to track it. This deficiency, however, can be compensated by fusing the output of multiple chemical sensors distributed in the observed space with kinematic track data of the objects moving in this area. In other words, an additional temporal dimension for processing chemical sensor attributes is spanned in tracking algorithms.

7.3.4 Varying Criteria and the Need for Flexibility

The criteria mentioned above may vary over time. For instance, we can distinguish bomber jets from fighter jets by means of their acceleration. However, new technical development has led to the introduction of more agile bombers, our criteria have to be changed accordingly. Another example would be that a former dirt road has been paved, so that the context data has to be updated. In both cases, it is desirable to have a flexible system that can be adapted to new needs without the necessity of reprogramming the whole system. In the next section we will see that this flexibility can be reached using a database management system.

7.4 Relational DBMSs for Processing Sensor Data

At a first glance, it seems to be an unusual idea to employ a conventional database system for processing sensor data. However there are many good reasons for processing sensor data directly in a database system: First, the amount of data to be processed must be taken into consideration. For instance, in radar systems, the amount of clutter will typically exceed the amount of true measurements and may become overwhelming. Therefore, radar data can be stored within a database which additionally provides a history control and a recovery mechanism. If data obtained from moving object sensors are to be tracked, tracking algorithms require the frequent retrieval of certain subsets of radar data which can be efficiently realized using the index mechanisms of a database.

Security aspects also have to be taken into account. As any computer system, sensor systems may be subject to system failures, due to hardware or software problems. In many tracking scenarios it is important that the complete tracking history can be reconstructed after a system failure, a functionality that is automatically provided by the recovery component of database management systems. Another issue which becomes more and more important is the fusion of multiple sensors, multiple platforms and multiple users. Data acquired from multiple sensors enhance the process results. In the field of moving object tracking, multiple view angles can be combined and false alerts are more easily distinguished. It is also possible to combine sensors of different types, like optical and radar sensors for example. Synchronizing these data streams is a difficult task, as there may be access violations, delays etc. This becomes even more likely if sensors are mounted on a mobile platform and the distribution of data is done using a wireless connection.

Another aspect is that multiple users should be able to retrieve the raw sensor data as well as the processed results. As database systems provide techniques for handling multiple user access and result distribution, they are well-suited for solving these synchronization problems. The persistent storage of raw data is of great importance when decisions must be justified against the logging history.

It is widely believed by now that conventional relational database systems are not well-suited for dynamically processing fast changing data. Therefore, various extensions for the structured query language (SQL) have been proposed [14, 28], as well as for stream processing engines [1, 2, 3, 20]. Some of them have even been designed as full-fledged commercial products (e.g, StreamBase [25]). We believe, however, that even conventional SQL queries can be efficiently employed for analyzing a wide spectrum of sensor data streams. In particular, we think that the application of incremental update propagation allows for evaluating continuous queries efficiently which involve high-frequent track data as well as static context data (see Section 7.5.2).

In the following section we will show how conventional SQL queries can be efficiently employed for analyzing a high frequent stream of radar data streams. In particular, we will show that the application of incremental update propagation considerably improves the efficiency of computing answers to continuous queries. A complete description of our approach can be found in [6].

7.4.1 Relational Databases and Relational Algebra

This section gives a brief introduction to *relational databases* and to the *relational algebra*. Today, most of the database management systems, e.g. Oracle, Microsoft SQL Server, IBM DB2, MySQL etc. rely on the relational model introduced by Edgar F. Codd in 1970 [10]. The basic idea is to use tables (relations) for storing data. A relation R is a subset of the Cartesian product of arbitrary sets:

$$R \subseteq A_1 \times A_2 \times \ldots \times A_n \tag{7.1}$$

Each element of relation is called *tuple*, the j-th component is called j-th attribute, the set A_j domain of attribute j. Relations are assumed to be duplicate-free, i.e., two tuples differ in at least one attribute value. Each attribute is given a name in order to distinguish it from other attributes. The structure and the name of a relation is called *schema*. The set of tuples representing a relation is called *state*. In practice, relations are represented by means of *tables* while tuples are given by means of *rows*. One important difference is that tables may represent multisets and thus are not always duplicate-free.

Databases are not only intended for storing data, but also for performing operations on stored data. These operations can be expressed using the *relational algebra* (RA). The relational algebra maps relations onto relations. The basic set operations, i. e. union ($\cup$), intersection ($\cap$) and set difference ($\setminus$ or $-$) can be directly transferred to the relational algebra with the limitation that both operands have to be union-compatible, i. e., both operands must have the same number of attributes and corresponding attributes must have the same domain.

The Cartesian product (also called cross product) is a bit redefined in comparison to its set correspondent, as it will not yield a pair of tuples but a new tuple. It is defined as follows

$$\{(a_1,a_2,\ldots,a_n)\} \times \{(b_1,b_2,\ldots,b_m)\} \mapsto \{(a_1,a_2,\ldots,a_n,b_1,b_2,\ldots,b_m)\} \quad (7.2)$$

Beside the set operators, there are RA specific operators. The *selection* σ select those tuples, that satisfy a given condition:

$$\sigma_c(R) = \{t \in R \mid c\} \quad (7.3)$$

where c is a predicate that may contain attributes, constants and logical operators. The *projection* $\pi_{a_1,a_2,\ldots,a_m}(R)$ removes all attributes from the relations that are not listed in the sub script $a_1,a_2,\ldots,a_m$. The *renaming* $\rho_{a\leftarrow b}(R)$ renames the attribute b to a. It may also be applied to relation names.

The composition of cross product and selection can be expressed using the *join* operator. There are various join variants and we will recall the most important ones, only. The *natural join* (symbol $\bowtie$) links two relations, by selecting those tuples from the cross product that have equal values in equal-named attributes. It is defined as follows

$$R \bowtie S = \pi_{A_1,\ldots,A_m,R.B_1,\ldots,R.B_k,C1,\ldots,Cn}\sigma_{R.B_1=S.B_1\wedge\ldots\wedge R.B_k=S.B_k}(R \times S) \quad (7.4)$$

where the attribute $B_1,\ldots,B_k$ represent the common attributes of relations R and S. Note that a projection is employed for removing doubled attributes from the result set. A Theta-Join (symbol $\bowtie_\theta$) works in a similar way by applying the selection $\sigma_\theta(\cdot)$ to the cross-product. A projection, however, is not implied. Semi joins ($\ltimes$, $\rtimes$) represent another join variant where the attributes of one input relation (hence there exist a left semi join and a right semi join) is projected out after performing a Theta-Join. The complement join (symbol $\overline{\bowtie}$) will only return tuples that do not have a matching partner.

In the following, we allow for defining equations of the form

$$\begin{aligned} C &= A \bowtie B \\ D &= \sigma_c(C) \\ \Leftrightarrow D &= \sigma_c(A \bowtie B) \end{aligned}$$

Relations that are derived by defining equations are known as *views*.

7.5 Expressing Anomalies in Relational Algebra

In order to implement anomaly detection using a sensor database system, the nature of anomalies has to be expressed in a relational way, using the relational algebra. We give now a few examples of RA expressions for anomaly detections. It is easy to see that only the definition of anomalies is specified, not the evaluation process for the detection process, which keeps the system flexible for modifications and adjustments. Due to the concept of logical data independence in relational database systems, necessary modifications are transparent to the user and can also be done while the system is running.

7.5.1 *Velocity/Acceleration Classification*

As an illustrative example, we present a relational algebra system that identifies the current position of potential threats. Assume that these threats consist in helicopters, military airplanes, and fighter jets that are additionally seen as a threat for airplanes. A possible system of RA equations could look like this:

$$\text{flightObjects} = \sigma_{\text{heightOverGround}>10}(\text{objectTracks}) \tag{7.5}$$

$$\text{civilAirplanes} = \sigma_{\text{velocity}<300 \wedge \text{velocity}\geq 10}(\text{flightObjects}) \tag{7.6}$$

$$\text{militaryAirplanes} = \sigma_{\text{velocity}\geq 300}(\text{flightObjects}) \tag{7.7}$$

$$\text{helicopters} = \sigma_{\text{velocity}<10}(\text{flightObjects}) \tag{7.8}$$

$$\text{fighterJets} = \sigma_{\text{acceleration}>3}(\text{militaryAirplanes}) \tag{7.9}$$

$$\text{wingsArms} = \sigma_{\text{acceleration}>6}(\text{fighterJets}) \tag{7.10}$$

$$\text{threats} = \pi_{\text{ID},x,y,z}(\sigma_{t=max(t)}((\text{helicopters}\cup \tag{7.11}$$

$$\text{militaryAirplanes}))) \tag{7.12}$$

$$\text{airThreat} = \pi_{\text{ID},x,y,z}(\sigma_{t=max(t)}(\text{fighterJets})) \tag{7.13}$$

In Equation 7.5, we filter out all objects that lie on the ground or fly lower that 10 m. In 7.6-7.8, we separate the flying objects into three classes, depending on their velocity. Equations 7.9 and 7.10 separate the objects according to their accelerations. The last two Equations 7.11 and 7.13 identify possible threats by providing the position and ID of tracks assigned to helicopters and military airplanes. In 7.13, threats to airborne objects, in this case fighter jets, are identified.

7.5.2 *Context Information and Several Sensors*

We consider now our example for rare events: a freight truck on a dirty road. Assume that we have a map of the surveillance area that contains the information "if a road is a dirt road" as context information and that we have a scalar function `onRoad` that is true if (and only if) a point is on a road. Then we could write

$$\begin{aligned}\text{rareSituation} = {} & \sigma_{\text{weight}>3500\wedge(t<6{:}00\vee t>22{:}00)}(\text{Cars}) \\ & \bowtie_{\text{onRoad(Cars.x, Cars.y, r.x1, r.y1, r.x2, r.y2)}} (\sigma_{\text{r.quality='Dirt'}}(\text{Roads})))\end{aligned} \tag{7.14}$$

A similar equation would return the off-road targets:

$$\text{offRoad} = \text{Cars} \,\overline{\ltimes}_{\text{onRoad(Cars.x, Cars.y, r.x1, r.y1, r.x2, r.y2)}} \text{Roads} \tag{7.15}$$

The next step in data fusion is the combination of measurements obtained from different sensor types. Let us reconsider the chemical sensor example for detecting CBRNE materials. Assume that we have a tracking algorithm that tracks persons in a field of view and that we have chemical sensors which measure the concentration of chemicals in the air. We could store the "field of smell" (FoS) for each chemical sensor in a table:

ID	x1	y1	x2	y2
1	0	0	1	1
2	4	4	3	2
...				

Then we combine the tracking information with the chemical detection of a critical substance in parts per million (ppm) to get the possible carriers *(possCarr)* of a substance:

$$\begin{aligned}\text{possCarr} = {} & \text{Tracks} \bowtie_{\text{tracks.x<fos.x1}\wedge\ldots\wedge\text{tracks.y>fos.y2}} \text{FoS} \\ & \bowtie_{\text{fos.id=Sensor.id}\wedge\text{tracks.t = Sensor.t}} (\sigma_{\text{ppm}>300}(\text{Sensor}))\end{aligned} \tag{7.16}$$

Now we can filter out all persons that have been detected as possible carriers at different times:

$$\begin{aligned}\text{suspects} = {} & \rho_{\text{pc1}\leftarrow\text{possCarr}}(\text{possCarr}) \\ & \bowtie_{\text{pc1.ID=pc2.ID}\wedge\text{pc1.t>pc2.t}} \rho_{\text{pc2}\leftarrow\text{possCarr}}(\text{possCarr})\end{aligned} \tag{7.17}$$

7.5.3 Incremental Evaluation of Relational Queries

In relational queries, the user does not specify how to answer the query. A query does only specify the answer relation. It is the responsibility of the underlying database system, to find an efficient query execution plan for determining the answer relation. However, database systems cannot always choose the most effective solution. As an example, consider the following anomaly view for detecting landings on airports

$$\text{landing} = \sigma_{\text{VertSpeed}<0\wedge\text{Flightlevel}<0}(\text{Tracks}) \bowtie_{\text{dist(Tracks.pos, Airports.pos);20km}} \text{Airports} \tag{7.18}$$

where the relation *Tracks* is constantly updated by new data, while the relation *Airport* is unchanged. A DBMS could choose the following evaluation plan:

1. Execute the selection operation on all tracks stored in the database.
2. Calculate the distance between all selected tracks from step 1 to all airports.
3. Select all combinations where the distance is below 20 km.

In a static database system, this plan is almost optimal, as the most selective restriction is moved to step 1. However, in constantly updated databases, this plan will suffer from the fact that the query is completely re-evaluated for every new entry in Tracks, although a small number of answer tuples may change, only.

To speed up the computation process, a straightforward solution is to store (materialize) the results and to compute the new answers, only. In order to compute new answer tuples, so-called update propagation (UP) methods [9, 15, 16, 21] can be employed. An update propagation process allows for incrementally computing new answer tuples and is initiated by the DBMS every time new stream data is recorded. The incremental evaluation is achieved by specialized RA expressions that describe the induced changes with respect to the original query. These expressions are known as *delta rules*. Delta rules can be automatically derived from the given operators within a query expressions. A list of transformation rules is shown in Table 7.1.

Table 7.1. Delta-Rules for basic RA operators

Join	$P = Q \bowtie R$	$P^+ = (Q^+ \bowtie R^0) \cup (Q^0 \bowtie R^+) \cup (Q^+ \bowtie R^+)$
		$P^- = (Q^- \bowtie R^0) \cup (Q^0 \bowtie R^-) \cup (Q^- \bowtie R^-)$
Intersection	$P = Q \cap R$	$P^+ = (Q^+ \cap R^0) \cup (Q^0 \cap R^+) \cup (Q^0 \cap R^0)$
		$P^- = (Q^- \cap R^0) \cup (Q^0 \cap R^-) \cup (Q^- \cap R^-)$
Union	$P = Q \cap R$	$P^+ = (Q^+ \cup R^+)$
		$P^- = ((Q^- \backslash R^0) \backslash R^+) \cup ((R^- \backslash Q^0) \backslash Q^+)$
Difference	$P = Q \backslash R$	$P^+ = ((Q^+ \backslash R^0) \backslash R^+) \cup (R^+ \cap Q^0) \cup (R^- \cap Q^+)$
		$P^- = Q^- \cup (R^+ \cap Q^0)$
Selection	$P = \sigma_c(R)$	$P^+ = \sigma_c(R^+)$
		$P^- = \sigma_c(R^-)$
Projection	$P = \pi_c(R)$	$P^+ = \pi_c(R^+)$
		$P^- = \pi_c(R^-)$

All relational algebra expressions can be transformed to incremental rules this way. In anomaly detection systems, it is handy to specialize the delta rules with respect to the relations that are expected to change frequently. In the above example, the delta rule for insertions into the *Tracks* relation relation would be:

$$\text{landing}^+ = \sigma_{\text{VertSpeed}<0 \wedge \text{Flightlevel}<0}(\text{Tracks}^+) \bowtie_{\text{dist(Tracks.pos, Airports.pos)} < 20\text{km}} \text{Airports}. \quad (7.19)$$

Obviously, this rule will perform faster if most data in the *Tracks* relation stays unchanged. In [6], we have shown how this technique can be applied to a complete tracking system which led to an improvement of the run time behavior from cubic to linear. This way, even a real-time analysis of sensor data streams with a relational database system becomes feasible.

7.6 Anomaly Detection for Improving Air Traffic Safety

The safety of air traffic is an issue that has become more and more difficult to handle for human air traffic controllers. The detection of anomalies in air space, such as critical encounters or planes deviating from their planned route, is one of the main tasks of Air Traffic Control (ATC). In recent history, many situations have been reported where controllers missed critical situations, like in the Ueberlingen accident [8]. Automatic anomaly detection can provide a solution to support human air traffic controllers. In the following, we want to give insights into a recently developed system for airspace monitoring [22, 23]

The "Airspace Monitoring System" (AIMS) monitors and analyzes flight data streams with respect to the occurrence of arbitrary, freely definable complex events. In contrast to already existing tools which often focus on a single task like flight delay detection, AIMS represents a general approach to a comprehensive analysis of aircraft movements, serving as an exemplary study for many similar scenarios in data stream management. In order to develop a flexible and extensible monitoring system, SQL views are employed for analyzing flight movements in a declarative way. Their definition can be easily modified, so that new anomalies can simply be defined in form of view hierarchies. The key innovative feature of AIMS is the implementation of a stream processing environment within a traditional DBMS for continuously evaluating anomaly detection queries over rapidly changing sensor data.

7.6.1 Problem Setting

The measurement of air traffic is mainly done by two means: primary and secondary radar. A primary radar measures the run-time (proportional to distance) and heading direction of electromagnetic waves that are reflected by the surface of flying objects. It is less accurate than secondary "radar" that emits an electromagnetic pulse and measures the active answer of a transponder. An antenna measures the direction and runtime of the answer pulse. Today, all aircrafts above 5.7 metric tons of mass and a possible true airspeed above 250 knots have to be equipped with a so-called *Mode S* transponder [11]. The answer message from this kind of transponder will also return the height and velocity as measured by the aircraft's instruments.

However, position measurement suffers from several drawbacks. First of all, radar measurements are usually incomplete as some signals may be missed. One reason is that some aircrafts head towards a direction in which the radar pulse is reflected weakly or diverted from the receiver antenna. Another problem arises from bad weather conditions which may lead to a high damping of the signal. In secondary radar, it may also occur that the transponder misses a request from the ground station or that it suffers from malfunction. Additionally, a radar system itself usually leads to clutter and false detections. Clutter results from reflections of the ground, humidity or other atmospheric effects as well as from the intrinsic noise added by the transmitter antenna. It will add noise to the "true" reflected signals and shows up as randomly distributed measured points (plots) over the FoV. False detections may arise from punctually disturbing sources like clouds, birds etc. Misdetection and false alerts/clutter are connected anti-proportionally as levering the amplification of the signal will also amplify the disturbing signals [24]. Another issue is that radar signals are not sharp in position. The measurement will be smeared out due to the nature of radar antennas. If multiple objects are in the FoV (which is likely in airspace), problems will arise from the connection between plot and plane.

To this end, tracking algorithms are deployed to construct tracks from the measurement data. In the past decade, many tracking algorithms have been proposed, e.g. [7, 27, 29], and this topic is still an active research area (at Fraunhofer FKIE in particular). It is also possible to incorporate tracking algorithms directly into sensor databases [6].

AIMS is aimed at supporting the analysis of aircraft movements based on the tracked radar observations and/or transponder signals to avoid situations where anomalies or critical events are detected too late or remain undetected at all. Many problems are caused by local deviations from flight plans which may induce global effects to air traffic. One effect is a considerable number of close encounters of planes in airspace occurring every day. Another less critical but increasingly important example is the violation of no-fly zones over cities for noise protection reasons, or over power plants for preventing terroristic attacks.

AIMS [22] is a prototype of a system for monitoring and analyzing local and global air traffic [1]. It has been developed at the University of Bonn in cooperation with Fraunhofer FKIE and EADS Deutschland GmbH. One aim of this prototype is the detection of anomalies within the movements of individual aircrafts (e.g., critical delays, deviations from flight plans or critical maneuvers). Based on that, a global analysis is supported (e.g., critical encounters, zones with high flight density, airport jams) which allows for adjusting air traffic influenced by local phenomena. Currently, the system is used to monitor the complete German airspace every 4 seconds with up to 2000 flights in peak times.

The key innovative feature of AIMS is the use of continuously evaluated SQL queries (stored in the DB as views, too) for declaratively specifying those situations that are to be detected. To this end, consistent tracks of individual planes have to be derived and complex events occurrences within these track data have to be found. Our general research aim is the development of efficient DBMS-based methods for

[1] `http://idb.informatik.uni-bonn.de/research/aims`

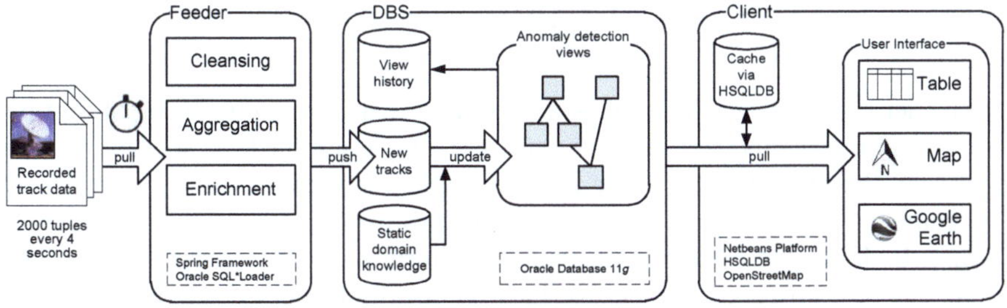

Fig. 7.2. The architecture of AIMS. The main data flow is from right to left, starting with the feeder process. The feeder cleanses the data and aggregates measurements which are assigned to the same time stamp. Afterwards, the data inserted into the DBS. The DB server combines measurements with context information and applies anomaly detection views. Consequently, the view results are displayed to the user using a client component which continuously fetches data from the DB server in regular intervals. In addition, the client manages a local cache for accelerating the access to the recent stream history.

real-time gathering and monitoring of streams of track data. In particular we want to answer the following questions with respect to the airspace data:

- Is it possible to detect aircrafts entering a critical situation like collision course, leaving the flight path or entering bad weather zones using SQL views?
- How can flights and their behaviour be classified using SQL views defined over track data?
- How can tracking of aircrafts be improved by using (derived) context information?

With respect to data stream management, the following research questions are addressed by AIMS:

- Which type of continuous query can be efficiently evaluated using a conventional relational DBMS?
- Up to which frequency/volume is it feasible to use a relational DBMS for evaluating continuous queries over a stream of track data?
- To which extent can an SQL-based analysis be used for airspace surveillance?

Using SQL queries as executable specifications has the advantage of being able to easily extend the system by additional criteria without having to re-program large amounts of code. In order to continuously re-evaluate the respective queries, data stream management systems (DSMS) like STREAM[3] or Aurora [1] could be used. However, DSMS do not provide all capabilities that are needed for a reliable in-depth analysis, like recovery control, multiuser access and processing of historical as well as static context data. In this book chapter, we show, how commercial database systems in combination with intelligent rule-rewriting can be used to processes a geospatial data stream. The proposed approach provides insights for the implementation and performance of related applications where geospatial or sensor data streams have to be processed. A graphical representation of the architecture of AIMS is shown in Fig. 7.2. The basic system consists of three main components:

1. A feeder component which takes a geospatial data stream as an input and periodically pushes its data into the database. This track feeder also continuously activates the re-evaluation of the anomaly detection views.
2. A graphical user interface programmed in Java using the NetBeans platform library. It shows the positions of analyzed aircrafts on an OpenStreetMap. This map can be configured in order to show the result of selected queries, only. Additionally, query results are displayed in tabular form.
3. An Oracle server which stores the stream data and performs the continuous evaluation of the user-defined anomaly detection views.
4. An in-memory database that works as a cache and stores the results of selected queries on the client side. This database system forms the basis for a time-shift and video recorder functionality.

7.6.2 View-Based Flight Analysis

Although there are various commercial implementations of flight tracking services (e.g., AirNav [26], FlightView [13] or FlightStats [12]), these services are often limited to a set of predefined tasks like delay detection or identification of basic flight states such as departing, approaching or cruising. In order to develop a flexible and extensible monitoring system, we have decided to used SQL views for analyzing our track data. The advantage of a view-based analysis is that the underlying definition can be easily recovered and modified while new anomalies can be simply defined in form of view hierarchies. In the following, we give two examples for this kind of analysis.

As a first example of an interesting event in airspace we consider landing flights. To this end, the plane must have a negative vertical speed (i.e., it is descending), it has to be below a certain flight level (like 3000 ft), and it must be in the vicinity of an airport, e.g. closer than 20 mi. These criteria can be expressed in relational algebra as follows:

$$\text{landing} = \sigma_{\text{VertSpeed}<0 \wedge \text{Flightlevel}<0}(\text{Tracks}) \bowtie_{\text{dist(Tracks.pos, Airports.pos)}} \text{Airports} \tag{7.20}$$

In a relational DBMS, however, queries are formulated in a high-level languages such as SQL rather than relational algebra expressions. Every algebra expression can be translated into SQL in a systematic way. For instance, the corresponding SQL query to the above statement looks as follows:

```
CREATE VIEW vwLanding AS
SELECT *
FROM TrackData t, airports a
WHERE t.vertSpeed<0
  AND t.flightLevel<3000
  AND dist(a.pos, t.pos)<20000;
```

Note that the function `dist` is a user-defined function (UDF) for calculating the Euclidean distance between two positions on the globe. The base table `airports` stores data about position and names of airports.

Another interesting anomaly is the determination of critical encounters where two planes undercut the prescribed distance of security. In AIMS a critical encounter is detected if two planes are closer than 300 ft. A corresponding statement in relation algebra may look as follows:

$$\text{encounter} = \rho_{\text{t}_1 \leftarrow \text{Tracks}}(Tracks) \bowtie_{\text{dist}(\text{t}_1.pos, t_2.pos) < 300 \wedge \text{t}_1.ID \neq \text{t}_2.ID} \rho_{\text{t}_2 \leftarrow \text{Tracks}}(Tracks) \tag{7.21}$$

The corresponding expression in SQL is given by:

```
CREATE VIEW vwEncounter AS
SELECT *
FROM TrackData t1, TrackData t2
WHERE dist(t1.pos, t2.pos)<300 AND t1.ID<>t2.ID;
```

Since the employed UDF-expression cannot be indexed, this implementation leads to a quadratic run-time. A better performance can be achieved by pre-selecting all those flights having approximate longitude and latitude values (e.g. ABS(t1.lat-t2.lat)<1 AND ABS(t1.lon-t2.lon)<1). This pre-partitioning can be performed fast because the values used are indexed. In Fig 7.3 a critical encounter of two flights is depicted. One can see the high number of radar plots used for determining the underlying tracks of the two involved planes.

The client of AIMS allows to freely add new user-defined anomaly detection views which are to be continuously evaluated by the system. For example, a user may combine landing flights and close encounters in order to define a new view for determining critical landings.

$$\text{criticalLandings} = \text{encounter} \bowtie_{\text{encounter.ID=landing.ID}} \text{landing} \tag{7.22}$$

The corresponding SQL statement looks as follows:

```
CREATE VIEW vwCritLandings
SELECT * FROM vwEncounter, vwLanding
WHERE vwEncounter.ID=vwLanding.ID
```

7.6.3 Enhancing Robustness and Track Precision

AIMS uses pre-processed radar data and associated transponder signals in order to determine position and heading at a high precision. To this end, a probabilistic multi-hypothesis tracker is employed for effectively combining various sensor signals. Consequently, every aircraft position is associated with an uncertainty ellipse indicating the probability density of position values. This approach can be used to predict new position values in case that no sensor data is received for an aircraft.

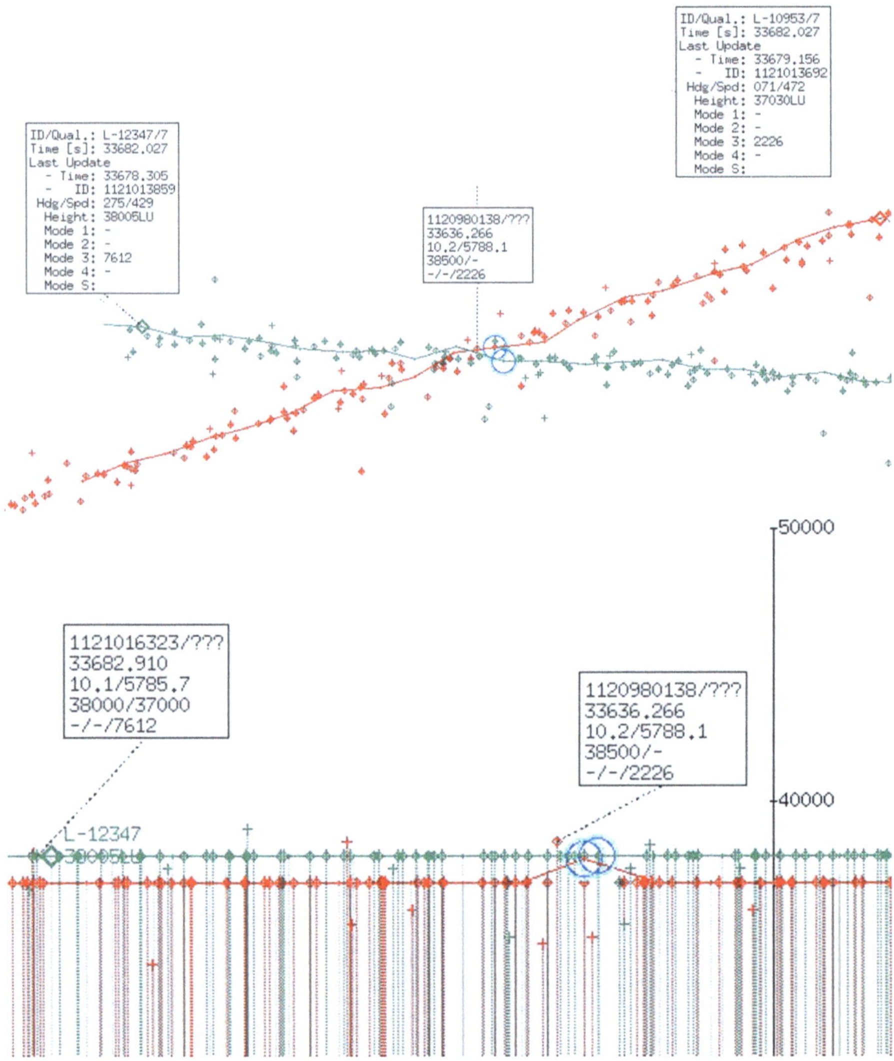

Fig. 7.3. Critical encounter of two airplanes as detected by the system. For testing purposes, the limit for a critical encounter was set to 1000 ft. In the upper image, the view from above is shown, the lower figure shows a height profile. The scale in the lower figure gives the altitude in feet. The lower part of the profile was cropped.

For improving the robustness of the underlying analysis, AIMS employs a "time to live" (TTL) approach. The reason is that given criteria in an anomaly detection view may be unsatisfied for a short time, leading to erroneous fluctuations in the result set. For example, planes may change their flight levels during their cruising period only because of noise in the underlying track data, while a landing plane may even gain altitude again for a short time in order to adjust its approach. In order to avoid such false alerts, we use the TTL approach: After a plane is no longer satisfying some given criteria, it is not deleted from the result set but rather an internal counter is decremented. A plane is deleted from the result set only if its TTL reaches the value zero. As soon as the underlying criteria are satisfied, however, the TTL is reset to its default value again. In this way, short term fluctuations are straightened a bit.

7.6.4 History Management

As soon as stream data have been processed they become historical data which have to be stored by the system. Historical data are necessary for provenance analysis or may be relevant for continuous queries involving long-term statistics. For example, the determination of the number of all landings in Frankfurt during the last hour could be done as follows:

```
SELECT count (*)
FROM (SELECT distinct track_id, callsign
      FROM landing_archive
      VERSIONS BETWEEN TIMESTAMP
        (systimestamp - interval '1' hour) AND
         systimestamp
WHERE airportname = 'Frankfurt Main');
```

This view employs the Total Recall facility of the Oracle DBMS for accessing time-stamped historical data. Total Recall is a technique that allows users to query historical versions of table data stored within a database archive. It turned out, however, that this elegant approach performs rather slow such that it is only suitable for purely historical queries (e.g., count all landings on Tuesday last week).

In case a statistical query with a sliding window is to be answered, AIMS employs an incremental approach where the new aggregate value is computed based on its former value and the new incoming stream data. For example, in the above query the count is increased by new landings detected in the current stream, and decreased by the number of landings which fall out of the one hour window.

For implementing a time slider functionality, neither Total Recall nor our incremental approach could be used. Therefore, we added a cache to the system which stores the recent continuous query results. To this end, the in-memory database system HSQLDB (Hyper Structured Query Language Database) has been used allowing to rewind to and replay from an arbitrary time point in the past. For example, the development of critical encounters could be comprehensively analyzed this way. In addition, the cached query results can be further analyzed using refined SQL statements.

For the incremental recomputation of anomaly detection views, a special update propagation method called Magic Updates is applied [5]. It reduces the size of intermediate joins by dynamically generated selection conditions. For example, consider the determination of delayed flights which are involved in a critical landing. For the determination of delayed flights we need to access the usually very large table of flight plans. However, the small number of critical landings can be used as an additional selection criterion for determining relevant flight plans, only. To this end, the determined critical landings are used to form so-called magic subqueries which are joined with the table of flight plans first. Using Magic Updates speeds up the recomputation of our anomaly detection views dramatically, improving the run-time from cubic to linear behaviour, w.r.t. the size of a stream package [6].

7.6.5 Experiences

AIMS could successfully identify critical situations like close encounters or deviations from the flight plan. It was interesting to notice the high number of serious deviations (sometimes more than 50 miles). In addition, the high number of critical approaches was very surprising, as there were far more close encounters than expected. We could also show violations of no-fly zones and determine zones with a critically high number of aircraft movements. Currently we are working on the determination of abnormal landing approaches.

Another result is that our incremental evaluation of SQL views indeed provides a suitable approach for analyzing this real world stream scenario. AIMS is capable of monitoring the entire German airspace by processing $\approx$1400 tuples (12 attributes) every 3 - 4 seconds. For performance measurements, the system has been tested with recorded data which were periodically fed into the system with increasing update frequency. Most of the monitoring tasks could be solved in less than 1 second (on a standard PC with Windows XP 32 bit and 4 GB of RAM), showing the feasibility of our approach.

7.7 Future Work and Conclusion

In the future, the system shall be expanded to compute more statistics about the observed air traffic, mainly in order to improve the organization of air traffic flow and to straighten out "hot spots", i.e. regions with abnormally high air traffic. It is also planned to add a component in order to simulate the consequences of a increase in traffic volume. We also plan to add further anomaly detection views for discovering blind areas where radar data is typically not available. In addition, the prediction capabilities of our system ought to be employed for improving the dynamic models used in our tracking software. For example, the knowledge about air traffic routes can be used to predict a flight curve even if no radar data is available. Another application may be used in meteorology: By analyzing the measured air speed (measured by Pitot Tubes) and comparing it to the detected ground speed (measured by radar and GPS), we are able to measure the wind speed at many points and to create a detailed wind map.

The field of anomaly detection has a wide range of application and can be useful in other domains than moving object monitoring, too, e.g. in the monitoring of machine states, stock exchange prices or in health care applications. The usage of database systems allows for a full insight into anomaly detection processes for historical evaluation and provableness. Moreover, the proposed concept can be seen as a special case of descriptive programming, i. e., the user specifies what to detect, but does *not* specify the concrete implementation of the underlying evaluation. In future applications, this way of functionally reactive programming will become more and more important as the complexity of systems and the need for flexibility increases.

References

1. Abadi, D., Carney, D., Çetintemel, U., Cherniack, M., Convey, C., Erwin, C., Galvez, E., Hatoun, M., Maskey, A., Rasin, A., Singer, A., Stonebraker, M., Tatbul, N., Xing, Y., Yan, R., Zdonik, S.: Aurora: a data stream management system. In: SIGMOD 2003, p. 666. ACM, New York (2003)
2. Abadi, D.J., Lindner, W., Madden, S., Schuler, J.: An integration framework for sensor networks and data stream management systems. In: VLDB 2004, pp. 1361–1364. VLDB Endowment (2004)
3. Arasu, A., Babcock, B., Babu, S., Datar, M., Ito, K., Nishizawa, I., Rosenstein, J., Widom, J.: Stream: the stanford stream data manager (demonstration description). In: SIGMOD 2003, p. 665. ACM, New York (2003)
4. Bar-Shalom, Y., Li, X.R., Kirubarajan, T.: Estimation with Applications to Tracking and Navigation, 1st edn. Wiley-Interscience (June 2001)
5. Behrend, A., Manthey, R.: Update Propagation in Deductive Databases Using Soft Stratification. In: Benczúr, A.A., Demetrovics, J., Gottlob, G. (eds.) ADBIS 2004. LNCS, vol. 3255, pp. 22–36. Springer, Heidelberg (2004)
6. Behrend, A., Manthey, R., Schüller, G., Wieneke, M.: Detecting Moving Objects in Noisy Radar Data Using a Relational Database. In: Grundspenkis, J., Morzy, T., Vossen, G. (eds.) ADBIS 2009. LNCS, vol. 5739, pp. 286–300. Springer, Heidelberg (2009)
7. Blackmann, S.S., Populi, R.: Design and Analysis of Modern Tracking Systems. Artech House, Boston (1999)
8. Bundesstelle für Flugunfalluntersuchung. Investigation Report AX001-1-2/02 (May 2004)
9. Ceri, S., Widom, J.: Deriving production rules for incremental view maintenance. In: VLDB 1991, pp. 577–589 (1991)
10. Codd, E.F.: A relational model of data for large shared data banks. Communications of the ACM 13(6), 377–387 (1970)
11. Eurocontrol. Mode S EHS - Requirements for the Carriage and Operation of Mode S Transponders (Online; Last validation March 1, 2010)
12. Flightstats. (2010), `http://www.flightstats.com/` [(Online, retrieved December 2010)]

13. FlightView (2010), `http://www.flightview.com/` [(Online, retrieved December 2010)]
14. Golab, L., Özsu, M.T.: Issues in data stream management. SIGMOD Rec. 32(2), 5–14 (2003)
15. Griffin, T., Libkin, L.: Incremental maintenance of views with duplicates. In: ACM SIGMOD 1995, pp. 328–339 (1995)
16. Gupta, A., Mumick, I. (eds.): Materialized Views: Techniques, Implementations and Applications. The MIT Press (1999)
17. ICS. The digital universe is still growing (2009), `http://www.emc.com/digital_universe` [(Online as of August 21, 2009)]
18. Koch, W.: Target tracking. In: Stergiopoulos, S. (ed.) Advanced Signal Processing Handbook – Theory and Implementation for Radar, Sonar, and Medical Imaging Real-Time Systems, ch. 8, CRC Press (2008)
19. Liggins, M.E., Hall, D.L., Llinas, J.: Handbook of Multisensor Data Fusion: Theory and Practice, Electrical Engineering and Applied Signal Processing, 2nd edn. CRC, Boca Raton (2008)
20. Madden, S., Franklin, M.J.: Fjording the stream: An architecture for queries over streaming sensor data. In: ICDE 2002, pp. 555–566 (2002)
21. Manthey, R.: Reflections on some fundamental issues of rule-based incremental update propagation. In: DAISD 1994, pp. 255–276 (1994)
22. Schüller, G., Behrend, A., Manthey, R.: AIMS: an SQL-based system for airspace monitoring. In: ACM SIGSPATIAL IWGS 2010, pp. 31–38. ACM, New York (2010)
23. Schüller, G., Saul, R., Behrend, A.: In-Memory Caching for Fast Stream History Access. In: ACM SIGSPATIAL IWGS 2011, pp. 31–38. ACM, New York (2011)
24. Skolnik, M.I.: Introduction to Radar Systems. McGraw-Hill (1981)
25. Stonebraker, M., Cetintemel, U.: "One Size Fits All": An Idea Whose Time Has Come and Gone. In: ICDE 2005, pp. 2–11. IEEE Computer Society, Washington, DC (2005)
26. Systems, A.: Airnavsystems (2010), `http://www.airnavsystems.com/` [(Online, retrieved December 2011)]
27. Van Keuk, G.: MHT extraction and track maintenance of a target formation. IEEE Transactions on Aerospace and Electronic Systems 38(1), 288–295 (2002)
28. Wang, H., Zaniolo, C., Luo, C.R.: ATLAS: a small but complete SQL extension for data mining and data streams. In: VLDB 2003, pp. 1113–1116. VLDB Endowment (2003)
29. Wieneke, M., Koch, W.: On sequential track extraction within the PMHT framework. EURASIP Journal on Advances in Signal Processing 2008, 1–13 (2008)
30. Wieneke, M., Koch, W.: Combined person tracking and classification in a network of chemical sensors. International Journal of Critical Infrastructure Protection 2(1-2), 51–67 (2009)

Chapter 8
Hierarchical Clustering for Large Data Sets

Mark J. Embrechts, Christopher J. Gatti, Jonathan Linton, and Badrinath Roysam

Abstract. This chapter provides a tutorial overview of hierarchical clustering. Several data visualization methods based on hierarchical clustering are demonstrated and the scaling of hierarchical clustering in time and memory is discussed. A new method for speeding up hierarchical clustering with cluster seeding is introduced, and this method is compared with a traditional agglomerative hierarchical, average link clustering algorithm using several internal and external cluster validation indices. A benchmark study compares the cluster performance of both approaches using a wide variety of real-world and artificial benchmark data sets.

8.1 Introduction

Hierarchical clustering is a clustering technique that can provide information regarding the structural relationships between the data within the clusters. This clustering method has several unique visualization tools including the dendrogram and the bicluster plot. Several cluster validation indices will be introduced in this chapter and the influence of cluster seeding on cluster validation indices will be shown for several common benchmark data sets.

The presentation of hierarchical clustering will begin with the basic algorithm and details on its implementation. This will be followed by a discussion of the selection of a distance metric and a linkage criterion for the algorithm. The distance metrics

Mark J. Embrechts · Christopher J. Gatti
Rensselaer Polytechnic Institute, Troy, NY
e-mail: {embrem, gattic}@rpi.edu

Jonathan Linton
University of Ottawa, Canada
e-mail: linton@telfer.uottawa.ca

Badrinath Roysam
University of Houston, Houston, TX
e-mail: broysam@central.uh.edu

P. Georgieva et al. (Eds.): Adv. in Intelligent Signal Processing and Data Mining, SCI 410, pp. 197–233.
springerlink.com

that will be presented include the Euclidean distance, the Manhattan distance, and the maximum distance, as these are simple to understand and are commonly used in practice. The discussion of the linkage criteria will include the complete (maximum), single (minimum), and average linkage criteria. The selection of the distance metric and the linkage criterion have an impact on the results of the clustering, and the implications of these choices will be shown through the use of examples.

A new agglomerative hierarchical clustering strategy will be introduced for large data sets. Performing hierarchical clustering on large data sets poses a challenge, because at least $N \times \frac{(N+1)}{2}$ memory allocations are required (corresponding to the cluster distances), where N is the number of patterns. For this purpose we will seed hierarchical clustering with initial clusters obtained from K-means clustering and proceed from that point.

An important detail of cluster analysis is the evaluation of the quality of the clustering. This evaluation is typically done using various cluster validation indices. Cluster validation indices are introduced for unlabeled data (internal validation metrics) and labeled data (external cluster validation indices). Commonly used internal cluster validation metrics include the Davies-Bouldin index, Dunn index, cluster silhouette width and the GAP index. Some common external cluster validation indices include the Rand Index, Hubert and Arabie's adjusted Rand Index, the Jaccard Index and the Fowlkes-Marlowe index. These validation indices will be explained and evaluated on 16 commonly used benchmark data sets. While cluster seeding can significantly speed up agglomerative hierarchical clustering, it will also affect the cluster quality, and thus the validation indices as well. Extensive benchmarks show that the impact of cluster seeding is often rather small.

This chapter aims to accomplish four things: 1) provide a thorough discussion of agglomerative hierarchical clustering and associated visualization tools; 2) suggest an elegant way to speed up agglomerative clustering; 3) introduce and discuss different cluster validation indices that can be used to evaluate the quality of clustering; and 4) demonstrate how cluster seeding affects cluster quality using a wide variety of common benchmark data sets. In addition to the tutorial aspects and innovation in speeding up hierarchical clustering with cluster seeding, this chapter will discuss the details of leaf (i.e., data) reordering for dendrograms (which is rarely discussed in the literature) and introduces the cluster cartoon plot as a way to represent clustering results for labeled data. The overall message of this chapter is that the results from clustering are far from unique and that while cluster validation indices are helpful to assess cluster quality, such indices can only serve as a rough guideline.

8.2 Introduction to Clustering

There is an ever expanding amount of data that can be collected online in an automated fashion that can help us gain insights to current trends in finance, health, technology, safety, and many other fields. Often these data are either completely or partially unlabeled. The term labeling is introduced here to indicate a category, state, or some integer index or continuous response associated with a data record. The vast amount of these data challenges the human ability to analyze and extract knowledge from them. It is therefore necessary to use automated computational tools to extract useful knowledge from these data. Such tools are ideally able to analyze large amounts of data and return useful and comprehensive knowledge. This process is commonly called *knowledge discovery in databases*, or simply KDD [20].

Clustering is an ideal preliminary tool used to look at objects. Clustering is intuitive to humans. When we have a collection of objects the first questions that come to mind are:

- What do we have here?
- Do all of these objects look the same?
- Is there one group or are there multiple groups of objects?
- Are the objects in such groups similar to each other?
- How many groups of patterns (i.e., data archetypes) are there?
- Do some patterns look different or very different (outliers), or are there patterns that do not clearly belong to a single group?
- Do the groups of patterns look very different from each other?
- Do the groups have very different numbers of patterns?
- Is there something we can say about the pattern density amongst different groups (same, vary)?
- Are the patterns scattered according to some underlying distribution?
- Do some pattern groups have distinct subgroups?
- Are there groups that distinguish themselves from the other groups?
- Is there a structure with substructures in the clusters or groups?

As an example of clustering objects we can look at Fig. 8.1 and consider objects such as trees, bushes, buffalos, and elephants and try to answer some of the questions posed above to see whether this helps us to better understand the picture.

Clustering is more complicated than it looks at first sight, and just answering the straightforward question of how many clusters are present in a data set is not as straightforward as it may seem. Looking at Fig. 8.2, most people would agree that the 20 data can be partitioned into 2 clusters. One could argue that some clusters seem to have subclusters and that it is also possible that there are 4 or more partitions possible for these data. This ambiguity is already indicative that, in practice, it is not always easy to determine a definitive and exact number of clusters, and that partitioning data into clusters is not a task that has a definitive and unique answer. As another example, Lego blocks could be partitioned by color, by size, or just by using equal random selections of blocks. In other words, more than one type of partitioning can be applied to the same data set and, without further specifications or directions, several of these partitioning results could be valid. Furthermore, it is not necessarily the case that one partition is correct and another partition is incorrect. Sometimes just presenting a subset of possible clusters helps us to understand more about a particular data set.

Fig. 8.1. Nature picture with different clusters of objects (bushes, trees, elephants, and buffalos).

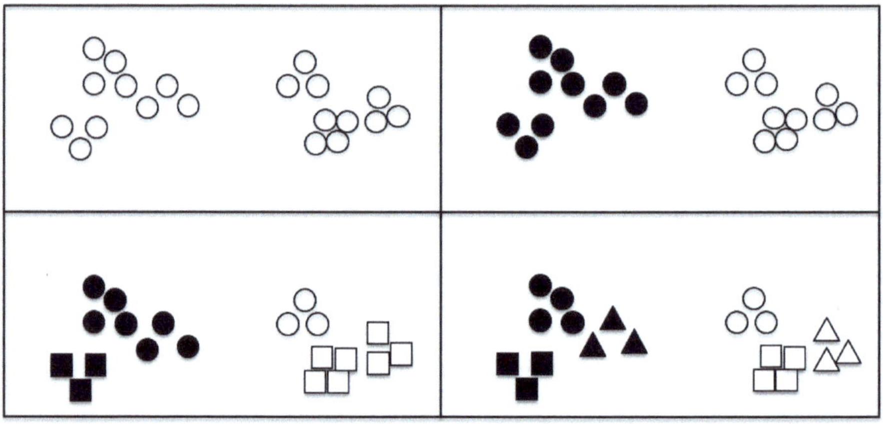

Fig. 8.2. Data set of objects with different types of partitioning in 2, 4, and 6 clusters.

Data clustering refers to an automated partitioning process of data, where the data objects are lumped into groups of similar objects and there is a dissimilarity between these groups. Data clustering is one of the key preliminary data analysis tools for a variety of reasons: (1) clustering can operate on labeled, partially labeled, or unlabeled data data; (2) clustering can be used to partition large data sets into smaller and more manageable sets of data; and (3) clustering can provide some preliminary visuals for a first insight into the data. There may be a variety of additional reasons at to why data clustering is desirable, but the key emphasis here is on preliminary data analysis.

Partitional clustering algorithms can be classified into *fuzzy* and *crisp*. A clustering algorithm is crisp when each object belongs to a single group. A clustering algorithm is fuzzy if each data object can belong to multiple clusters at once, i.e., each data object has a degree of membership to more than one cluster.

Following Xu and Wunsch [66, 67], consider the matrix $\mathbf{X}_{NM}$ which contains a set of N pattern vectors, $\mathbf{x}_M$, each with M attributes (also named variables, descriptors or features):

$$\mathbf{X}_{NM} = \left[\mathbf{x}_1,\ ...,\ \mathbf{x}_i,\ ...,\ \mathbf{x}_N\right]^T \text{ where } \mathbf{x}_i = \left(x_1^i,\ ...,\ x_j^i,\ ...,\ x_M^i\right) \in \Re^M$$

Crisp clustering produces a partition of $\mathbf{X}_{NM}$ into K non-overlapping subgroups $\{C_1,...,C_K\}$ such that:

$$\mathbf{X} = \{C_1,...,C_K\} \ (K \le N), \text{ s.t.}$$

$$\begin{aligned}
&1)\, C_l \neq \varnothing,\ l = 1,...,K\\
&2)\, \mathbf{X} = \cup_{l=1}^{K} C_l\\
&3)\, C_l \cap C_m = \varnothing\ (l,\ m = 1,...,K \text{ and } l \neq m)
\end{aligned}$$

There are many different clustering algorithms, and each algorithm has pros and cons. A nice survey of clustering and a discussion and comparison of the various algorithms can be found in Xu and Wunsch [66, 67], Jain [31, 32], and Tan [60]. There may be many different motivations for clustering that can influence the selection of a particular clustering algorithm. Common motivations for clustering can be: (i) data exploration, (ii) dividing large data sets to create smaller, more manageable sets for specialized machine learning algorithms, (iii) data visualization, (iv) outlier detection, (v) image segmentation, and (vi) application specific tasks such as gene categorization in bioinformatics, the construction of phylogenetic trees (or phonetic trees) in biology, similarity searches in medical images, document classification in text mining and web mining, and many others. A brief taxonomy for cluster analysis is shown in Fig. 8.3.

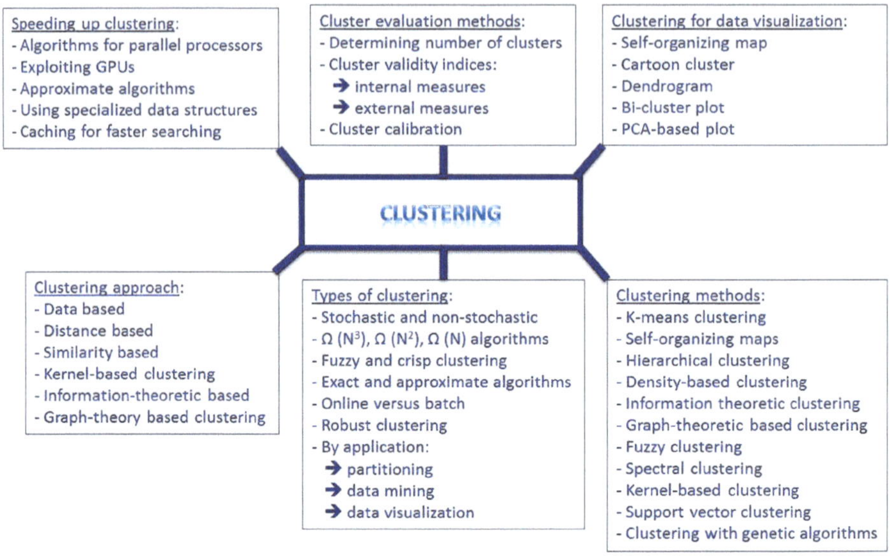

Fig. 8.3. Taxonomy for cluster analysis.

The taxonomies described above involve different algorithmic approaches for clustering (e.g., based on distances between data versus similarities between data, kernel-based clustering, etc.), different types of clustering (e.g., fuzzy, non-fuzzy, algorithms that scale well either in memory or speed), different clustering methods (e.g., K-means clustering, hierarchical clustering, self-organizing maps, support vector clustering, information theoretic clustering, graph-theory-based clustering, etc.), methods for clustering visualization, methods for speeding-up clustering and cluster evaluation methods. Some of these issues will be addressed in more detail in this chapter.

8.3 Hierarchical Clustering

The most popular clustering algorithms are the traditional K-means clustering algorithm and hierarchical clustering algorithms [32, 67]. The aim of the K-means algorithm is to represent the clusters by their centroid and to assign the data, based on an iterative procedure, to the clusters with the nearest centroid [66]. Hierarchical clustering was first proposed by Johnson [34]. Some excellent reviews of hierarchical clustering can be found in Day [14], Jain [30], and Anderberg [3]. Hierarchical clustering algorithms can be either agglomerative or divisive. In agglomerative hierarchical clustering each data point starts out as an individual cluster and clusters are built from the bottom up; in divisive hierarchical clustering all of the data starts out as a single cluster that is successively divided into multiple clusters. Agglomerative hierarchical clustering produces a tree-like, nested partition of the data ($\mathbf{X}_{NM}$) into non-overlapping subgroups H_l such that:

$$\begin{aligned}
&\mathbf{X}, H = \{H_1, ..., H_Q\} \ (Q \leq N), \text{ s.t.} \\
&1)\ H_q \neq \varnothing,\ q = 1, ..., Q \\
&2)\ H_Q = \mathbf{X} \\
&3)\ \text{if } C_l \in H_q \text{ and } C_m \in H_r,\ q > r \Rightarrow \\
&\quad C_l \subset C_m \text{ or } C_l \cap C_m = \varnothing\ \forall l, m \neq l, \text{ and } m, l = 1, ..., Q
\end{aligned}$$

This chapter will concentrate on agglomerative hierarchical clustering because agglomerative clustering leads to some interesting data visualization methods, and because there is still room for improvement in agglomerative clustering considering that these methods generally do not scale well in computing time with a growing number of data objects.

The most basic algorithm for hierarchical clustering is SAHN [59], which is an acronym for Sequential Agglomerative Hierarchical Non-overlapping clustering. The algorithm begins with each of the N patterns as its own cluster. Objects are then grouped in $N-1$ successive steps (into $< N$ sets), ending in a single cluster with N objects. SAHN is a pairwise-group method where, at each iteration, the two closest clusters are grouped. There is a sequence of $[H_1, ..., H_Q]$ partitions of objects, where H_1 is the disjoint partition, H_Q is the conjoint partition, and H_i is a refinement of H_j for all $1 \leq i \leq j \leq Q$. The SAHN algorithm is sequential in the sense that the same algorithm is used iteratively to generate $H_i + 1$ from H_i for all $0 \leq i < Q$. This algorithm is illustrated in Fig. 8.4.

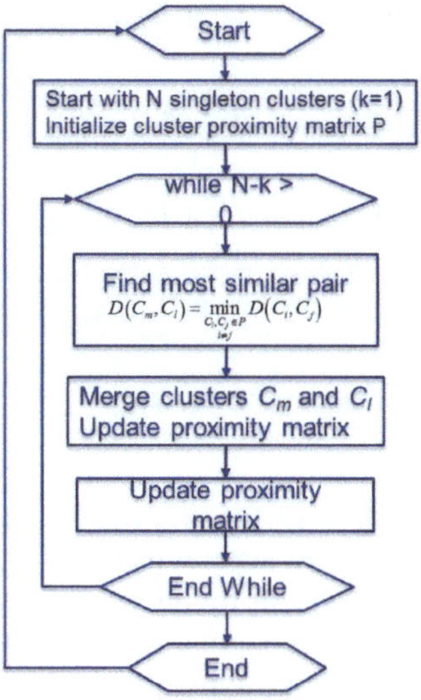

Fig. 8.4. SAHN algorithm for agglomerative hierarchical clustering [59].

The SAHN algorithm is based on either a distance matrix or a similarity matrix between the data points and can come in several flavors depending on: a) which distance (similarity) measure is used in the distance (similarity) matrix, and b) the linkage criterion that defines the 'closest clusters' to be merged in each iteration step.

Hierarchical clustering can proceed starting from just a distance matrix (or, alternately a similarity matrix) between the patterns. The distance matrix contains the distances between all the data points. These distances can be expressed in a variety of available metrics (e.g., Euclidean distance, Manhattan distance, Mahalanobis distance, Minkowski distance, Chebychev distance, etc.). Generally speaking, the distances are symmetric, and the distance matrix needs to be precalculated and stored in memory. Hierarchical clustering can also be based on a similarity matrix which instead contains similarities between data points (rather than distances). A common way to express a similarity between data vectors is the correlation coefficient, however, there are many other similarity measures that can be used instead. Some popular similarity measures are the cosine similarity measure and the Jaccard-Tanimoto similarity measure. Many of these distance and similarity metrics are listed below.

$$\text{Euclidean distance:} \quad \|\mathbf{a}-\mathbf{b}\|_2 = \sqrt{\sum_{i=1}^{M}(a_i-b_i)^2}$$

$$\text{Squared Euclidean distance:} \quad \|\mathbf{a}-\mathbf{b}\| = \sum_{i=1}^{M}(a_i-b_i)^2$$

$$\text{Manhattan distance:} \quad \|\mathbf{a}-\mathbf{b}\|_1 = \sum_{i=1}^{M}|a_i-b_i|$$

$$\text{Mahalanobis distance:} \quad D(\mathbf{a},\mathbf{b})_{Mahalanobis} = \sqrt{(\mathbf{a}-\mathbf{b})^T S^{-1}(\mathbf{a}-\mathbf{b})}$$

$$\text{Minkowski distance:} \quad D(\mathbf{a},\mathbf{b})_{Minkowski} = \left(\sum_{i=1}^{M}|a_i-b_i|^p\right)^{\frac{1}{p}}$$

$$\text{Chebychev distance:} \quad D(\mathbf{a},\mathbf{b})_{Chebychev} = \max_i(|a_i-b_i|)$$

$$\text{Cosine similarity measure:} \quad S(\mathbf{a},\mathbf{b})_{cosine} = \frac{\mathbf{a}\cdot\mathbf{b}}{\|\mathbf{a}\|\,\|\mathbf{b}\|}$$

$$\text{Jaccard-Tanimoto similarity measure:} \quad S(A,B)_{Jaccard} = \frac{|A\cap B|}{|A\cup B|}$$

The second issue is the definition of the closest cluster. In a first instance hierarchical clustering can be divided into several different algorithms depending on the interpretation of what exactly is meant by 'closest clusters': i.e., the so called link or amalgamation criterion. The most popular linkage criteria are: the single link, the average link, and the complete link, as illustrated in Fig. 8.5. In the single link algorithm, the cluster distance is determined by the closest pair of two points belonging to two different clusters. In the complete link algorithm, the distance between two clusters is determined by the two most distant points between two different clusters. For the average link case, the distance between two clusters is averaged over all the distances between all possible point combinations between two clusters. These three different clustering modes or link criteria will generally produce very different clusters and each have pros and cons.

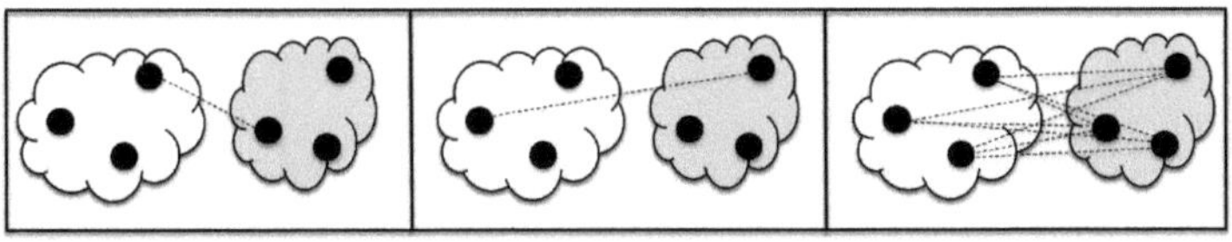

Fig. 8.5. Illustration of three popular link criteria for hierarchical clustering: (a) single link (left figure), (b) complete link (center figure) and (c) average link hierarchical clustering (right figure).

The selection of the most appropriate linkage criterion is application and data dependent. Note that in addition to the three linkage criteria mentioned above, other amalgamation criteria exist as well, including Ward's method, UPGMA (unweighted pair group method with arithmetic averages), WPGMA (weighted pair group method with arithmetic average, WPGMC (weighted pair group method with centroid average), and WPGMS (weighted pair group method with Spearman's average) [14].

The results of hierarchical clustering can be visualized with a dendrogram, which is illustrated in Fig. 8.6. Dendrograms that are application-specific and that illustrate biological evolution are commonly referred to as phylogenetic trees. Note that the preferred type of distance metric and cluster agglomeration type is often application dependent. For most examples in this chapter, the Euclidean distances are applied and the average link criterion is applied to determine which clusters to agglomerate.

Hierarchical clustering has several advantages over other clustering methods, which include the following:

- The algorithm has no stochastic elements;
- The algorithm can be made recursive;
- Data can be structured into a dendrogram that often correlate with some time-dependent characteristic for a particular data set;
- Different metrics can be exploited for a particular application;
- The algorithm is nonparametric in the sense that, aside from choosing a distance metric and an agglomeration method, there are no other parameters involved in the algorithm.

However, hierarchical clustering has several drawbacks as well, as follows:

- The algorithm scales poorly in both memory and computing time with increasing data size;
- The clustering can be very different depending on the distance (similarity) metric and cluster agglomeration metric used (Figs. 8.7 and 8.8);

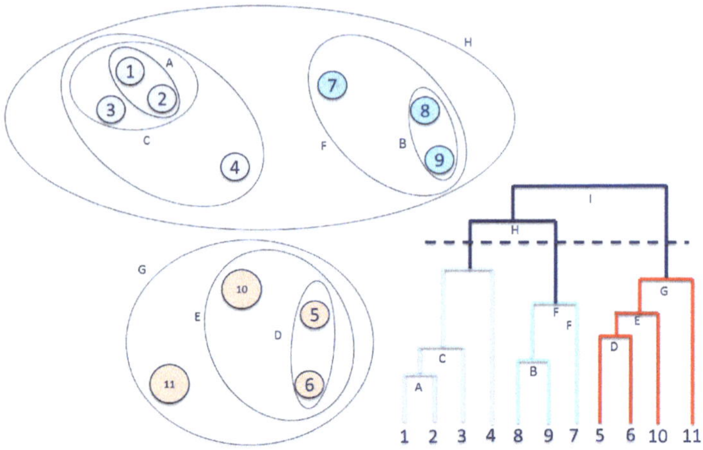

Fig. 8.6. Illustration of the concept of hierarchical clustering and the resulting dendrogram. Each pattern starts out as a single cluster (i.e., a singleton) and the clusters that are closest together are successively merged into larger clusters.

- The algorithm does not always lead to the most obvious clusters;
- There is no going back in the algorithm; in other words, once two clusters are merged, all of the data points within these clusters are dissolved.

It should also be noted that the different linkage criteria used in hierarchical clustering have advantages and disadvantages. Specifically, algorithms based on the single link and complete link criteria are generally much faster than those that use the average link criterion. The complete link algorithm may be very sensitive to the presence of outliers.

Figures 8.7 and 8.8 show different dendrograms for hierarchical clustering of Fischer's iris data [21] for various distance metrics and linkage criteria. While the Euclidean distance metric and average link criterion can be considered the default modes for hierarchical clustering, the optimal combination of distance metric and linkage criterion is often application-dependent and of course depends on how one defines *optimal*.

Figure 8.9 presents a taxonomy for hierarchical clustering. In this case we divide hierarchical clustering algorithms depending on: (i) whether they are agglomerative or divisive; (ii) whether they operate on distances or similarity matrices; (iii) according to distance or similarity metric; (iv) according to linkage criteria; and (v) according to the applied procedure for speeding up the algorithm. Some popular and free software packages include Orange, Spicy Cluster, CRAN and ELKI [64].

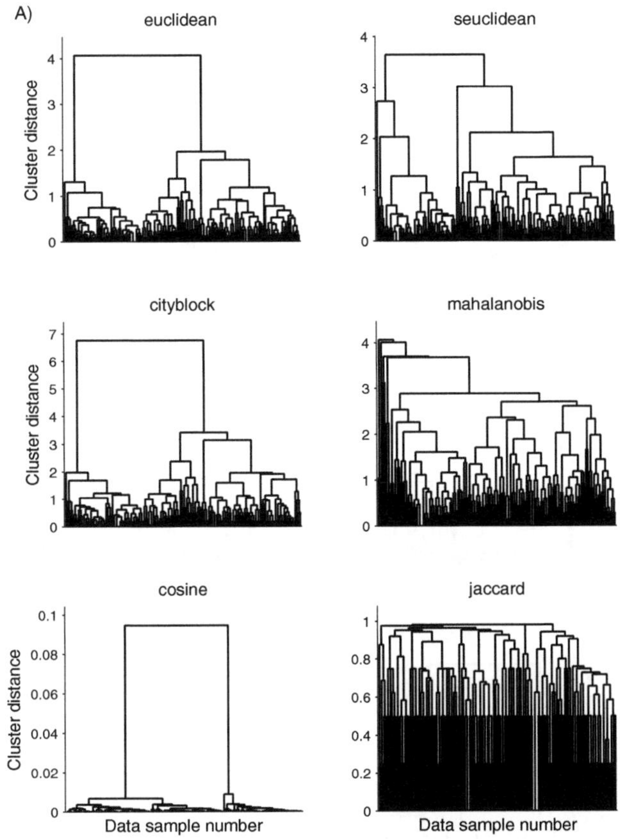

Fig. 8.7. Different dendrograms are produced with different distance metrics in hierarchical clustering (average link mode) applied to Fischer's iris data.

8.4 Displaying Hierarchical Clustering with Dendrograms

The results of hierarchical clustering can be stored using a number of different data formats. A particularly useful data format that concisely contains the entire hierarchical cluster tree is a 3-column matrix used by Matlab [43] (Table 8.1). In accordance with the hierarchical clustering algorithm, each of the n data points are initially considered as individual clusters, and cluster numbers 1, ..., n represent the original data points in this format. All clusters are represented by the cluster numbers $n+1$, ..., $2n-1$. This 3-column data format uses the first 2 columns to represented which data are merged at each iteration of the clustering process where the clustering iterations proceed down the 3-column matrix. The 3rd column represents the distance at which the clusters in the corresponding row were clustered. A 4th column can be implicitly constructed, containing the cluster numbers $n+1$, ..., $2n-1$ corresponding to the merged clusters in columns 1 and 2.

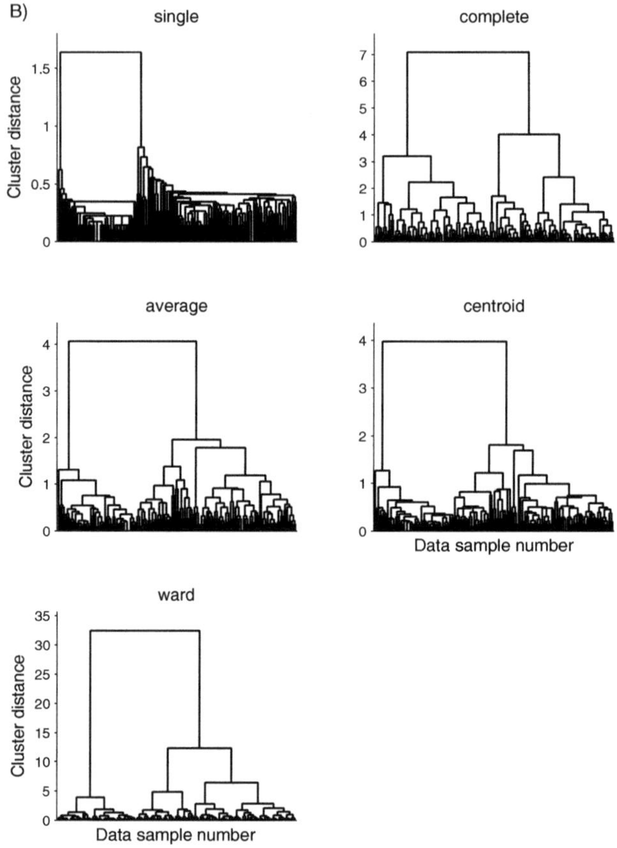

Fig. 8.8. Different dendrograms are obtained for different hierarchical clustering linkage criteria (Euclidean distance metric) applied to Fischer's iris data.

Table 8.1. Example of hierarchical clustering representation.

$Merger_1$	$Merger_2$	Distance	Implicit cluster number
2	3	1.42	7
1	5	2.23	8
7	8	3.61	9
4	6	4.00	10
9	10	5.10	11

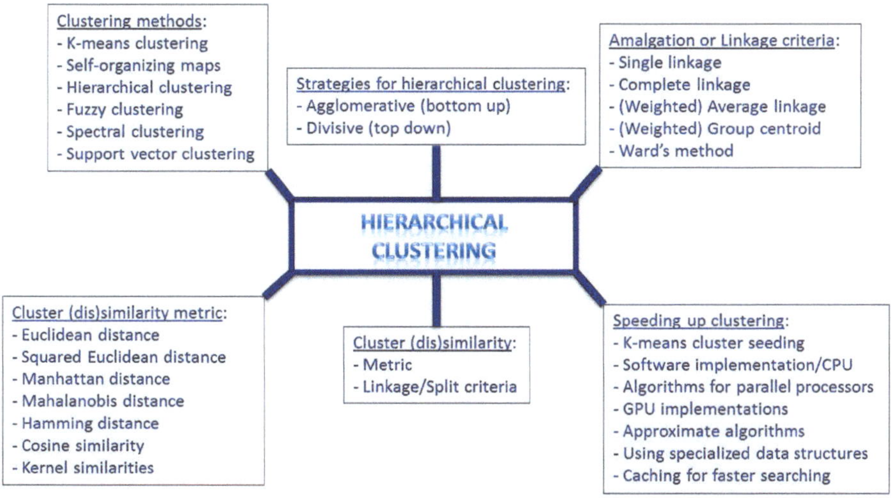

Fig. 8.9. Taxonomy for hierarchical clustering distinguishing between agglomerative or divisive, distance metric, linkage criteria, distance or similarity based, and speeding up methodology.

8.4.1 Data Reordering

The most frequently used method to display the results of a hierarchical cluster is the dendrogram, also known as a phylogenetic tree. The dendrogram essentially displays how the data were clustered during the clustering process and it does this by showing which data were merged and the distance at which the merging occurred. An example dendrogram for the hierarchical clustering in Table 8.1 is shown in Fig. 8.10.

Figure 8.10 shows a dendrogram with branches that do not overlap each other, which is how dendrograms are traditionally displayed. For most of the linkage methods used in hierarchical clustering (excluding the centroid and mediam linkage methods), the data can be reordered along the x-axis such that the branches do not cross, resulting in a monotonic dendrogram. The seriation of the data, or the ordering of the leaf nodes, is not immediately evident nor is it given from the output of hierarchical clustering, such as that in Table 8.1. Furthermore, there is little literature that discusses the topic of how to order the data such that branches do not cross. If the dendrogram is drawn with the data in its original order along the x-axis however, the branches will often cross over each other (Fig. 8.11(**a**)). The reordering of the data can be accomplished by starting at the top of the tree, walking down through each branch, and picking up and storing only the leaf nodes as each branch is walked. This method repeatedly uses the same procedure of walking down each descendant branch until leaf nodes are reached and this makes it amenable to a short recursive algorithm.

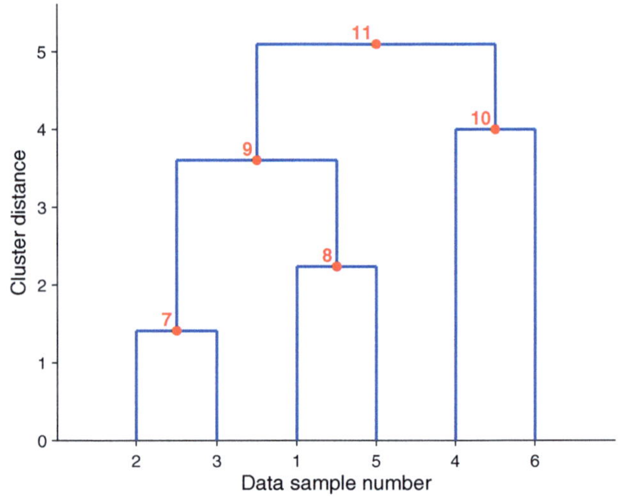

Fig. 8.10. An example dendrogram for the hierarchical clustering shown in Table 8.1. The original data (numbered 1, ..., 6) are shown on the x-axis and the connecting lines show how data are combined. Vertical lines extended upward to the distance at which two clusters were merged corresponding to the height on the y-axis, and the horizontal lines connect which clusters are merged. Note that this figure also shows the higher numbered clusters ($n+1$, ..., $2n-1$) corresponding to the implicit cluster numbers in Table 8.1 however, these numbers are not typically included in the dendrogram.

8.4.2 *Leaf Reordering*

The dendrogram shown in Fig. 8.11(b) shows, for n data, only one of the possible 2^{n-1} different data orderings that can produce a consistent dendrogram. The purpose of reordering the data beyond merely producing a dendrogram without crossing branches would be to order the data in such a manner that reveals additional information regarding the relationships between adjacent data samples. *Information* in this context is based on some metric that is descriptive of the relationship between clusters including, for example, correlation, Euclidean distance, etc. Thus, for each of these different metrics, there is an *optimal* data sample ordering such that the ordering maximizes (or minimizes, depending on the metric) the particular metric. Optimizing the ordering of the data samples while also achieving a dendrogram that does not have crossing branches may seem like a multi-objective optimization problem. However, the goal of producing a dendrogram without crossing branches can be achieved by exploiting the structure of the hierarchical tree, and this essentially becomes a constraint which limits the number of possible permutations of the data samples. One leaf ordering algorithm that is frequently used for biclustering (hierarchical clustering on both data and features) is the algorithm developed by

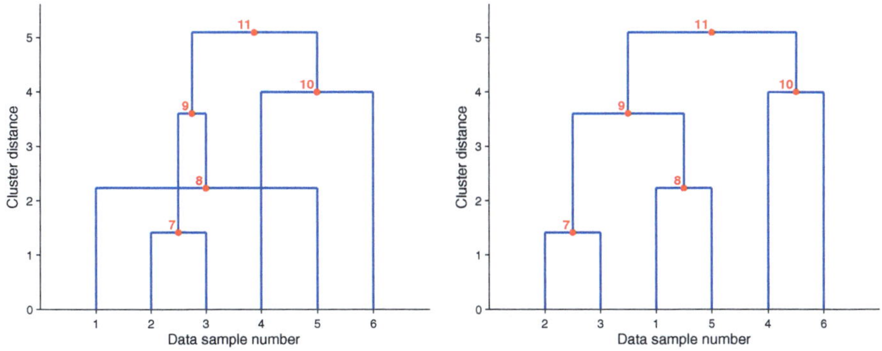

(a) Dendrogram with crossing branches. (b) Dendrogram without crossing branches.

Fig. 8.11. This figure shows the same dendrogram from hierarchical clustering displayed with the data in its original order (a) and the data reordered (b) so that the branches do not cross over each other.

Bar-Joseph et al. [4]. This algorithm has a time complexity of slightly faster than $O(n^4)$ and is thus computationally demanding for large data sets.

We will not further discuss methods of optimizing the order of the data samples in the dendrogram, rather we will explain just one method to simply rearrange the leaf nodes of the dendrogram. The particular method described here relies on the fact that each set of leaf nodes has a single common parent node; that is, a node that lies above it in the dendrogram. For example, the dendrogram Fig. 8.12(**a**) shows that leaf nodes #2 and #3 have a common parent of node #7; similarly, clusters #7 and #8 have a common parent of node #9. We can first isolate a common parent node and then identify all of its leaf nodes. Once these have been identified, the order of these leaf nodes within the complete ordering of the data samples can be flipped.

As an example, suppose we would like to flip the leaf nodes that comprise cluster #9. The leaf nodes can be obtained by simply walking down each branch of the parent node #9 and picking up an storing the leaf nodes (in the same manner which was done to obtain the complete data sample ordering). The leaf nodes of cluster #9 are then (in order) [2, 3, 1, 5], as shown in Fig. 8.12(**a**). These leaf nodes reside in indices 1–4 in the data sample ordering. If the order of the leaf nodes at these indices is simply flipped, resulting in [5, 1, 3, 2], the dendrogram can be redrawn with a new data sample ordering that again does not contain crossing branches (Fig. 8.12(**b**)).

8.5 Data Sets

For studying the influence of cluster seeding on computation time savings and cluster quality, 16 benchmark data sets will be considered (Table 8.2; note that 'V'

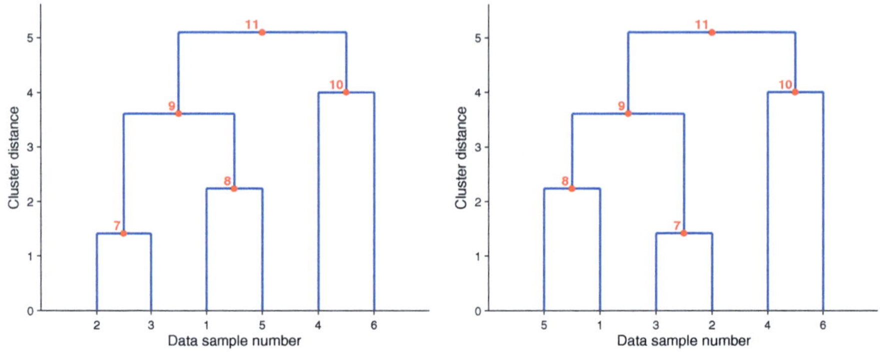

(a) Original dendrogram. (b) Dendrogram with flipped leaves.

Fig. 8.12. This figure shows the same dendrogram in two different configurations. The dendrogram in (**a**) is in its original form whereas the dendrogram in (**b**) has the leaves of cluster 9 (with leaves [2, 3, 1 5]) flipped.

means there is a variable number of samples or classes). This table lists the number of data samples, the number of attributes or features, and the most commonly assigned number of classes. These benchmark data sets are chosen because they exhibit different characteristics that can be associated with typical data sets and most of these data sets are frequently cited in the literature. Additionally, some of the author's or collaborator's benchmark data are introduced here to contribute to the repository of benchmark data, and these data sets include the Santos 2-D clustering data [56], Linton's *Journal of Business Ethics* (JBE) data [41], and Karen Smith's Microglia data [58].

The 16 benchmark data sets considered are (a) Italian olive oil data [22, 65]; (b) Fischer's iris data [21]; (c) Joaquim Marques de Sá Portuguese Rock data [15]; (d) Tobacco data [33]; (e) Wieland's intertwined spiral data [19]; (f) Kohonen's animal data [37, 5]; (g) two overlapping Gaussians; (h) several mostly non-overlapping Gaussians; (i-j) Linton's *Journal of Business Ethics* data sets (one based on a one-word dictionary and one based on a two-word dictionary) [41]; (k-n) four of Santos' 2-D clustering data sets (clock, anvils, Bermuda and beans) [56]; (o) Duke Leukemia data [25], and (p) Karen Smith's Microglia data [58].

These data were selected because they exhibit a wide variety of characteristics that are typical for real-world data sets: (a) some data sets are multi-class (e.g., Iris, Rock, many Gaussians, some Santos data sets); (b) some data sets are easy to classify (e.g., Olive Oil, Iris, Tobacco, Animal, Leukemia); (c) some data sets can be difficult to classify (e.g., Wieland's intertwined Spiral data, overlapping Gaussians, Microglia); (d) some data sets can have mixed classes (e.g., overlapping Gaussians, some of the Santos data sets); (e) some clustering data sets have no unique classes (e.g., Animal, Santos); (f) some data sets can be large in the number of data records (e.g., many Gaussians, Microglia); (g) some data sets can be large in the number

Table 8.2. Benchmark data sets for studying cluster seeding for hierarchical clustering.

Data set name	# data samples	# attributes	# classes
Italian olive oil	572	8	9
Fischer's iris data	150	4	3
JMDS Portuguese rock data	134	18	6
Tobacco	26	16	2
Wieland's spiral data	194	2	2
Kohonen's animal data	16	16	V
Two Gaussians	V	2	2
Many Gaussians	V	2	V
Linton's JBE data (1 word)	29	340	4
Linton's JBE data (2 word)	29	521	4
Leukemia	38	7129 (2529)	2
Microglia	1772	300 (80)	3
Santos 2-D clustering data			
clock	126	2	3
anvils	201	2	2
bermuda	182	2	4
beans	140	2	2

of features (e.g., Leukemia data, Karen Smith's Microglia data, and Linton's text mining data sets); (h) some data sets are real-world data (e.g., medically-related data such as the Leukemia and the Microglia data), other data sets are synthetic (e.g., overlapping spirals and Gaussians), while other data sets represent a clear timeline (e.g., Linton's text mining data).

Figure 8.13 represents a subset of the Santos 2-D clustering data sets. A complete overview all of the Santos 2-D clustering data sets is shown in Appendix A. These data sets form an interesting collection of benchmark data for clustering because they exhibit many features commonly seen in real-world data sets such as overlapping clusters, outliers, subclusters, variations in density, nonlinearly separable cases, etc. Santos performed experiments with these data sets in which children of different grade levels were asked to indicate the clusters in these data. Depending on the point of view of the data, different clusterings are possible for several of these data sets.

8.6 Cluster Plots

One of the goals of clustering is often to present different ways of visualizing data as a first analysis tool for data mining [17]. A prime motivation for data visualization is the fact that it is challenging to describe large multi-dimensional data sets using a single picture. The purpose of these visualizations is often to get an initial idea of the structure of the data and to help answer some of the following questions: (a) Is there structure in the data? (b) Are there outliers? (c) Do the data exhibit obvious groups or classes? (d) Do the data follow a timeline trend? Common data visualization methods are the heatmap, principal component plot, starplots, box plots,

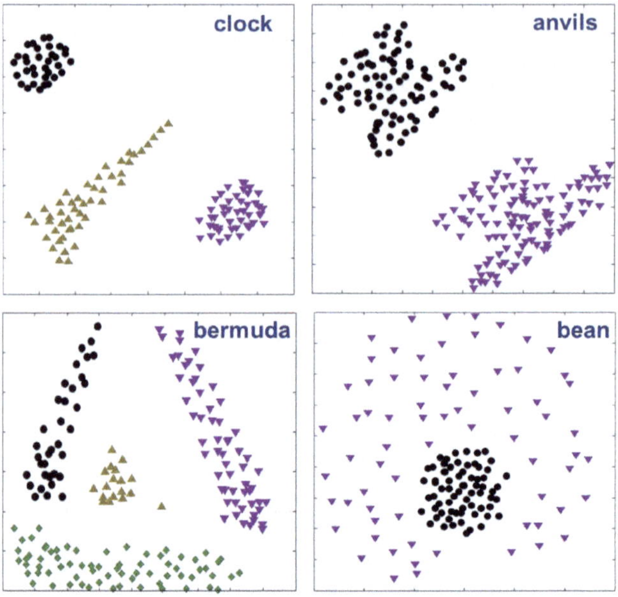

Fig. 8.13. Subset of 4 of the Santos 2-D benchmark clustering data sets (clock, anvils, Bermuda and beans).

and parallel coordinate plots [17]. Several types of plots are specific to clustering such as dendrograms (described above), the cartoon cluster plot, bicluster plots, and Self-Organizing Maps (SOMs) [37].

8.6.1 Cartoon Cluster Plot

An interesting and novel way to represent clustering results for labeled data (i.e., data for which the classes are known) is the cartoon cluster plot (Fig. 8.14). The cartoon cluster plot can be considered as a 2-D representation of multi-dimensional data, where the identification numbers for each pattern are shown within the clusters (randomly positioned within the cluster) and the different classes are represented by different colors. Such a plot allows one to visually evaluate how well a particular clustering result corresponds to the assigned classes and whether the clusters are mostly homogeneous with respect to the classes.

Figure 8.14 shows a cartoon cluster plot for Linton's one-word text mining data of the *Journal of Business Ethics* [41]. These 29 data contain 4 classes (for indicating 5 or 4 year groups) that follow a timeline. One purpose of this data set is to study the evolution of key words in publications of the journal from 1982 through 2010. The corresponding confusion matrix (Table 8.3) shows that there is a total of 7 misclassifications (out of 29 data records). These misclassifications are in contingent classes and such errors can be expected from timeline data.

Fig. 8.14. Cartoon cluster plot for labeled data for Linton's one-word *Journal of Business Ethics* text mining data.

Table 8.3. Confusion matrix for Linton's text mining data.

	Class #1	Class #2	Class #3	Class #4
Class #1	6	2	0	0
Class #2	0	6	1	1
Class #3	0	1	5	1
Class #4	0	0	2	5

8.6.2 *Timeline Analysis with Principal Components Analysis*

An interesting application for clustering can emerge when data follow a clear timeline history. Such a data set could be the works of Shakespeare, where it might be of interest to date a specific Shakespeare manuscript relative to the portfolio of Shakespeare manuscripts. Similar applications might relate to dating paintings. In our case, we consider Linton's *Journal of Business Ethics* text mining data set, where a word count table is made according to a one-word dictionary of the 340 most relevant words of the 29 years of journal data by year. Hierarchical clustering this data leads to the dendrogram displayed in Fig. 8.15. The dendrogram shows a clear grouping by time. For this data set, four classes were assigned to the 29 data records according the publication year.

More revealing for Linton's text mining data is a principal component plot, where the data records are plotted as a function of the two first principal components [17] as shown in Fig. 8.16. The timeline in this figure clearly shows that there is an evolutionary trend in time. A detailed analysis of the journal indicates that sudden changes in the timeline direction correspond to the time spots where the journal either changed editor or merged with other journals.

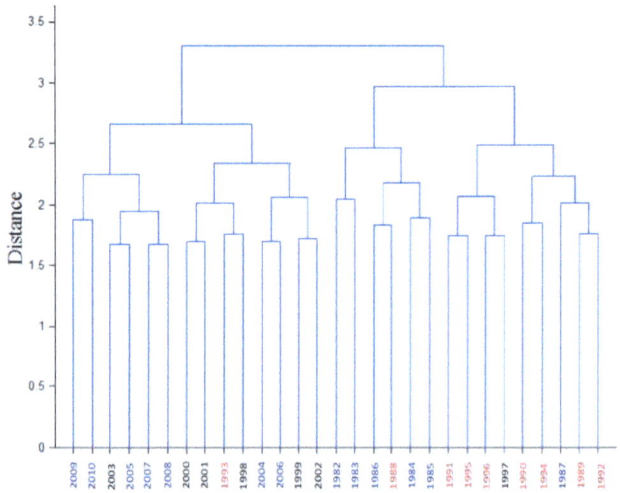

Fig. 8.15. Dendrogram obtained with the SAHN algorithm on Linton's one-word *Journal of Business Ethics* data [41].

8.6.3 Bicluster Plots

A unique visualization tool associated with hierarchical clustering is the bicluster plot. A bicluster plot can be obtained by clustering the attributes as well as the data records, and then by reordered both the data and feature hierarchical trees individually (cf. the leaf reordering procedure for displaying dendrograms). A heatmap of the rearranged data structure is then combined with both dendrograms in a single biclusterplot. Figure 8.17 shows the biclusterplot that was obtained from Linton's single-word text mining data showing that there is a clear structure in the data.

Biclusterplots are a common visualization tool for analyzing gene expression array data [7, 9, 42, 51, 55, 61]. The purpose is to discover internal structure in the data in either a supervised or an unsupervised way. Current research related to bicluster plots focuses on developing fast methods for representations that show better structure in the data with new optimized leaf-ordering algorithms [4].

8.7 Assessing Cluster Quality with Cluster Evaluation Indices

One of the common questions in clustering is how to identify the natural number of clusters present is in a particular data set. In other words, can we identify the best partition between several partitioning alternatives for a data set? A related question concerns how to compare different clusterings for a particular data set to each other. The answer to these questions can be quite involved, especially when there is no ground truth available (e.g., pre-determined class labels). Even when the data are labeled, different types of class labels may be possible for the same data set

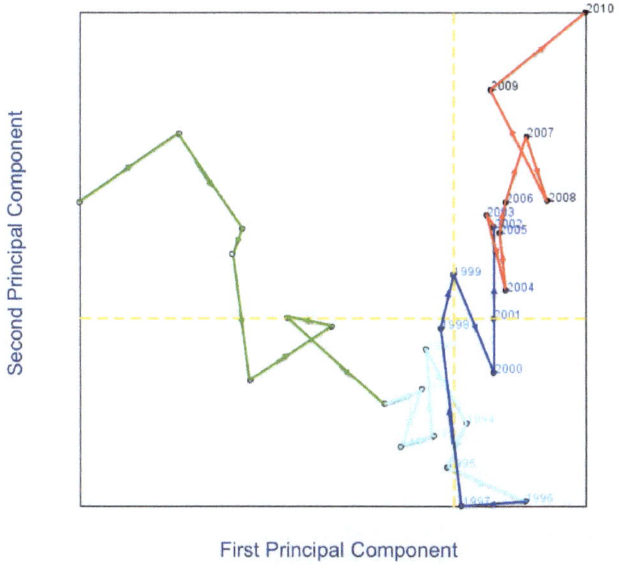

Fig. 8.16. Principal component plot for Linton's *Journal of Business Ethics* data with a timeline connecting successive journal years (ranging from 1981 through 2010).

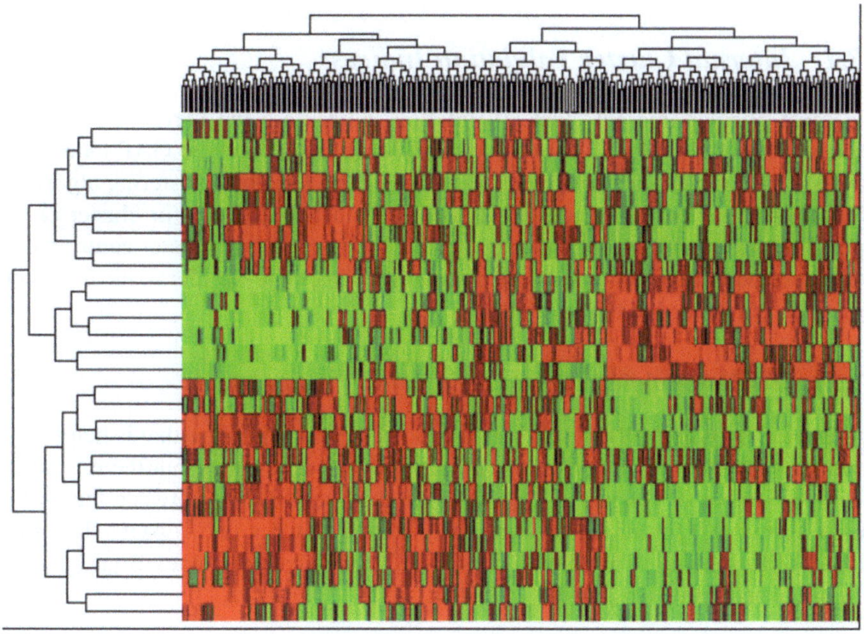

Fig. 8.17. Biclusterplot for Linton's one-word *Journal of Business Ethics* text mining data, which shows clear structure in the data.

and the classes may not necessarily represent an absolute gold standard or ground truth. Many real-world data from medical fields can contain errors in the assigned labels, and, as is often the case in biological data, there are always special cases and exceptions.

There are two types of indices to compare clusterings to each other: internal cluster validation indices that do not rely on external class labels, and external cluster validation indices that assume there are class labels that represent a ground truth labeling. Given that there are several different cluster validation indices, one is often tempted to try to determine the best index. We avoid answering this question by taking a holistic view: i.e., several different cluster validation indices should be interpreted as a fingerprint to describe a particular clustering for a particular data set. We will expand on this holistic view when comparing clusterings for different data sets in the next section. In this section, only the most commonly used cluster validation indices are introduced and we refer the reader to the literature [36, 67] for a more complete overview. For the internal cluster validation, indices include the Davies-Bouldin index, the Dunn index, the Silhouette width validity index, and the GAP statistic. The literature also introduces the C index [27], the Goodman Kruskal index [24], and the isolation index [50]. As for external cluster validation, the Rand index, the adjusted Rand index (ARI), the Jaccard coefficient, and the Fowlkes-Mallows index will be discussed. Sledge et al. generalized cluster validation indices to fuzzy clustering [57].

8.7.1 *Internal Cluster Validation Indices*

The most popular internal cluster validation indices are the Davies-Bouldin Index, the Dunn Index, the Cluster Silhouette Width Index, and the GAP index. The Davies-Bouldin Index was originally defined for the K-means clustering algorithm where there are clearly defined cluster centers, cluster centroids, or cluster prototypes. Because hierarchical clustering typically does not have cluster centroids, an alternate Davies-Bouldin Index which is more appropriate for hierarchical clustering is also introduced in this section.

8.7.1.1 Davies-Bouldin Index

The Davies-Bouldin (DB) cluster validity index [13] is a measure that is helpful for estimating the ideal number of clusters in a data set. It assumes that there is a prespecified number of clusters and is based on the average ratio between the within cluster scatter (S_l) for all clusters and the distance between two clusters. The ideal number of clusters is the number of clusters (K) for which the Davies-Bouldin index is minimal according to:

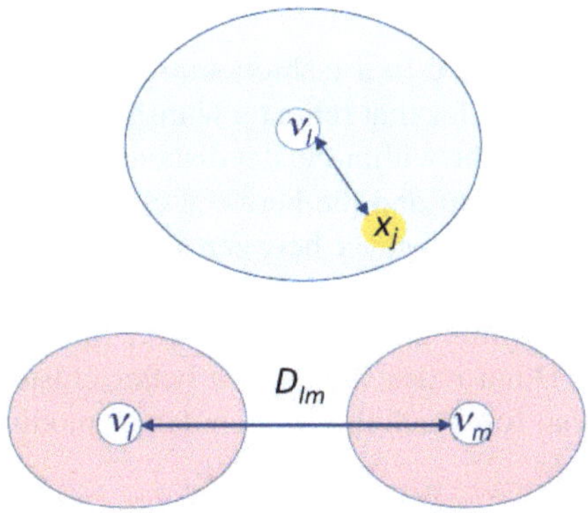

Fig. 8.18. Illustration of some of the metrics used for defining the Davies-Bouldin index, based on ratios containing within cluster scatter and between cluster distances. The optimal number of clusters is assigned to the number of clusters with a minimal Davies-Bouldin index.

$$DB = \frac{1}{K}\sum_{l=1}^{K} R_l$$

$$R_l = \max_{l\in K, l\neq m} R_{lm}$$

$$R_{lm} = \frac{S_l + S_m}{D_{lm}}$$

$$D_{lm} = \sqrt{\frac{1}{N_l N_m}\sum_{l\in C_l}\sum_{m\in C_m} \|\mathbf{x}_l - \mathbf{x}_m\|^2}$$

$$S_l = \sqrt{\frac{1}{N_l}\sum_{j=1} \|\mathbf{x}_j - v_l\|^2}$$

where S_l is the within cluster scatter for cluster l, v_l is the cluster centroid for cluster l, and x_l represents an individual data point in cluster l. D_{lm} represents the distance between two clusters and R_{lm} is the joint cluster scatter over the distance between the clusters l and m. N_l is the number of data in cluster l and N_m is the number of data in cluster m. Figure 8.18 highlights the concept of the terms used in the Davies-Bouldin index. Note that in this case the concept of a cluster centroid was used. In K-means clustering the centroid is usually defined as the average pattern within a cluster. In hierarchical clustering, it is common to redefine the Davies-Bouldin index such that the concept of a centroid is avoided altogether. In this case the sum of the distances between centroids is replaced by the average distance between all possible pairs of points in two clusters. The within-cluster distance is replaced by the average distance between all point pairs within a single cluster. This alternate Davies-Bouldin index is indicated with the abbreviation (DB_A) in this chapter.

8.7.1.2 Dunn Index

The Dunn index [16] is defined in a fashion similar to the Davies-Bouldin index and contains elements and ratios that reflect a within-cluster distance and between-cluster distance. In this case the within-cluster distance ΔS is the maximum distance between two points in a cluster and the between-cluster distance $d(K,L)$ between clusters K and L is the smallest distance between any pair of points where each point belongs to a different cluster (K and L). The Dunn index (Fig. 8.19) is defined as the ratio of the minimal within-cluster distance over the maximal between-cluster distance, and a higher Dunn index indicates a better cluster quality. The optimal number of clusters is that for which the Dunn index is maximized.

$$Dunn = \frac{\min(\Delta S)}{\max(\delta(S,T))}$$

$$\Delta S = \max_{x,y \in S}(d(x,y))$$

$$\delta(S,T) = \min_{x \in S, y \in T}(d(x,y))$$

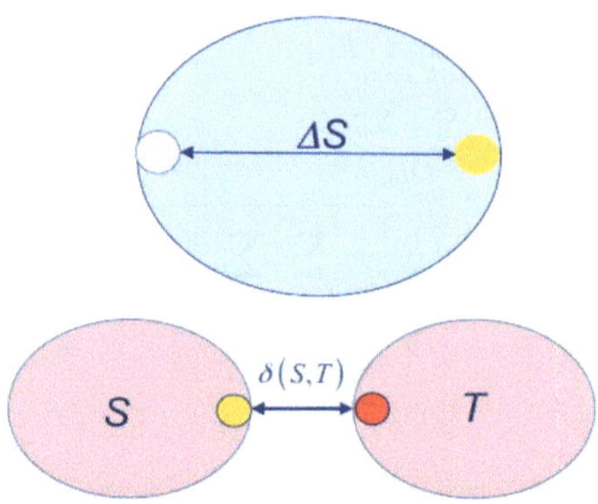

Fig. 8.19. Illustration of some of the metrics used for defining the Dunn index based on ratios containing the smallest within cluster distances ΔS and the largest between cluster distances $\delta(S,T)$. The optimal number of clusters is that which maximizes Dunn index.

8.7.1.3 Cluster Silhouette Width Index

The cluster silhouette width (SHW) validation index [54] is based on the silhouette width for each sample, the average silhouette width for each cluster, and the overall average silhouette width for all of the data. This index is a measure that compares cluster tightness and cluster evaluation (Fig. 8.20). The optimal number of clusters maximizes the cluster silhouette width index according to:

$$SHW = \frac{1}{N}\sum_{k=1}^{N_C}\frac{1}{A}\sum_{l=1}^{A_l} S_l$$

$$S_l = \frac{b_l - a_l}{\max(b_l, a_l)}$$

$$a_l = \frac{1}{A_l}\sum_{i,j\in A_l} d(x_i, y_j)$$

$$b_l = \frac{1}{B_l}\sum_{i\in A_l, j\in B_l} d(x_i, y_j)$$

where N_C is the number of clusters, A_l is the number of data in cluster l, a_l is the average within-cluster distance, and b_i is the minimal average between-cluster distance between two clusters (including the cluster for which a_l was calculated). A silhouette value close to unity for a sample indicates that the sample is assigned to the correct cluster, a sample silhouette width close to 0 means that that sample could be assigned to another close cluster as well, and a silhouette with close to -1 is indicative of a misclassified sample.

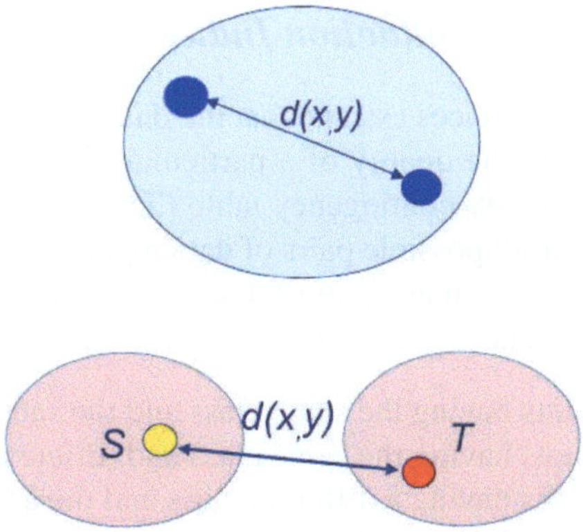

Fig. 8.20. Illustration of some of the metrics used for defining the Cluster Silhouette Width Validation index. The optimal number of clusters is that which maximizes the silhouette width index.

8.7.1.4 GAP Statistic

The GAP statistic is used to estimate the number of clusters in a data set [26, 62]. The GAP statistic compares the cluster dispersion obtained for a clustering algorithm to the cluster dispersion obtained with the same algorithm for a random distribution of the same number of data with the same number of (now uniformly random) attributes. Depending on the algorithm, cluster dispersion can be defined in different ways. In the case of hierarchical clustering, cluster dispersion is defined by the sum over all the clusters of the sum of the pairwise distances between all possible point pairs within a cluster. More formally:

$$Gap(k) = \mathbf{E}\{\log(W_k)\} - \log(W_k) \;\text{ where }\; W_k = \sum_{r=1}^{k} \frac{1}{2n_r} D_r$$

where $\mathbf{E}\{\cdot\}$ is the expectation operator for the reference (i.e., random) distribution, k is the number of clusters, W_k is the pooled within-cluster sum of squares around the cluster means, and D_r is the sum of all pairwise distances in cluster r. The GAP statistic Gap_k therefore computes the difference between the values of $\log(W_k)$ for the real data and the reference data. Tibshirani introduces some additional logarithmic ratios and calibration elements for the exact definition of the GAP statistic. The GAP statistic can be computed for different numbers of clusters k on a single data set to obtain the GAP statistic as a function of the number of clusters. The number of clusters within a data set based on the GAP statistic is defined at that for which there is a sudden jump in the difference of the cluster dispersions between the actual data and the random gauge data. For a more comprehensive description of this metric, the reader is directed to the original work found in [62].

8.7.2 External Cluster Validation Indices

External cluster validation indices assume that the data have a class label and utilize this class label to evaluate the quality of a particular clustering. These indices are generally based on a two-way contingency table (Table 8.4) generated by tallying the concurrency states for all possible pairs of data points. These concurrency states are based on the consideration that a pair of data points can belong to four states: A, B, C and D, where these states have the following meaning:

A: # pairs of data points having the same class and the same cluster
B: # pairs of data points having the same class and different clusters
C: # pairs of data points having a different class and the same cluster
D: # pairs of data points having a different class and different clusters

Table 8.4. Contingency table for comparing all pairs of data points.

	Same cluster	Different cluster
Same class	A	B
Different class	C	D

In this section the Rand Index, Hubert and Arabie's Adjusted Rand Index, the Jaccard Coefficient and the Fowlkes-Mallows Index will be defined. These indices can be applied either to compare the quality of two different clusterings or indicate how well a particular clustering corresponds to pre-assigned labels or classes.

8.7.2.1 Rand Index

The Rand index is a powerful method to assess how well two different clusterings compare to each other. In a similar fashion this index can also be used to indicate how well a particular clustering corresponds to pre-assigned labels or classes. In this case one of the clustering indices in the contingency table is replaced by the class labels. Based on the contingency table for the two clusterings or the contingency table for a particlar clustering and class labeling set, the Rand Index (*RI*) [53] is defined as:

$$RI = \frac{A+D}{A+B+C+D}$$

The Rand Index ranges between 0 and 1. It can generally be observed that a Rand Index < 0.5 corresponds to poor clustering, while a Rand Index > 0.5 indicates a meaningful or plausible clustering. An index that is derived from the Rand Index, but more specific for a meaningful clustering, is the Adjusted Rand Index (*ARA*), introduced by Hubert and Arabie [28].

8.7.2.2 Adjusted Rand Index

The Adjusted Rand Index (*ARI*) [28, 38] ranges between 0 and 1, where any $ARI > 0$ indicates a meaningful clustering. The Adjusted Rand Index can also be derived from a contingency table and, similar to the Rand Index, it can be applied to compare two different clusterings to each other or to indicate how well a particular clustering corresponds to pre-assigned class labels. The Adjusted Rand Index is computed as follows:

$$ARI_{HA} = \frac{\binom{N}{2}(A+D) - [(A+B)(A+C)+(C+D)(B+D)]}{\binom{N}{2}^2 - [(A+B)(A+C)+(C+D)(B+D)]}$$

8.7.2.3 Jaccard Coefficient

A frequently used external cluster validation index is the Jaccard coefficient [29], which is also based on a cluster contingency table. The Jaccard coefficient can either be used to measure the clustering quality between two different clusterings or to indicate how well a particular clustering corresponds to pre-assigned labels or classes.

$$Jaccard\ Coefficient = \frac{A}{A+B+C}$$

8.7.2.4 Fowlkes-Mallows Index

The Fowlkes-Mallows index [23] can be considered as an alternate to the Jaccard coefficient and is also defined based on a cluster contingency table, according to:

$$\mathit{Fowlkes\text{-}Mallows} = \frac{A}{\sqrt{(A+B)(A+C)}}$$

8.8 Speeding Up Hierarchical Clustering with Cluster Seeding

Average link hierarchical clustering with the traditional SAHN algorithm does not scale well in computing time. This section discusses the scaling of the SAHN algorithm in detail and introduces a hybrid K-means cluster seeding approach to mitigate the poor scaling behavior with regard to computing time and memory requirements

8.8.1 Scaling of Hierarchical Clustering in Memory and Time

Hierarchical clustering has a serious drawback compared to several other clustering methods. One drawback is that the algorithm is based on a distance matrix that has to be kept in memory. Considering that the distance matrix is symmetric, this means that the required memory scales as $\frac{N(N-1)}{2}$. In addition, average link clustering scales as N^3 in time, because for each cluster agglomeration, the algorithm searches through $\frac{N(N-1)}{2}$ cluster dissimilarities in order to determine the pair of most similar clusters to merge, and the algorithm works through $N-1$ iterations. Figure 8.21 shows benchmark comparisons for the SAHN algorithm for memory requirements and computing time on the clustering of 2 Gaussian distributions versus the number of data points with 256 features (254 features are uniform random). The number of data points ranges from 100 to 10,000. Calculations were performed on a 1 GHz MacBook Air (UNIX gnu compiler) and a 3 GHz Red Hat Linux workstation equipped with an Intel Core i5-2500 CPU, using the gnu compiler (g++ 4.4.5) and the ICC compiler 12.0.4 with a -O3 optimization flag. The -xhost flag was used in addition for the ICC compiler. As expected, the comparisons show that the 3 GHz workstation performs at three times the speed of the 1 GHz MacBook Air. Memory scales as N^2 once the number of data becomes large enough to mask the additional fixed memory requirements for data storage and additional coding overhead in the algorithm. As expected, the computing time scales as N^3.

While these benchmark results show no real surprises, an interesting observation can be made. For the larger number of data, the turn around time is on the order of hours however, the real limitation to clustering the larger data sets is the memory requirement. Indeed, while the algorithm would be excessively slow on 100,000 data points, the real challenge is to find computer equipment that can satisfy the memory requirements. In other words, in spite of the fact that the computing time scales poorly (N^3), the real bottleneck on large data sets is the memory requirement.

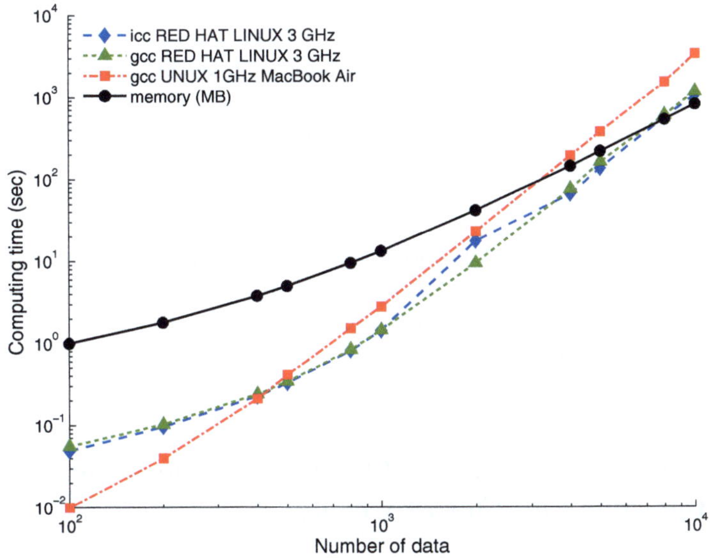

Fig. 8.21. Benchmark comparison of memory requirements and computing time for the SAHN algorithm on two overlapping Gaussians with 256 attributes as a function of the number of data.

It also should be noted that even when the SAHN algorithm scales as N^3, a fast code implementation can still offer a tremendous improvement, and for that reason the iterative cluster amalgamation scheme is usually based on *updating* the cluster dissimilarities rather than recalculating all dissimilarities de novo.

8.8.2 Speeding Up Hierarchical Clustering

This section will focus on methods for speeding up the SAHN algorithm for the average link case. The literature mentions three popular approaches used to accelerate the hierarchical link algorithms: (a) using parallel processors; (b) using a cache memory storage of the previous search results for the closest dissimilarities with a tree like data structure; and (c) using clever graph-based algorithms with a quadtree-like data structure. In addition, these three approaches can be combined to yield performance improvements.

The above-mentioned cubic scaling in time is only for the SAHN (average link) algorithm. It should be noted that the single link algorithm can utilize efficient minimum spanning tree [1, 35, 48] algorithms that scale with N. However, one needs to keep in mind that each of the $N-1$ cluster amalgamations requires the construction of a new (or updated) spanning tree. Li [40] shows that an efficient single link algorithm with p parallel processors scales as $O\left(\frac{N^2}{p}\right)$. It is also mentioned that on a single processor the best scaling for a single link algorithm is $O\left(N^2\right)$ [10, 49].

The literature points out that most of the parallel implementations for hierarchical clustering are impractical, mainly due to the large number of parallel processors with shared memory that are required, and these methods are therefore only of theoretical relevance [2, 8, 12, 52]. Epstein introduces an efficient and interesting implementation for hierarchical clustering with a quadtree data structure for dynamic updating of the closest pairs [18]. While Epstein discusses an improved scaling performance for each iteration in detail, it is not clear what the final scaling in time would be for a real-world data set. In addition, it is not clear whether the updating is approximate or whether the clustering is fully equivalent to the SAHN algorithm. The exploitation of nearest neighbor searches and the dendrogram microstructure for hierarchical clustering is further discussed by Dash and Weiss [11, 63]. Murthag discusses several clever and fast algorithmic implementations for hierarchical clustering [44, 45, 46, 47].

8.8.3 Improving the Scaling of Computing Time for the SAHN Algorithm with Cluster Seeding

An obvious way to improve the scaling of average link hierarchical clustering in time is to seed the clustering with the cluster results obtained from a different and faster clustering algorithm such as K-means or single or complete link-based algorithms. For this purpose, it is worth noting that standard K-means clustering implementations scale as N^2 in time and that faster implementations are possible as well. The drawback of such a scheme is that the resulting final clustering may be different than the clustering that would be obtained by implementing the standard SAHN algorithm. Such a hybrid clustering scheme was first mentioned by Kwon and Han [39], however, in their paper they seed the clustering with hierarchical clustering and then continue with a traditional K-means clustering scheme in order to improve and stabilize the cluster performance for this particular application rather than minimize computing time.

In order to study the influence of K-means seeding on the cluster performance of hierarchical clustering, the standard SAHN algorithm was compared with a SAHN algorithm using K-means cluster seeding using two different seeding experiments on 16 data sets: one experiment seeds with $\frac{N}{2}$ clusters obtained from K-means clustering and the other experiment seeds with $\frac{N}{4}$ clusters obtained in a similar way. The time savings are obvious; due to the cubic scaling, half the number of data points should cluster in 1/8th of the computing time, while one fourth of the data points should be 64 times faster. Because the K-means algorithm scales quadratically at most in computing time, the required time for cluster seeding with the K-means algorithm can be ignored, even with multiple seeds. In this case, 100 different initializations where typically used for seeding the K-means clustering algorithm and the clustering initialization corresponding to the smallest cluster dispersion was then selected. In order to evaluate the impact of cluster seeding on the cluster performance, the optimal number of clusters obtained with various internal (Davies-Bouldin, Dunn, Cluster Silhouette width, and GAP index) and external cluster validation indices (Adjusted Rand Index, Jaccard coefficient, and Fowlkes-Mallows index) are tabulated in Table 8.5. The optimal number of clusters with the two cluster seeding experiments is listed as well if that number deviates from the initial hierarchical clustering result.

From Table 8.5 it is obvious that the various cluster validation indices for the standard SAHN algorithm vary widely and are only consistent for clearly separated clusters such as 5 non-overlapping Gaussians. It is interesting to note that most studies in the literature that introduce and discuss cluster validation indices only consider this type of data, and then proceed to point out the advantages of a new cluster index on a particular data set for which the new index now shows a clear advantage in estimating the number of clusters. Therefore, in spirit of Sledge et al. [57], a more holistic approach is proposed for comparing cluster validation indices: such that all indices should be considered together to serve as a fingerprint for the cluster performance and cluster structure for a particular data set [6]. As a justification for this approach, consider the example where there are two clusters of individuals based on descent (e.g., Asians and Europeans), and that both clusters also have subclusters consisting of male and female members for the Asians and the Europeans. Clustering these data does not necessarily lead to a unique result for the final number of clusters and finding two or four clusters are both plausible answers in this case. In addition, the two prime clusters could be very different as well, because they could either distinguish between male and female or between Asian and European.

Table 8.5. Benchmark studies comparing cluster performance for hierarchical clustering with K-means cluster seeding.

	# samp.	# att.	# class	GAP	DB	Dunn	SHWI	ARI	Jaccard	Fowlkes
Italian Olive Oil	572	8	9	3/3/2	4/5/6	13/2/2	9/9/6	5/5/4	5/5/4	5/5/4
Fischer's Iris Data	150	4 (E)	3	2	2	2	2	3	2	2
Portugues Rock Data	134	18	6	3	5	4/2/2	4/5/5	5/4/4	5/4/4	5/4/4
Tobacco	26	16 (E)	2	2	2	4/2/2	12/12/7	2	2	2
Wieland's Spiral Data	194	2 (E)	2	3	10/11/11	2/12/4	5/10/3		2	2
Kohonen's Animal Data	16	16	V	2	12/8/3	2	12/8/4	3	3	2
Gaussians										
2 overlapping	300	2 (E)	2	4/4/3	6	2	6	3/3/4	2/3/2	2/3/2
5 non-overlapping	1500	2 (E)	5	5	5	5	5	5	5	5
Linton's JBE (1 word)	29	340	4	5/2/2	2/3/3	3/3/3	3	7/6	7/6/4	7/6/4
Linton's JBE (2 word)	29	521	4	5/2/2	12/12/7	2/2/2	12/12/7	4/6/4	4/2/2	4/2/2
Leukemia	38	7129 (2529)	2	3/2/2	12/12/12	2/2/2	12/12/12	3/2/3	2	2
Microglia	1772	300 (80)	3	5/2/2	2/3/3	6/6/2	6/6/6	3/5/4	2	2
Santos 2-D data										
clock	126	2	3	3	2	2	3	3	3	3
anvils	201	2	2	2	2	2	2	2	2	2
bermuda	182	2	4	3	7/8/8	6/2/2	6/8/6	7/6/6	7/6/6	7/6/6
beans	140	2	2	4	3/5/4	2/2/4	11/4/4	3	2/3/3	2/3/3

With these remarks in mind, one can conclude from Table 8.5 that cluster seeding leads to different clustering results for hierarchical clustering, but that the differences in estimating the number of clusters in the data vary more depending on the particular cluster validation index rather than the cluster seeding. Thus, while cluster seeding changes the clustering result as far as the number of clusters is concerned, such changes may prove to be acceptable and valid depending on the particular data set and application. The improved scaling in time of a hybrid hierarchical clustering scheme needs little elaboration. In this study it was observed that the signature plot including different cluster indices and a different number of clusters is relatively stable for each data set. As an illustration the cluster signatures for the clock data and the microglia data sets as shown in Figure 8.22. For the clock data several cluster validation indices (Davis-Bouldin, Dunn_A, Cluster Silhouette Width, and Adjusted Rand Index) are shown versus the number of clusters for a full hierarchical clustering and for hierarchical clustering with a K-means cluster seeding with 31 clusters. Similarly, these same indices are shown for the microglia data where the hierarchical clustering utilized the original 1772 data and a K-means cluster seeding with 443 data (being one fourth of the original data). It can be seen that the clock data exhibit a very different fingerprint or signature for the behavior of the cluster validation indices versus the number of clusters than the microglia data. These fingerprints are almost identical for the same data set regardless of the number of cluster seeds that was used to accelerate the hierarchical clustering. In other words, even when the optimal number of clusters can be influenced by cluster seeding, the typical clustering signature of the cluster validation indices for a particular data set remains very similar.

8.8.4 Improving the Scaling of Memory for the SAHN Algorithm with a Divide and Conquer Approach

It deserves to be repeated here that, in practice, the real bottleneck in clustering large data sets relates to both memory requirements and computing time. Most cluster algorithms scale with N^2 in memory, because the data distance matrix is of size $\frac{N(N-1)}{2}$. An obvious way to overcome the memory bottleneck is to take a divide-and-conquer approach where the initial data are split up into several subsets of data at random, and each subset is initially independently clustered. The clustering of these subsets can either proceed with or without K-means cluster seeding. The hierarchical clustering for the subsets is now halted at a sufficiently large number of clusters so that the combined cluster distance matrix can still fit into computer memory. At this point, one can proceed with hierarchical clustering from the combined (but reduced) cluster distance matrix. It is not demonstrated here explicitly that such a divide-and-conquer approach would lead to a similar profile of the cluster validation indices, and this aspect deserves further attention and study.

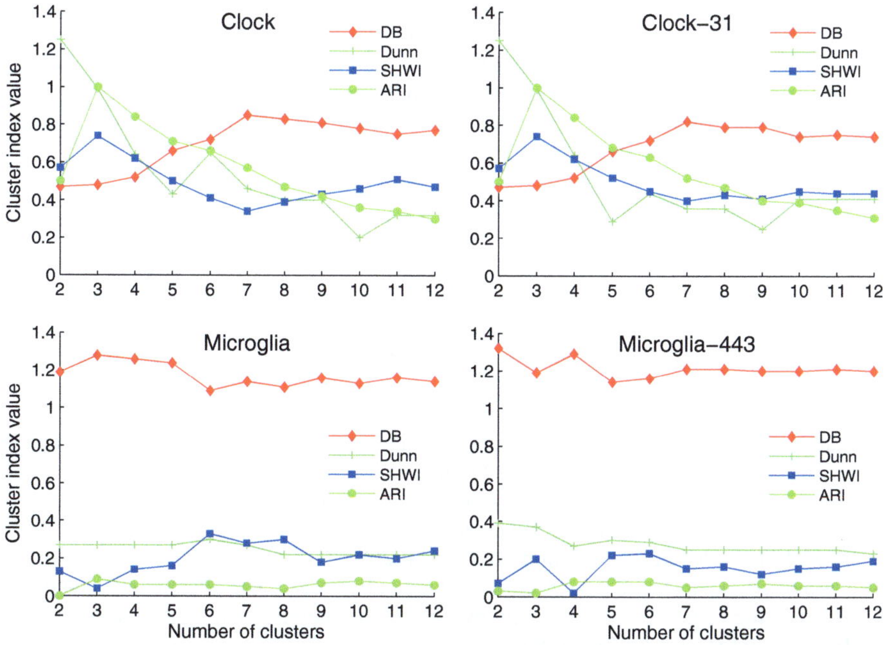

Fig. 8.22. Illustration of clustering signatures (i.e., cluster evaluation indices versus number of clusters) for the Santos clock data and the microglia data for hierarchical clustering with (right figures) and without (left figures) cluster seeding. Even when the optimal number of clusters can vary as a consequence of cluster seeding the clustering signature remains very similar.

8.9 Conclusions

This chapter serves as a review on hierarchical clustering and several related data visualization techniques such as the cartoon cluster plot, the dendrogram and the bicluster plot. A novel hybrid clustering approach is introduced to mitigate the poor scaling of the computing time for the average link hierarchical clustering algorithm using K-means cluster seeding or any other available clustering method. The cluster performance of this hybrid approach was evaluated on 16 benchmark data sets and evaluated using a fingerprint or cluster signature based on several internal and external cluster validation indices. The scaling in computing time of this hybrid algorithm is now governed mostly by the scaling of the algorithm used for cluster seeding. In a similar way, the memory limitation for large data sets can be overcome by a divide-and-conquer approach where the data is split up into several smaller data sets that fit into memory. While such hybrid approaches can affect the optimal number of clusters found for the individual cluster evaluation indices, the overall clustering signature (i.e., plot of the behaviour of several cluster indices versus the number of clusters) remains relatively stable.

Acknowledgements. The authors would like to than Min-Nan Chen for preparing the *Journal of Business Ethics* data sets. We wish to acknowledge the support of the Natural Sciences and Engineering Research Council (NSERC) in conducting this research. This work was partially sponored by the Defence Advanced Research Projects Agency (DARPA) Microsystems Technology Office under the auspices of Dr. Jack Judy through the Space and Naval Warfare Systems Center, Pacific Grant No. N66001-11-1-4015.

References

1. Akl, S.G.: An adaptive and cost-optimal parallel algorithm for minimum spanning trees. Comput. 3, 271–277 (1986)
2. Akl, S.G.: Optimal parallel merging and sorting without memory conflicts. IEEE Trans. on Comput. 36(11), 1367–1369 (1987)
3. Anderberg, M.R.: Cluster Analysis for Applications. Academic Press (1973)
4. Bar-Joseph, Z., Gifford, D.K., Jaakkola, T.S.: Fast optimal leaf ordering for hierarchical clustering. Bioinforma. 17(S1), 22–29 (2001)
5. Behbahani, S., Ali Moti Nasrabadi, A.M.: Application of SOM neural network in clustering. Journal of Biomedical Science and Engineering 2, 637–643 (2009)
6. Bezdek, J.C.: Private conversation. In: IJCNN (2010)
7. Carmona-Saez, P., Pascual-Marqui, R.D., Tirado, F., Carazo, J.M., Pascual-Montano, A.: Biclustering of gene expression data by non-smooth non-negative matrix factorization. BMC Bioinforma. 7(1), 78 (2006)
8. Chen, G.: Design and analysis of parallel algorithm. Higher Education Press (2002)
9. Cheng, Y., Church, G.M.: Biclustering of expression data. In: Proc. Int. Conf. Intell. Syst. Mol. Biol., vol. 8, pp. 93–103 (2005)
10. Dahlhaus, E.: Parallel Algorithms for Hierarchical Clustering and Applications to Split Decomposition and Parity Graph Recognition. Algorithms 36, 205–240 (2000)
11. Dash, M., Lui, H., Scheuermann, P., Tan, K.L.: Fast Hierarchical clustering and its validation. Data Knowl. and Eng. 44(1), 109–138 (2003)
12. Datta, A., Soundaralakshmi, S.: Fast parallel algorithm for distance transform. IEEE Trans. Sys., Man, and Cybern. 33(5), 429–434 (2003)
13. Davies, D.L., Bouldin, D.W.: A cluster separation measure. IEEE Trans. Pattern Analysis Mach. Intell. 1, 224–227 (1971)
14. Day, W.H.E., Edelsbrunner, H.: Efficient algorithms for agglomerative hierarchical clustering methods. Classif. 1, 7–24 (1984)
15. Marques de Sá, J.P.: Applied Statistics using SPSS, STATISTICA and MATLAB and R. Springer (2007)
16. Dunn, J.: Well separated clusters and optimal fuzzy partitions. Cybern. 4, 95–104 (1974)
17. Eibe, F., Hall, M.A.: Data Mining: Practical Machine Learning Tools and Techniques. Morgan Kaufmann (2011)
18. Epstein, D.: Fast hierarchical clustering and other applications of dynamic closest pairs. Exp. Algorithms 5, 1–10 (2000)
19. Fahlman, S.E.: Faster-Learning Variations on Back-Propagation: An Empirical Study. In: Proceedings of the 1988 Connectionist Models Summer School. Morgan Kaufmann (1989)
20. Fayyad, U.M., Pietetsky-Shapiro, G., Smyth, P., Uthurusamy, R. (eds.): Advanced in Knowledge Discovery and Data Mining. MIT Press (1996)
21. Fisher, R.A.: The Use of Multiple Measurements in Axonomic Problems. Ann. Eugen. 7, 179–188 (1936)

22. Forina, M., Armanino, C.: Eigenvector projection and simplified nonlinear mapping of fatty acid content of Italian olive oils. Ann. Chem. 72, 125–127 (1981)
23. Fowlkes, E.B., Mallows, C.: A method for comparing two hierarchical clusterings. Am. Stat. Assoc. 78, 553–569 (1983)
24. Goodman, L., Kruskal, W.: Measures of associations for cross-validations. Am. Stat. Assoc. 49, 732–764 (1954)
25. Golub, T.R., Slonim, D.K., Tamayo, P., Huard, C., Gaasenbeek, M., Mesirov, J.P., Coller, H., Loh, M.L., Downing, J.R., Caligiuri, M.A., Bloomfield, C.D., Lander, E.S.: Molecular classification of cancer: class discovery and class prediction by gene expression. Monitoring Science 15, 531–537 (1999)
26. Hastie, T., Tibshirani, R., Friedman, J.: The Elements of Statistical Learning: Data Mining, Inference, and Prediction. Springer (2009)
27. Hubert, L., Schultz, J.: Quadratic assignment as a general data-analysis strategy. Br. J. Math. Stat. Psychol. 29, 190–241 (1976)
28. Hubert, L., Arabie, P.: Comparing partitions. J. Classif. 2, 193–218 (1985)
29. Jaccard, P.: Etude comperative de la distribution florale dans une portion des Alpes et des Jura. Bull. de la Socíété Vaudoise des Sciences Naturelles 37, 574–579 (1901)
30. Jain, A.K., Dubes, R.C.: Algorithms for clustering data. Prentice-Hall (1988)
31. Jain, A.K., Murthy, M.N., Flynn, P.J.: Data Clustering: A Review. ACM Comput. Surv. 31(3), 264–323 (1999)
32. Jain, A.K.: Data clustering: 50 years beyond K-means. Pattern Recognit. Lett. 3(8), 651–666 (2010)
33. Jiang, J.H., Wang, J.H., Chu, X., Ru-Qin, R.Q.: Neural network learning to non-linear principal component analysis. Analytica Chemica Acta. 336, 209–222 (1996)
34. Johnson, S.C.: Hierarchical clustering schemes. Psychometrika 32(3), 241–254 (1967)
35. Jun, M., Shaohan, M.: Efficient Parallel Algorithm s for Some Graph Theory Problems. Comput. Sci. Technol. 8(4), 362–366 (1993)
36. Kaufman, L., Rousseeuw, P.: Finding Groups in Data. Wiley Interscience (1990)
37. Kohonen, T.: Self-Organizing Maps. Springer (1995)
38. Krieger, A.M., Green, P.E.: A Generalized Rand-Index method for consensus Clustering of Separate partitions of the Same Data Base. Classif. 16, 63–89 (1999)
39. Kwon, S., Han, C.: Hybrid clustering method for DNA microarray data analysis. Gene Inform. 13, 258–259 (2002)
40. Li, Z., Li, K.-L., Xiao, D., Yang, L.: An Adaptive Parallel Hierarchical Clustering Algorithm. In: Perrott, R., Chapman, B.M., Subhlok, J., de Mello, R.F., Yang, L.T. (eds.) HPCC 2007. LNCS, vol. 4782, pp. 97–107. Springer, Heidelberg (2007)
41. Linton, J., Chen, M.-N.: Working paper: Analysis of the Evolution of the Field of Business Ethics through Text Mining. University of Ottawa (2011)
42. Madeira, S., Oliveira, A.: Biclustering algorithms for biological data analysis: a survey. IEEE/ACM Trans. Comput. Biol. Bioinforma. 1(1), 24–25 (2004)
43. The MathWorks, Natick, MA
44. Murthag, F.: Expected-time complexity results for hierarchic clustering algorithms which use cluster centres. Inform. Process. Lett. 16, 237–241 (1983)
45. Murthag, F.: A survey of recent advances in hierarchical clustering algorithms. Comput. J. 26, 354–359 (1983)
46. Murthag, F.: Complexities of hierarchical clustering algorithms: State of the art. Comput. Stat. Q 1(2), 101–113 (1984)
47. Murthag, F.: Comments on 'Parallel algorithms for hierarchical clustering and cluster validity. IEEE Trans. Pattern Analysis Mach. Intell. 14(10), 1056–1057 (1992)

48. Nath, D., Maheshwari, S.N.: Parallel algorithms for the connected components and minimal spanning tree problems. Informa. Process. Lett. 14(1), 7–11 (1982)
49. Olson, C.F.: Parallel algorithms for hierarchical Clustering. Parallel Comput. 21, 1313–1325 (1995)
50. Pauwels, E.J., Frederix, G.: Finding salient regions in images: nonparametric clustering for image segmentation and grouping. Comput. Vis. Underst. 75, 73–85 (1999)
51. Prelic, A., Bleuer, S., Zimmermann, P., Wille, A., Bhlmann, P., Gruissem, W., Hennig, L., Thiele, L., Zitzler, E.: A systematic comparison and evaluation of biclustering methods for gene expression data. Bioinforma. 22(9), 1122–1129 (2006)
52. Rajasekaran, S.: Efficient parallel hierarchical clustering algorithms. IEEE Trans. Parallel Distrib. Sys. 16(6), 497–502 (2005)
53. Rand, W.M.: Objective criteria for the evaluation of clustering methods. Am. Stat. Assoc. 66, 846–850 (1971)
54. Rousseeuw, P.J.: Silhouettes: A Graphical Aid to the Interpretation and Validation of Cluster Analysis. Comput. Appl. Math. 20, 53–65 (1987)
55. Santamaría, R., Therón, R., Quintales, L.A.M.: A Framework to Analyze Biclustering Results on Microarray Experiments. In: Yin, H., Tino, P., Corchado, E., Byrne, W., Yao, X. (eds.) IDEAL 2007. LNCS, vol. 4881, pp. 770–779. Springer, Heidelberg (2007)
56. Jorge Manuel Fernandes dos Santos: Data Classification with Neural Networks and Entropic Criteria. Ph. D. Dissertation (School of Engineering, University of Porto FEUP (2007)
57. Sledge, I.J., Havens, T.C., Bezdek, J.C., Kelleher, J.M.: Relational cluster validity. In: Aranda, J., Xambó, S. (eds.) Plenary and Invited Lectures of the 2010 World Congress on Computational Intelligence, Barcelona, Spain, pp. 151–185 (2010)
58. Smith, K.: Private Communication (2011)
59. Sneath, P.H.A., Sokal, R.R.: Numerical Taxonomy. W.H. Freeman (1973)
60. Tan, P.N., Steinbach, M., Kumar, V.: Introduction to Data Mining. Addison-Wesley (2005)
61. Tanay, A., Sharan, R., Shamir, R.: Biclustering algorithms: A survey. In: Handbook of Computational Molecular Biology. Chapman and Hall (2004)
62. Tibshirani, R., Walther, G., Hastie, T.: Estimating the number of clusters in a dataset via the Gap Statistics. J. R. Stat. Soc. B 63, 411–423 (2001)
63. Weiss, S.F.: A probabilistic algorithm for nearest neighbor searching. In: Oddy, R.N., Robertson, S.E., Van Rijsbergen, C.J. (eds.) Information Retrieval Research, Butterworths, pp. 325–333 (1981)
64. Wikipedia, `http://en.wikipedia.org/wiki/Hierarchical_clustering` (last accessed August 4, 2011)
65. Zapan, J., Gasteiger, J.: Neural Networks in Chemistry and Drug Design, 2nd edn. Wiley VCH (1999)
66. Xu, R., Wunsch II, D.: Survey of clustering Algorithms. IEEE Trans. Neural Netw. 16(3), 645–678 (2005)
67. Xu, R., Wunsch II, D.: Clustering. IEEE Press Series on Computational intelligence. Wiley (2008)

Appendix A: Santos 30 2-D Benchmark Clustering Data Sets

The Santos two-dimensional clustering data sets [56] are an interesting collection of benchmark data for clustering because they have features of many of the real-world data sets and are shown in Fig. 8.23. These data can be easily visualized because they are two-dimensional, and they exhibit interesting characteristics such as overlapping clusters, outliers, several clusters, nonlinearly separable cases, etc. Santos performed experiments with these data sets where children of different grade levels were asked to indicate the clusters in these data. Depending on the point of view, different clusterings are possible for several of these data sets. Also indicated on the figure are suggested clusterings in different colors.

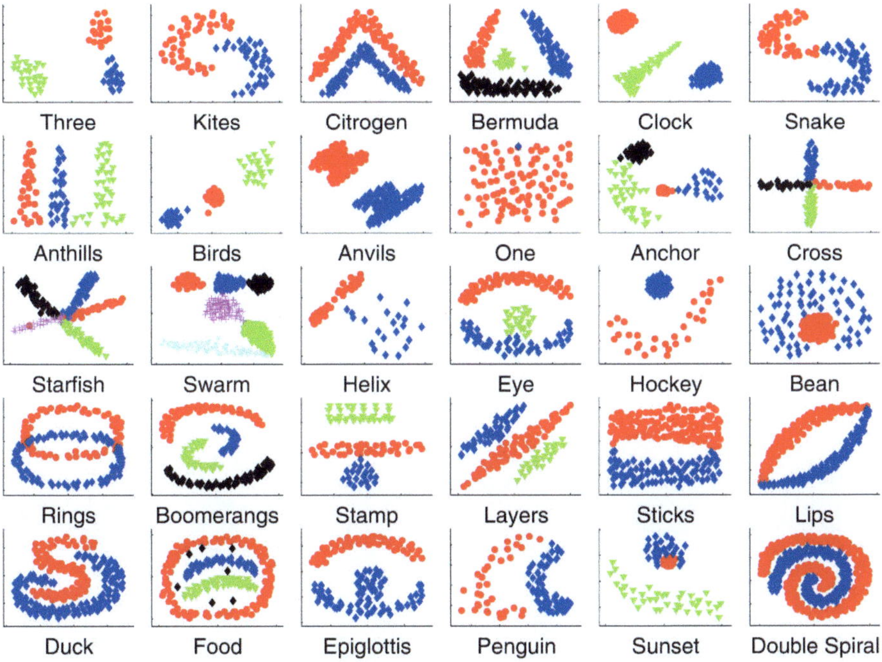

Fig. 8.23. Santos 2-D benchmark clustering data sets with suggested clusterings [56].

Chapter 9
A Novel Framework for Object Recognition under Severe Occlusion

Stamatia Giannarou and Tania Stathaki

Abstract. This work introduces a novel technique for the automatic identification of real world objects in complex scenes. The identification problem requires the comparison of assemblies of image regions with a previously stored view of a known prototype. Shape context representation and matching are employed for recovering point correspondences between the image and the prototype. Assuming that the prototype view is sufficiently similar in configuration with an object in the complex scene, the correspondence process will succeed. However, a number of ambient conditions such as partial object occlusion and contour distortion, may affect the performance of the matching process and consequently the identification result. A novel multistage type of clustering of suspicious image locations is applied in a novel fashion to enable the identification of regions of interest on the complex scene, based on a set of density and figural continuity metrics. In order to increase the robustness of the identifier to mismatches and reduce the computational cost of the process, a selection of the initial suspicious regions is applied. The performance of the identifier has been examined in a great range of complex image and prototype object selections.

9.1 Introduction

Object recognition is a highlighted field in computer vision and image processing. Its objective is to enable computers to recognize image patterns without human intervention. The task of object recognition is usually divided into two stages, namely the 'low-level' vision and 'high-level' vision. The first stage involves the extraction of significant features from an image, such as object boundaries, and usually the segmentation of the image into separate objects. The goal of high-level vision is to recognize these objects by finding effective and perceptually important shape

Stamatia Giannarou · Tania Stathaki
Imperial College London
Exhibition Road, SW7 2AZ, London, UK

P. Georgieva et al. (Eds.): Adv. in Intelligent Signal Processing and Data Mining, SCI 410, pp. 235–258.
springerlink.com

features. This work is concerned with the latter task and in particular with shape analysis and identification techniques.

Two basic approaches to shape analysis are the shape representation and description. Shape representation methods yield a non-numeric representation of the shape, for example a graph. On the other hand, shape description refers to methods that describe a shape numerically by generating a feature vector that uniquely characterizes it. However, shape analysis is not an easy task since a shape in an image is often corrupted with noise, defects, various and arbitrary distortion and occlusion. A shape descriptor with good retrieval accuracy should be tolerant to the above deformations. Furthermore, shape description methods should be invariant against possible transformations of an image caused by moving an object so as to change its perceived position, size (scale) and orientation in the image. A performance evaluation study of various local shape descriptors is found in [19].

Furthermore, shape identification refers to methods for comparing shapes using feature vectors generated from a pre-specified mathematical shape descriptor, in order to determine whether these shapes are the same or not. A query shape is matched to one of the model (prototype) shapes stored in a database by comparing their feature vectors using a similarity metric. An extensive survey of shape matching techniques in computer vision can be found in [28].

This work describes a new prototype-based method for identifying multiple objects in a complex scene. In our approach, prior knowledge of the target object's shape is described by a prototype template which consists of the object's representative features. These features are derived by employing the Shape Context descriptor that uses local histograms of the edge map of the image [1]. Matching between the contour points of the prototype and the complex image is investigated after applying a shape context similarity procedure. A novel multistage type of clustering of the detected matched points is employed for the identification of suspicious regions on the complex scene. At first stage an initial group of clusters is estimated. What follows is a series of processes, each of which restricts the previous group of clusters to a sub-group based on certain properties of the clusters. The success of the proposed identifier is verified by experimental results on a variety of complex images and prototypes.

9.2 Prior Work on Shape Analysis and Identification

The problem of shape analysis has been pursued by many authors resulting in a great number of various approaches. A rich literature has been published on the state of the art on shape analysis techniques [16, 32]. These methods can be classified according to various criteria. In general, shape representation and description techniques are categorized into contour-based and region-based methods, according to whether shape features are extracted from the contour only or from the whole shape region, respectively. A further classification is into global and structural methods, based on whether the shape is represented as a whole or by segments, called primitives. Another categorization can be made on the basis of whether the shape features are calculated in the spatial or transformed domain, resulting in space and transform domain techniques, respectively.

In cases where the boundary information is not available, region-based methods are preferred for shape analysis. Common techniques of this category use a specific type of the so called moment invariants, first introduced by Hu [11]. These moments, known as geometric moments, are derived from a nonlinear combination of lower order moments and they have the desirable properties of being invariant under translation, rotation and scaling. The algebraic moments [25] is another kind of moment invariants, computed from the first m central moments. Furthermore, the orthogonal Zernike moments were first proposed by Teague in [26] and use the Zernike polynomial as basis function. A comparison of moment-based descriptors is given in [24]. Another efficient region-based descriptor is the generic Fourier descriptor (GFD) [31,30]. It is obtained by applying the two-dimensional Fourier transform on a polar-raster sampled image shape.

Although region-based methods are robust and can be used in various applications, contour-based descriptors are more popular in the literature. This is mainly because humans are supposed to discriminate shapes based on their contour features and also because usually the shape interior content does not bear significant information. A representative example is the shape signature approach, which is a comprehensive way of representing the essence of the shape. The signature is a one dimensional function derived from the boundary points of the shape. Common shape signatures include centroid distance, tangent angle, curvature and area [7], [21]. However, shape signatures are undesirable since they are sensitive to noise and their matching is computationally complex. Stochastic models and in particular autoregressive models [13], is a different approach of contour-based methods. This technique is based on the stochastic modeling of a one dimensional function similar to the shape signature mentioned above. The main drawbacks of this descriptor are its computational cost and the difficulty to associate it with any physical meaning.

One of the most widely used contour-based shape description methods is Fourier descriptors [5]. These descriptors are obtained by applying Fourier transform on shape boundary which is usually represented by a shape signature (e.g centroid distance). The Fourier transformed coefficients are called Fourier descriptors of the shape. This method is noise insensitive and only a small number of coefficients are enough to capture the overall shape features.

A more recent approach to shape analysis is the curvature scale space (CSS), described in [20]. The CSS descriptor aims at tracking the position of curvature zero-crossings (inflection points) in a shape boundary filtered at varying smoothing scales. Thus, a final CSS contour map is obtained indicative of the location of the zero-crossing points and the scale at which they occurred. The peaks of this map are then used as the CSS descriptors to index the shape. The powerfulness of this approach stems from its ability to capture the location and the degree of convexity (or concavity) of curve segments on the shape boundary, features which are very important even to human perception in recognizing objects. Since the work of [20], improved and more efficient versions of this approach have been designed [14].

Another leading approach to shape analysis is the one proposed by Lateki et al. [15]. The uniqueness of this method is the novel process of digital curve evolution that is applied to contours prior to the calculation of the similarity measure in order to eliminate the effect of digitization noise and segmentation errors. Object contours are represented by a simple closed polygonal curve. The best possible correspondence of the visual parts of different contours is estimated and then the similarity between the corresponding parts is computed based on the L_2 distance of their tangent space representation (turn angle representation).

A significant contribution to shape analysis is the scale invariant feature transform (SIFT) proposed by Lowe [18]. This technique actually combines a scale invariant region detector and a descriptor based on the gradient distribution in the detected regions. The descriptor is represented by a three dimensional histogram of gradient locations and orientations while the contribution of each of these bins is weighted by the gradient magnitude. Another method, similar to the SIFT descriptor is the so called Shape Context descriptor [1]. This method yields a two-dimensional histogram with axes the log-distance and polar angle. The descriptor is applied on the edge points only and describes the edge distribution in the surrounding region of each contour point. With the appropriate modifications, shape context becomes invariant to rotation, shift and scale changes of the target object. In general, Shape Context descriptor is a simple, rich descriptor that enforces good shape matching and recognition and for these reasons it is employed in this work.

The object identification task is concerned with the decision on whether two observed objects are in fact one and the same object. This is a challenging problem in computer vision mainly due to the nonrigid nature of the objects in most of the applications. A number of different methods have been proposed in the literature for object identification. Harris [10] and Lowe [17] use a 3D model to track rigid objects. Snakes [27] have also been applied to track the contour of deformable objects, while deformable templates [33] are very appealing in tracking tasks because of their capability and flexibility.

This Chapter is organized as follows. The shape context representation and matching technique is analysed in the next section. In Section 9.4 we explain the clustering method used in this work. A novel identification method is proposed in Section 9.5. Experimental results produced from the application of the identifier on various images and prototypes are presented in Section 9.6. We discuss the performance of the identifier and conclude with Section 9.7.

9.3 Shape Context Representation and Matching

9.3.1 Shape Context Descriptor

The so called shape context descriptor for object contour representation was first introduced by Belongie [1]. In order to extract this descriptor, the edge map (collection of edge pixels) of the image of interest is required. This can be easily obtained by applying an edge detection process (as for example the Canny edge detector [3]) to the image.

Any edge pixel can be associated with a shape context. Let an object P represented by a set of contour (edge) points (pixels) $\{p_1 \dots p_m\}$. The notation $P \equiv \{p_1 \dots p_m\}$ may be used. Throughout this chapter we will keep this notation wherever is possible, for reasons of simplicity. However, it is important to stress out that each pixel is explicitly associated with the 2-D vector of its spatial coordinates and therefore, a more formal notation could be $p_i \equiv (x_i, y_i)$. The shape context of a point p_i is the two dimensional histogram that describes the distribution of edge pixels in a pre-specified discretised neighborhood of the point p_i and it is computed as:

$$H_K(p_i) = (h_1(p_i), h_2(p_i), \dots, h_K(p_i))$$

where $h_k(p_i)$ expresses the number of contour points in the kth bin, mathematically defined as:

$$h_k(p_i) = \#\{p_j \neq p_i \mid p_j \in P, p_j \in \text{bin}(k)\}$$

In this work, the above mentioned neighborhood is a circle centered at the edge point of interest with radius R. The histogram bins are obtained from the uniform partition of that circle in the log-polar space (r, θ). The log-polar space is preferred to the Cartesian space since it provides a contour descriptor that gives more emphasis to the closest neighbors of the point of interest. The resolution of the circle (number of bins) is controlled by the sampling rate of the logarithmic distance r from the center of the circle and the sampling rate of the polar angle θ. If we assume N bins for the logarithmic distance $r = \{r_i = \frac{Ri}{N}, i = 1, \dots N\}$ and M bins for the polar angle $\theta = \{\theta_i = \frac{2\pi i}{M}, i = 1, \dots M\}$ then $K = NM$ and we may rewrite the shape context descriptor as a function of the three parameters R, N, M as $H_{R,N,M}(p_i)$. The shape of the object P is described by the set of shape contexts that correspond to its contour points as:

$$SC_P \equiv \{H_{R,N,M}(p_i) \mid p_i \in P, i = 1, \dots, m\}$$

In order to obtain good identification results, the size of the examined neighborhood around contour points must be appropriately defined. A neighborhood, large enough to capture the configuration of the entire shape relative to a reference contour point, leads to a global descriptor which is effective when we are concerned with the retrieval of whole prototype objects from complex scenes. The key characteristic of the global descriptor is the underlying assumption that objects that belong to the same class have overall similar shape. Intuitively, the performance of the global descriptor degrades in cases of distortion or partial occlusion of the prototype object in the test image. To overcome this weakness, a local descriptor would be more efficient especially for the identification of partially visible or distorted objects in the complex scene since, in that case, local shape properties are more representative features of the class of the object. Therefore, the challenge in shape context representation is to define the size of the descriptor aiming at the extraction of shape features that are most distinctive for retrieval under any ambient condition.

The shape context descriptor turns out to be a powerful tool, enabling easy and robust correspondence recovery. By construction, the descriptor is invariant under translation of the test object within the image. Rotation invariance is achieved by considering the positive x-axis for the local coordinate system to be the tangent direction at each point. Uniform scale invariance is achieved if the radial distances in the log-polar representation are normalized by the mean distance between all pairs of points in the shape. Obviously this approach to scale invariance is of practical use only when the entire object under consideration is present in the image and furthermore, the minimum size circle that includes the entire object does not contain parts of other objects so that the global shape descriptor can be applied. We propose to guarantee scale invariance by estimating for each detected edge point multiple shape context descriptors with varying radii. The optimum scale is the one that gives the minimum total matching cost when estimating the correspondences between the points on the input image object and the points on the prototype object, as it is described later on.

Assume the object P introduced above and a complex scene represented also as a collection of edge pixels $S \equiv \{s_1 \dots s_n\}$. The similarity between two shapes is related to the similarity between their sets of shape contexts. We can then set up the problem of corresponding (matching) contour points between a prototype shape and a complex scene as that of finding pairs of shape contexts with the maximum similarity. The cost of matching two points is equal to the χ^2 distance between their shape contexts. More specifically, the cost of matching a point s_i on the scene S to a point p_j on the shape P is denoted by C_{ij} and is defined as:

$$C_{ij} = \frac{1}{2} \sum_{k=1}^{K} \frac{[h_k(p_j) - h_k(s_i)]^2}{h_k(p_j) + h_k(s_i)}$$

where $h_k(\cdot)$ stands for the number of contour points in the kth bin and K is the total number of bins defined previously. The above expression yields an $n \times m$ matrix C, where m, n are the number of edge points on P and S, respectively. Given the set of costs, C_{ij}, between all the pairs of points, we can proceed to estimate the optimum set of corresponding pairs, $\{s_i, p_{\alpha(i)}\}$, between the scene and the target. The function $\alpha(\cdot)$ stands for the point mapping $\alpha : S \rightarrow P$, defined later on.

9.3.2 *Many-to-One Edge Point Matching*

When we tackle the problem of matching edge points from a complex real life scene to edge points of a prototype (target) object image we face the issue that the former image usually contains additional objects, textured surfaces, noise and other types of signals which produce "irrelevant" edge points. Provided that the local shape context model used in this work is a robust model, the irrelevant edge points should not be matched to points on the prototype object's contour. Furthermore, part of the target object's contour might be absent in the complex image, due to obscuration of the object in the scene, or other variability. As a result, several contour points on the prototype object will have no matches on the scene. Finally, there are situations where the scene might contain multiple objects similar and/or identical to the prototype and in that case, a target contour point can be matched to multiple scene points. Based on the above problem specifications, the assumption of one-to-one correspondence between the target and the scene points which has solely been examined in the available literature [12], would be a very restrictive one, failing to identify multiple targets in the scene. Hence, we seek not a one-to-one matching, but rather a many-to-one correspondence between the scene points and the target.

Let us assume the $n \times m$ non-negative cost matrix C_{ij}, where the rows and columns correspond to points from the sets S and P, respectively. The set E of point pairs of size $n \times (m+1)$ is introduced, including all the possible matches between S and P expressed in terms of edge point index pairs, where $(i, j) \in E$ implies that $i \in S$, $j \in P$. Our objective is to match specific scene points to points on the prototype object image by selecting a subset of the set E according to certain optimization criteria which will be explained later. In our approach we initially assume that all points in the complex scene, including the irrelevant edge points, will be matched to a point in the prototype image. For that reason we introduce a so called "dummy" point to the set P. This additional point extends automatically the $n \times m$ cost matrix C_{ij} to a new one of size $n \times (m+1)$ by padding the new column with a "dummy" cost ϵ_d. A point will be matched to the dummy point whenever there is no real match available at cost smaller than ϵ_d.

In the proposed approach, our main concern is to find a set $A \subset E$ of size n of corresponding point pairs $A = \{(s_i, p_{\alpha(i)}), i \in [1, n]\}$ that minimizes the total cost of matching $\sum_{i=1}^{n} C_{i\alpha(i)}$, subject to the constraint that each scene point that belongs to A is matched to exactly one target point and each target point in A is matched to at least one scene point. Therefore, an many-to-one assignment problem can be formulated as finding a function $\alpha : \{1, \ldots, n\} \to \{1, \ldots, m+1\}$ such that values in $\{1, \ldots, m+1\}$ can be matched more than once. Here, $\alpha(i) = m + 1$ indicates that the scene point s_i is not matched with a target contour point but with its virtual "dummy" point.

The extended cost matrix is the input to the assignment problem and the resulting assignment, excluding pairs of the form $(i, m+1)$, is the output. The many-to-one assignment problem under study can be mathematically modeled as a straightforward modification of the one-to-one assignment problem [12], given below:

$$\min \sum_{ij} C_{ij} x_{ij}$$

$$\text{s.t} \sum_{j} x_{ij} = 1, \forall i \tag{9.1}$$

$$\sum_{i} x_{ij} \geq 1, \forall j \tag{9.2}$$

$$x_{ij} = 0 \text{ or } 1, \forall i, j$$

where $x_{ij} = 1$ if scene point i is assigned to prototype contour point j and equal to 0, otherwise. The set of constraints in (9.1) ensures that every scene point is assigned to only one prototype point. The set of constraints in (9.2) ensures that a prototype point can be assigned to more that one scene points. In practise, the above problem can be solved by finding for each row i of C the column index $\alpha(i)$ associated with the minimum C_{ij} value of row i, mathematically defined as:

$$\alpha(i) = \text{argmin}\{C_{ij} : j = 1, \ldots, m+1\}$$

The main concern in this work is to locate multiple targets in an input image. Let an image Ic representing a complex scene (input image) and a second one Ip containing a single prototype object. The edge map of each image can be easily obtained by applying an edge detection process to the image. Thus, the point sets $\{c_1, \ldots, c_K\}$ and $\{p_1, \ldots, p_L\}$ (where usually $L < K$) emerge, corresponding to contour points of the represented objects on the Ic and Ip image, respectively. The Shape Context representation technique is applied on both images, independently. Different points on object contours will be represented by different shape contexts. However, homologous points on similar shapes will have similar shape contexts. The proposed one-to-many assignment problem is employed to find for each scene point c_i the prototype contour point $p_{\alpha(i)}$ with the most similar shape context. In the following section, it will be proven that the location of the matched points on the input image provides significant information for the identification of the target in the scene.

9.4 Clustering of the Matched Points on the Complex Scene

With reference to the above analysis, the shape context matching process results in the extraction of a set of edge points $\{c_i \mid i = 1, \ldots, J, \text{ where } J \leq K\}$ on the input image which have the best correspondence to the contour points $\{p_{\alpha(i)} \mid i = 1, \ldots, J\}$ on the prototype shape. The aim of this section is to identify regions of interest on the input image which could indicate the presence of the target object in the scene.

Ideally, if the whole prototype object exists in the input image, the matched points will lie within a localized region of the image being indicative of the location of the prototype object on the scene. However, this in not always the case since the presence of a distorted or occluded version of the prototype in the complex scene may affect the performance of the matching process. For instance, in the case of occlusion, a contour point, $p_{\alpha(i)}$, that belongs to a part of the prototype shape which does not appear in the input image might be corresponded to an edge pixel, c_i, with a similar shape context which, however, may lie on a test object dissimilar to the prototype one. Likewise, in the case of distortion, a contour point on the prototype may be mismatched because the shape context of its actual homologous point on the complex image is not the most similar one. These mismatched edge points are randomly distributed along the scene.

During retrieval, if the prototype or a transformed (rotated or partially occluded) version of it appears on the complex scene, the matching process will form regions on the complex scene with dense and/or sparse distributions of matched points. If part of the prototype object is present in the complex scene without being significantly distorted we should expect at least one localized dense distribution of matched points. Intuitively, dense regions have greater possibility of being regions of interest, while sparse regions are likely to correspond to objects which exhibit weak similarities with the prototype object. Furthermore, isolate matched points are probably mismatches that should be ignored and removed. In this contribution, we propose the use of data clustering in a novel fashion for the partitioning of the matched points on the complex scene into groups (clusters) of neighboring points followed by a multistage type of elimination of the "irrelevant" clusters. The algorithm is divided in three stages. Initially, the clusters are estimated using a standard technique. In the second stage, we seek to eliminate among the estimated clusters these which consist of matched points that form (almost) straight lines. The idea behind this process is that straight lines form parts of a large variety of objects and therefore, are poor identifiers of a specific object. Finally, from the remaining clusters we seek to discard the clusters with sparse distributions of matched points estimated using a density metric that is described later.

Generally, the aim of clustering is the partitioning of a data set into groups such that the minimum similarity between points within a group is greater than the maximum similarity between points among different groups. The main categorization of clustering methods is based on whether the number of clusters is known a priori or not. K-means [2] and Fuzzy C-means [23] techniques belong to the first category while Mountain [29] and Subtractive Clustering [6] algorithms treat the number of clusters as one of the problem unknown parameters and estimate the clusters sequentially. In this work, since the appropriate number of clusters is not known in advance, a method that belongs to the latter category is employed and in particular, Subtractive Clustering [6] is chosen.

Subtractive clustering has evolved from the need to overcome Mountain method's inefficiency in cases of high dimensional data [29]. According to the former method, each data point is considered as a potential cluster centroid and the likelihood that each point would define a cluster centroid is measured, based on the density of the surrounding data points. In subtractive clustering, the density measure is calculated on data points rather than on grid points as in the Mountain clustering [29]. Let us consider a data set $\mathbf{X} = \{\mathbf{x}_1, \mathbf{x}_2, \ldots, \mathbf{x}_n\}$ in the 2-dimensional space $\mathbb{R}^2$ where x_{uv} will be the vth coordinate of the uth data point, for $u = 1, \ldots, n$ and $v = 1, \ldots, 2$. The density function at a data point $\mathbf{x}_i$ is defined as:

$$D(\mathbf{x}_i) = \sum_{j=1}^{n} \exp\left[-\frac{\|\mathbf{x}_i - \mathbf{x}_j\|^2}{(r_a/2)^2}\right]$$

where r_a is a positive constant representing a neighborhood radius and $\|\mathbf{x}_i - \mathbf{x}_j\|^2$ is the Euclidean distance between $\mathbf{x}_i$ and $\mathbf{x}_j$. Clearly, the higher the number of neighbors around a data point, the higher the density value that corresponds to this point.

The first cluster centroid, $\mathbf{x}_{c_1}$, is chosen as the data point with the highest density value. Then, a modified density function is used to update the density value at each data point, $\mathbf{x}_i$. At the kth iteration of cluster formulation the density values are calculated as:

$$D^k(\mathbf{x}_i) = D^{k-1}(\mathbf{x}_i) - D^{k-1}(\mathbf{x}_{c_{k-1}}) \exp\left[-\frac{\|\mathbf{x}_i - \mathbf{x}_{c_{k-1}}\|^2}{(r_b/2)^2}\right]$$

where r_b is a positive constant. The objective of the above update is to eliminate the effects of the cluster centers that have already been identified. Thus, at each round, k, each data point is assigned a new density value which is equal to the density value assigned at the previous round, $k-1$, reduced by an amount inversely proportional to the distance of the data point from the newly found cluster center, $\mathbf{x}_{c_{k-1}}$. Therefore, at each round, the data points near the newly found cluster center will have significantly reduced density measures, so these data points can not become cluster centers afterwards. This process ends when a sufficient number of clusters has been attainted.

9.5 Object Identification

The main objective of the following analysis is to determine the actual regions of interest on the complex scene, i.e, regions that include a part of the prototype shape, based on the extracted corresponding points, $\{(c_i, p_{\alpha(i)}) \mid i = 1, \ldots, J\}$, between the scene and the prototype. The prototype object identification is based on the examination of the matched points, c_i, on the scene. By applying Subtractive Clustering, these points can be spatially partitioned into clusters $\{CS_w \mid w = 1, \ldots, W\}$, while the ones with no close neighbors remain ungrouped. Not all of the above clusters bear equally significant information for the tracing of the prototype on the scene. Therefore, a data mining process would be desired to facilitate the identification process.

9.5.1 Cluster Elimination Based on Cluster-Activity Estimation

As part of the data elimination, only the matched points that belong to a cluster will be considered for further examination, while the isolated matched points not included in any cluster, will be ignored. This is a realistic assumption since these points are not sufficient to indicate a region of interest in the scene as they are associated with isolated points on the prototype's contour rather than with parts of it and therefore, are likely to be mismatched outliers. On the contrary, a neighborhood of matched points belonging to the same cluster may correspond to a neighborhood of edge points on a single object's contour and therefore indicate a region of interest on the complex scene.

Further to the above point elimination, a cluster selection can be performed. In particular, we assume that clusters with low-activity do not bare significant information. A low-activity cluster is defined as the one which includes only a small number of points with high curvature on the boundary formed by the cluster points. Therefore, the boundary parts enclosed in low-activity clusters are edges with few corners, such as lines or low order curves. The idea for the above data mining procedure is sensible especially in the case where the derived feature vectors for the object description incorporate local rather than global shape information. Since low-activity regions include object boundary parts (lines or simple curves) which are common for a large number of shape contours, it is highly likely that when such a part is present on the prototype's boundary it may be matched with a similar boundary part on the contour of a scene object, different to the target, leading to mismatches.

The activity of a cluster is related to the number of corners on the boundary formed by the edge points of the cluster. This boundary is moulded by joining the points of the cluster using the so called Connected-Component Labeling theory described in [9]. The entire process consists of a number of processing steps applied on the cluster points. To begin with, lets assume the cluster CS_w on the scene, where the included points form a set of connected-components, CCI_{wi}, where w and i stand for the cluster and the component index, respectively. In this context a connected component is a set of edge points such that each point is connected to at least one other member of the set and the set is maximal with respect to this property. These components must be detected and if they are more than one they must be prolonged, in order to be joined if it is possible. The goal is to form in the cluster fewer and consequently longer connected-components and ultimately a single one, to facilitate further processing of the cluster points, as it is explained later on. As already mentioned, the technique described in [9] is employed for the segmentation of the cluster into separate components. During prolongation, following the technique proposed in [8], a new point is added to each endpoint of a connected-component, belonging to the best path issued from this endpoint. Depending on the endpoint configuration, three candidate pixels are determined for the prolongation, as shown in Figure 9.1. For each path leading from the endpoint to the candidate pixel, a cost function is computed. The candidate belonging to the best path is then marked as a component point. The process is iterated from this new endpoint until the endpoint of another component is reached or until a pre-defined fixed number of iterations is realized. The cost function used in this work is the distance between the candidate of the examined path and the endpoints of the other components in the cluster. The connected-components emerged after the prolongation process are represented by CCF_{wj}, where w and j stand for the cluster and the component index, respectively. In the already mentioned ideal case where the prolongation process yields a single connected component, the index j takes only the value of 1 and therefore it can be dropped. A representative example of applying the contour closing process is illustrated in Figure 9.2.

During corner point detection in clusters, each CCF_{wj} component is processed separately. In order to arrange the points of the CCF_{wj} component as an ordered sequence of adjacent points, the component becomes the input to a contour tracer.

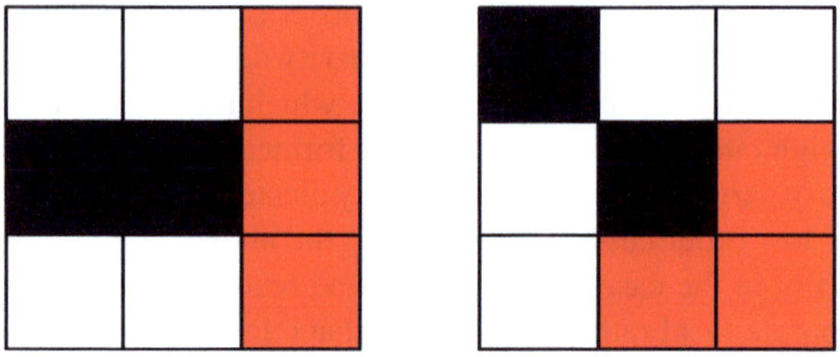

Fig. 9.1. Candidate points in prolongation for two different boundary configurations. The black pixels represent the component points and the red ones the candidate points.

The boundary tracing approach employed in this work is based on the work proposed in [4]. The output of the tracer is an array containing the co-ordinates of the component's points. Corners on the boundary of a component correspond to points of high curvature. Intuitively, curvature is the amount by which a geometric object deviates from being *flat*, but this is defined in different ways depending on the context.

For the curvature estimation, the boundary of the traced CCF_{wj} component can be represented by the set of its points' coordinates $\{(x_i, y_i) \mid i = 1 \ldots N\}$, where N is the number of the component's boundary points. According to [22], the curvature at a point i on the discrete boundary curve can be computed as:

$$Crv_i = \Delta x_i \Delta^2 y_i - \Delta y_i \Delta^2 x_i \tag{9.3}$$

where, Δ and Δ^2 stand for the first-order and second-order difference operator, respectively. The derivatives in (9.3) are defined as:

$$\Delta x_i = \frac{x_{i+1} - x_{i-1}}{\sqrt{(x_{i+1} - x_{i-1})^2 + (y_{i+1} - y_{i-1})^2}} \text{ and } \Delta y_i = \frac{y_{i+1} - y_{i-1}}{\sqrt{(x_{i+1} - x_{i-1})^2 + (y_{i+1} - y_{i-1})^2}}$$

$$\Delta^2 x_i = \frac{\frac{x_{i+1}-x_i}{\sqrt{(x_{i+1}-x_i)^2+(y_{i+1}-y_i)^2}} - \frac{x_i - x_{i-1}}{\sqrt{(x_i-x_{i-1})^2+(y_i-y_{i-1})^2}}}{\sqrt{[1/2(x_{i+1}+x_i) - 1/2(x_i + x_{i-1})]^2 + [1/2(y_{i+1}+y_i) - 1/2(y_i + y_{i-1})]^2}}$$

$$\Delta^2 y_i = \frac{\frac{y_{i+1}-y_i}{\sqrt{(x_{i+1}-x_i)^2+(y_{i+1}-y_i)^2}} - \frac{y_i - y_{i-1}}{\sqrt{(x_i-x_{i-1})^2+(y_i-y_{i-1})^2}}}{\sqrt{[1/2(x_{i+1}+x_i) - 1/2(x_i + x_{i-1})]^2 + [1/2(y_{i+1}+y_i) - 1/2(y_i + y_{i-1})]^2}}$$

To eliminate insignificant boundary details, both boundary coordinate functions are convolved with the discrete gaussian filter $g(n, \sigma)$ given by:

$$g(n, \sigma) = \frac{1}{2\pi\sigma^2} e^{\frac{-n^2}{2\sigma^2}}, \text{ for } n = -fw/2 \ldots fw/2$$

(a) (b)

Fig. 9.2. (a) Initial set of connected-components in a cluster (CCI_w) (b) The new connected-component set (CCF_w) yielded by applying the contour closing process on CCI_w.

where fw stands for the width of the filter and σ represents the standard deviation of the distribution. The corner detection result is sensitive to the smoothing parameter; a large smoothing scale will blur the corners, while a low scale will detect many false alarms.

The corner points on the boundary of a component are defined as the maxima of absolute curvature that are above a threshold value. The number of corners in a cluster is given by the total number of corner points on the boundary of all the connected-components in the cluster. Intuitively, the higher the number of corners detected in the cluster, the higher the activity of the cluster. The activity threshold, $activity\_threshold$, is introduced to distinguish the high-activity clusters from the low-activity ones. That means if the number of corner points in the cluster exceeds the activity threshold, the cluster is a high-activity one or of low-activity, otherwise.

9.5.2 Cluster Selection for the Identification of Suspicious Regions

The only points on the scene that can provide information to locate the prototype are those belonging to high-activity clusters. The object identification approach proposed in this work is based on the idea that a region on the scene, specified by the range of influence of a cluster's center, will enclose a part of the prototype object if the points in that cluster correspond to a dense neighborhood of points on the prototype shape. This idea stems from the fact that since these two sets of points are supposed to be homologous, representing the same part of the object, they should follow similar spatial distributions. However, the spatial distribution alone is not powerful enough to indicate the presence of a part of the prototype's boundary in a cluster. For that reason, the figural continuity constraint is incorporated in the identification process. The idea is that neighboring points on a boundary part included in a cluster on the test image should correspond to neighboring points on the prototype's contour.

As an indicative measure of how "crowded" a neighborhood of points is, the neighborhood density is introduced. Lets assume a cluster CS_w on the complex scene that includes a set of o points $CS_w \equiv \{c_i \mid i = 1, \ldots, o\}$, that correspond to

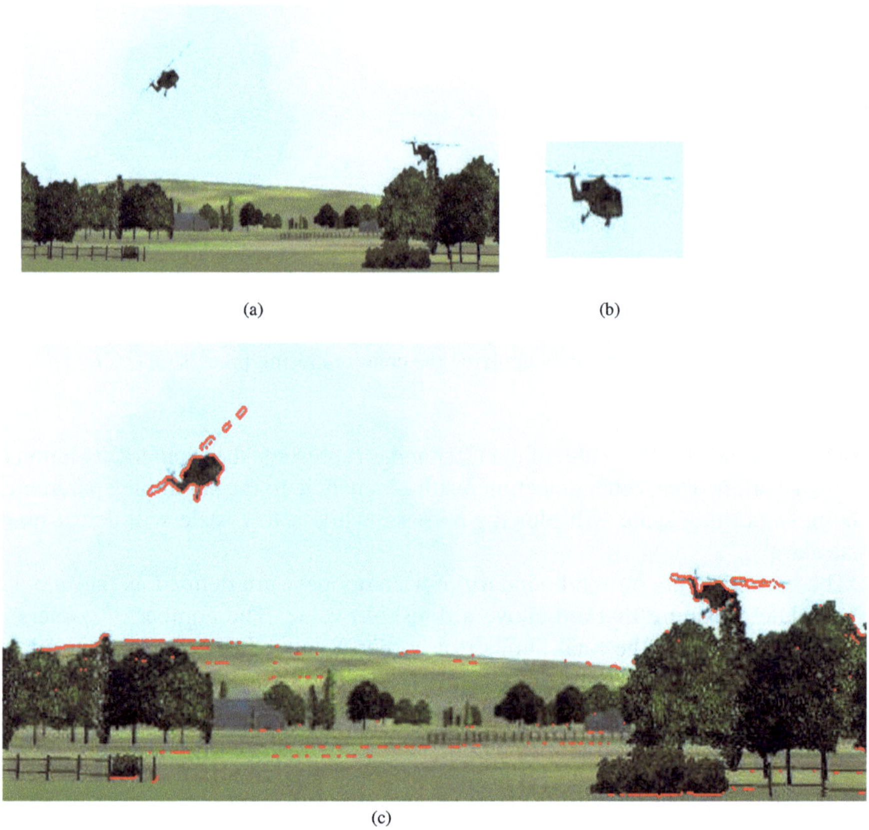

Fig. 9.3. (a) Terrain scene (b) Prototype helicopter (c) Matched points on the terrain scene highlighted in red.

another set of o points $Y \equiv \{p_{\alpha(i)} \mid i = 1, \ldots, o\}$ on the prototype shape. Since the matching between the target and the scene points follows an *many-to-one* matching approach, it is possible that two different points in the cluster are mapped to the same point on the prototype shape, resulting in point repetitions in the Y set. The point set $Y_{unq} \subseteq Y$ is introduced, including the unique elements of Y and represented as $Y_{unq} \equiv \{(x_1, y_1), (x_2, y_2), \ldots, (x_q, y_q)\}$, where $q \leq o$. The neighborhood density, D_{nbr}, of a cluster expresses the population of points in Y_{unq} per unit area and is mathematically defined as:

$$D_{nbr} = \frac{q}{A}$$

where, A is the maximum distance between the points in Y_{unq} and their center of mass, defined as:

$$A = \max_i \sqrt{(x_i - cm_x)^2 + (y_i - cm_y)^2}$$

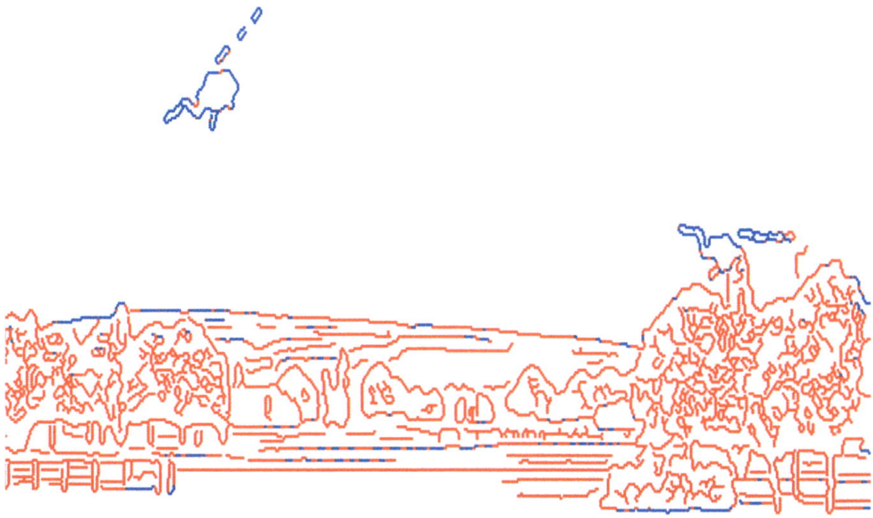

Fig. 9.4. Edge map of the terrain scene with the matched points highlighted in blue.

where, cm_x is the x-coordinate of the center of mass given by $cm_x = \frac{1}{q}\sum_{i=1}^{q} x_i$ and cm_y is the y-coordinate, similarly defined. Therefore, CS_w will be a candidate cluster for enclosing a part of the prototype object if the following expression is satisfied:

$$D_{nbr\,w} > density\_threshold$$

where, $density\_threshold$ is a constant, experimentally defined, as we will explain later on. Intuitively, the higher the neighborhood density, the more likely for the cluster to include a part of the prototype. The clusters that have survived the above thresholding process are further examined for figural continuity, in order to take the final decision for the presence of the target in the scene.

Let us consider the cluster CS_w introduced above, including a set of points that correspond to the point set Y on the prototype shape. The points in each of the above sets CS_w and Y form a number of connected-components represented by $\{CCS_{wj} \mid j = 1 \ldots n_j\}$ and $\{CCY_{j'} \mid j' = 1 \ldots n_{j'}\}$, respectively, where j, n_j and j', $n_{j'}$ stand for the indices and the total numbers of the components for the two sets. For the estimation of the figural continuity in the cluster each pair $(CCS_{wj}, CCY_{j'})$ of components is examined separately.

The first step when examining two components CCS_{wj}, $CCY_{j'}$ for figural continuity is to identify the set of corresponding pairs between their boundary points. Let us assume that a set MP of $k \leq n_j$ matched pairs exists between the two components, denoted as $MP \equiv \{(ccs^{wj}_{r(i)}, ccy^{j'}_{\alpha(r(i))})\}$ with $r(i)$ a function such that $i < j \Rightarrow r(i) < r(j)$. In this notation, $ccs^{wj}_{r(i)}$, $ccy^{j'}_{\alpha(r(i))}$ are points that belong to the sets CS_w and Y, respectively and $\alpha(.)$ is the mapping function $\alpha : CS_w \rightarrow Y$. From the property $i < j \Rightarrow r(i) < r(j)$ we observe that the index $r(i)$ follows the

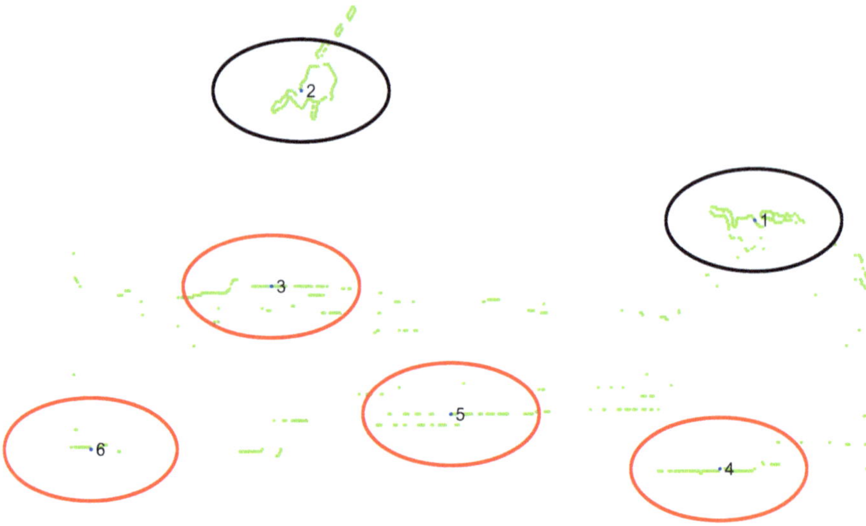

Fig. 9.5. Clustering of the matched points on the terrain scene.

tracing order of the points on the boundary of CCS_{wj}, where $ccs^{wj}_{r(i)}$, $ccs^{wj}_{r(i+1)}$ are not necessarily consecutive boundary points on the component. If for a pair of components the set MP is empty, that pair is not further considered for the estimation of the cluster's figural continuity.

The set of the boundary index differences of the matched points of CCS_{wj} is denoted as $BD_{CCS_{wj}} = \{(r(i+1)-r(i)), i = 1, \ldots, k-1\}$. The one dimensional signal that represents consecutive binary index differences $BD_{CCS_{wj}}(i) = r(i+1) - r(i)$ is formulated. For the component $CCY_{j'}$ the corresponding set and one dimensional signal are denoted by $BD_{CCY_{j'}} = \{(\alpha(r(i+1)) - \alpha(r(i))), i = 1, \ldots, k-1\}$ and $BD_{CCY_{j'}}(i) = \alpha(r(i+1)) - \alpha(r(i))$, respectively.

In order for the set of points on CCS_{wj} to follow the same figural distribution as on $CCY_{j'}$ the signals $BD_{CCS_{wj}}(i)$, $BD_{CCY_{j'}}(i)$ should be similar and ideally identical. Therefore, if we define a positive function of their difference $E_{CCS_{wj},CCY_{j'}}(i)$, the summation $\frac{1}{k-1}\sum_{i=1}^{k-1} E_{CCS_{wj},CCY_{j'}}(i)$ should be small and ideally zero. We call the restriction of this summation to a small value the *figural continuity* constraint. Hence, the figural continuity constraint for the entire cluster CS_w may be defined as the average of the figural continuity with respect to all component pairs $(CCS_{wj}, CCY_{j'})$, given below.

$$\frac{1}{n_j \cdot n_{j'}} \sum_{j=1}^{n_j} \sum_{j'=1}^{n_{j'}} \left[\frac{1}{k-1} \sum_{i=1}^{k-1} E_{CCS_{wj},CCY_{j'}}(i) \right] < continuity\_threshold$$

The parameter *continuity_threshold* should have a small value. The choice of *continuity_threshold* is decided after exhaustive experimental results.

9.6 Experimental Results

In this work, the objects are represented by a set of edge points detected by the Canny operator [3]. In existing literature [1], the contour is approximated by picking a subset of the initial set of edge pixels using uniform sampling. In our approach, all the emerged points from the internal and external contours of the objects are preserved to contribute to the shape representation, in order to enable a more detailed boundary information. The shape context descriptor was formed by taking 5 bins for the log distance r and 12 bins for the polar angle θ, as suggested in [1], resulting in a total number of bins equal to $K = 60$. In the current experimental work, a local shape context descriptor is used where the examined neighborhood around each contour point is experimentally set to be a circular region of $R = 26$ pixels radius. In order to ensure the descriptor's rotation invariance, the tangent direction at each point is used as the positive x-axis for the local coordinate system in the shape representation.

During the detection of regions of interest on the complex scene, the scene points that have been matched with target points are grouped into clusters by employing subtractive clustering. The clusters' size is an input variable to the clustering process which should be appropriately defined. The range of influence of a cluster depends on the application. Intuitively, the higher the scale at which the prototype object appears on the complex scene relative to the size of the image, the bigger the cluster's radius should be. In this work, we are concerned with small target identification in military applications and therefore a small cluster would be more appropriate for accurate identification results. In our experimental results we propose that each cluster center will have an ellipsoidal neighborhood of influence with major and minor radii set equal to the 20% of the width and the height of the data (matched points) space on the input image, respectively. If the minor and major radius are equal, the neighborhood of influence becomes a circle.

In the proposed identification approach a set of thresholds was set and evaluated as explained in this section. During the cluster processing three thresholds are introduced, namely the $activity\_threshold$, the $density\_threshold$ and the $continuity\_threshold$ which are experimentally evaluated and set to $activity\_threshold = 5$, $density\_threshold = 0.80$ and $continuity\_threshold = 3$, respectively. During the matching between the scene and the prototype points the dummy point cost, ϵ_d, was introduced which plays the role of a threshold that distinguishes the true matches from the false matches (points assigned to the dummy point) in the scene. Lets introduce the set $Sc \equiv \{Sc_i \mid i = 1 \ldots n\}$, where n is the total number of scene points, which includes the minimum costs of matching each scene point to a point on the prototype shape, sorted in ascending order. An initial estimation of the cost ϵ_d is $\epsilon_d = Sc_m$, where m is the total number of target points and usually $m < n$. The above cost evaluation is based on the idea that if the target is present on the scene, ideally the first m point-matchings with the minimum cost will correspond to the points of the test object on the scene which is most similar to the target. In order to efficiently handle cases where the target may appear in the

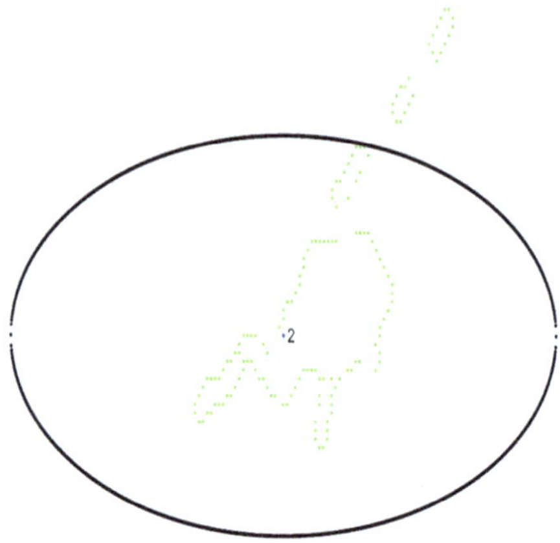

Fig. 9.6. Zoom in the matched points on a part of the terrain scene.

scene partially occluded or distorted (cases where the matching cost increases), the value of ϵ_d is defined as:

$$\epsilon_d = \alpha \cdot Sc_m$$

where α is a constant, greater than 1. In this work, α was set equal to 1.5.

The proposed framework is quite general and can be applied on a number of different identification tasks. For example, in military applications target object identification is of significant importance. This is as well the scenario we assume in this piece of work. We have applied the proposed object identification algorithm on a large number of images. The first simulation demonstrates the rotation invariance of the identifier and its ability to handle partial occlusion by illustrating the identification of a prototype helicopter in a terrain scene which contains both a rotated and an occluded version of it. The robustness of the identifier to distortion is verified with the identification of an airplane in a synthetic image in the presence of boundary noise.

The identification of a helicopter in a terrain scene is presented in Figures 9.3 and 9.5. Figures 9.3(a)-(b) illustrate the original terrain image and the prototype helicopter. A rotated and an occluded version of the target are present in two different locations in the complex scene. The outcome of the shape context matching is presented in Figure 9.3(c) where the matched points are shown on the complex scene, highlighted in red. The edge map of the terrain scene is illustrated in Figure 9.4 where the points on the image correspond to edge pixels on the terrain scene and the matched points are highlighted in blue. As it is observed, several mismatches are present, mainly on the ground of the scene. The subtractive clustering process results in 6 clusters, shown in Figure 9.5, where the circles mark the limits of the interest regions whose centroids are highlighted in blue. For a clearer view of the clustering result on the terrain image a closer view of the second cluster and the included points

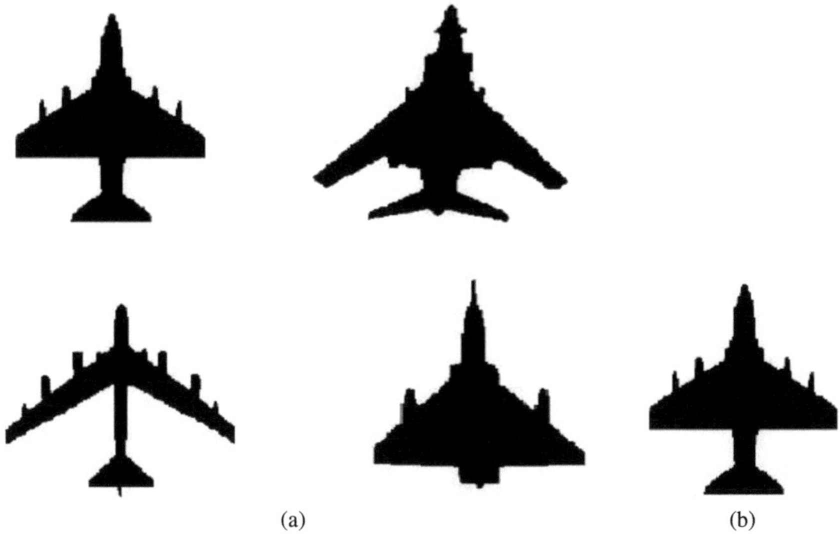

Fig. 9.7. (a) Synthetic image with four different aircraft models (b) Target aircraft.

is presented in Figure 9.6. Based on the extracted cluster activity values ($Activity = [7, 7, 3, 3, 2, 0]$), only the first two clusters have sufficiently high activity that exceeds the predefined activity threshold and should be further considered for the target identification in the scene. The derived clusters correspond to point neighborhoods with density values given by $D_{nbr} = [2.29, 1.67, 0.80, 0.72, 0.42, 0.41]$ and the figural continuity of the included boundary parts is determined by the values $CNT = [2.35, 1.51, 5.64, 2.48, 3.11, 1.91]$. According to the predefined thresholds, only the first two clusters survive the activity, the density and the figural continuity thresholding processes. Hence, these two clusters (points in the black circles) are the only regions on the terrain scene which include a part of the target's contour, which agrees with the actual scenario.

The experimental results presented in Figures 9.7-9.8 illustrate the ability of the proposed identifier to recognize a particular aircraft model in a set of aircrafts of different type. The aircraft in Figure 9.7(b) is selected to be the target which is compared against a complex scene (Figure 9.7(a)) which contains 4 different aircraft models, including the target. The test objects participating in this experiment are aircrafts of different type, bearing similar major characteristics and differing in small details on their boundary. So, the challenge in this experiment is the identification of the target aircraft model on the complex scene which includes object parts common in all the aircrafts such as, the nose of the aircraft, the tail and the outward curvy part of the wings. Figure 9.8(a) presents the point matching result with the red and blue points having the same role as in the previous simulations. A closer view of the matched and unmatched points on the boundary of one of the test aircraft objects is presented in Figure 9.9. The subtractive clustering result of the matched

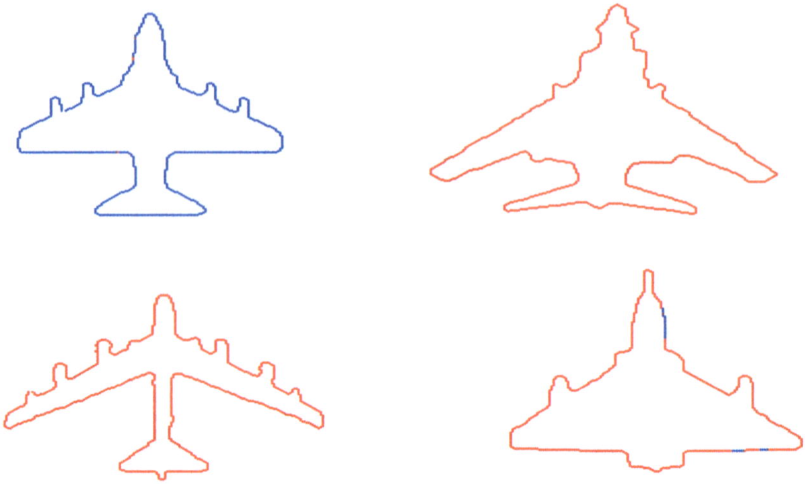

(a)

(b)

Fig. 9.8. (a) Matched points on the complex image (b) Clustering of the matched points.

points in the complex image is illustrated in Figure 9.8(b), where the red and black circles mark the limits of the interest regions whose centroids are highlighted in blue. Based on the cluster activity results, $Activity = [10, 28, 0]$, the third cluster is rejected as a low-activity cluster. The two remaining high-activity clusters (points in the black circle) survive the density and the figural continuity thresholding processes ($D_{nbr} = [3.46, 3.55, 0.25]$ and $CNT = [0, 0.07, 0]$) and therefore enclose part of the target, which agrees with the actual scenario.

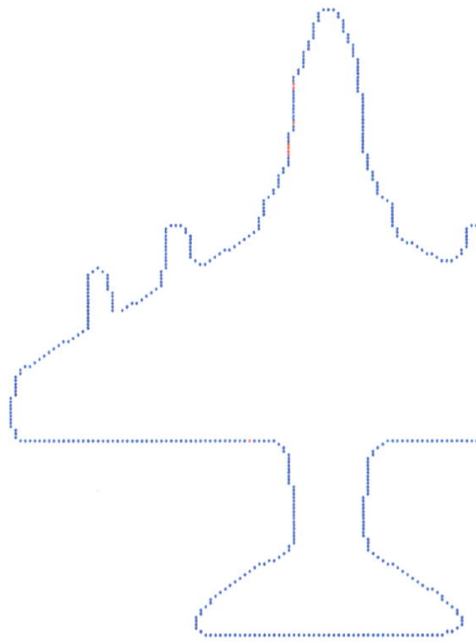

Fig. 9.9. Zoom in the matched points on a part of the complex image.

(a) (b)

Fig. 9.10. (a) Synthetic image with four different aircraft models corrupted by noise (b) Target aircraft.

The robustness of the identifier to distortion is verified by repeating the previous simulation in the presence of boundary noise. The boundaries of the objects in the complex scene are corrupted with Gaussian white noise, giving Signal-to-Noise ratio equal to 30dB, as shown in Figure 9.10(a). In this simulation we aim at locating a target aircraft (Figure 9.10(b)) in an input image which includes a distorted version of the target. The results of the point matching and subtractive clustering processes

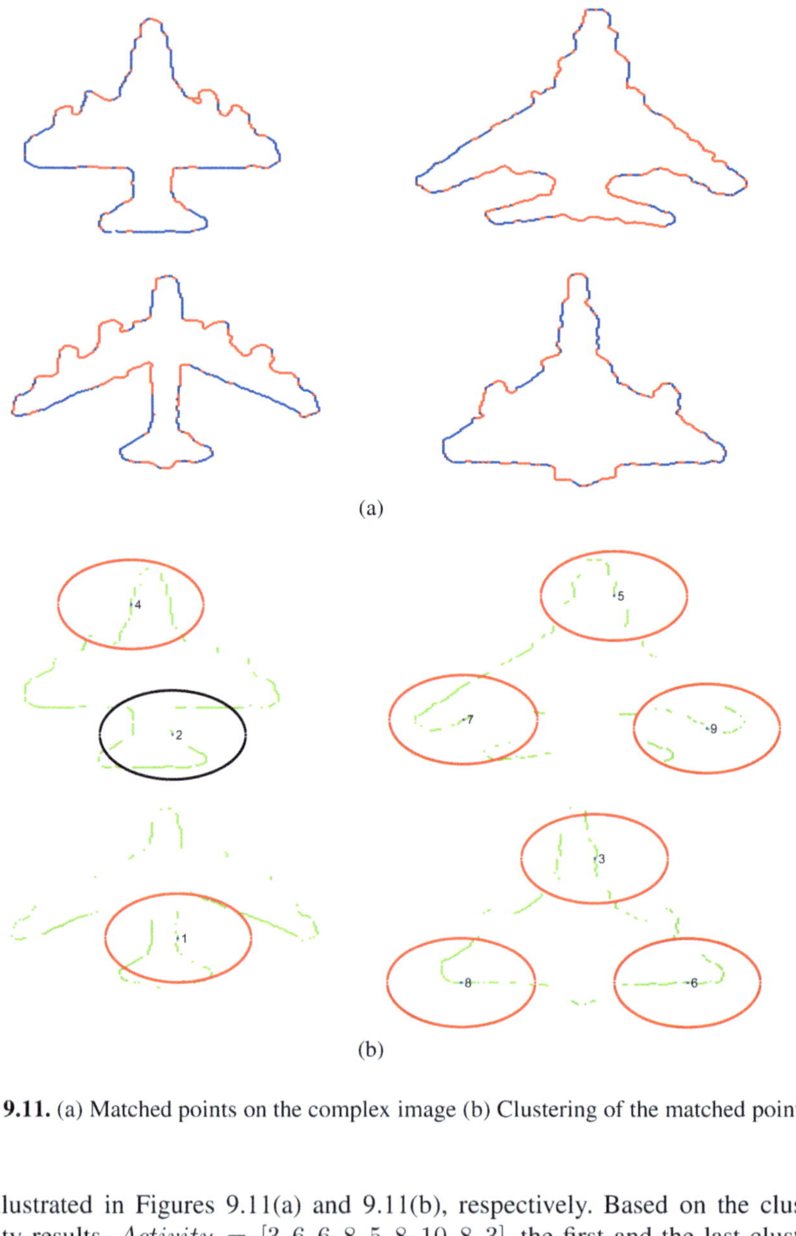

Fig. 9.11. (a) Matched points on the complex image (b) Clustering of the matched points.

are illustrated in Figures 9.11(a) and 9.11(b), respectively. Based on the cluster activity results, $Activity = [3, 6, 6, 8, 5, 8, 10, 8, 3]$, the first and the last clusters are low-activity clusters. Among the remaining high-activity clusters only the second cluster (points in the black circle) survives the density and the figural continuity thresholding processes ($D_{nbr} = [0.56, 0.81, 0.35, 0.36, 0.40, 0.38, 0.38, 0.52, 0.39]$ and $CNT = [1.80, 1.03, 1.89, 1.58, 1.52, 0.86, 1.42, 2.07, 1.33]$) and therefore encloses part of the target, which agrees with the actual scenario.

9.7 Discussion

In this work, a novel approach to prototype-based object identification is presented. Shape context representation and many-to-one matching is employed to extract the point correspondence between a complex scene and a target object. The novelty of this work is the use of subtractive clustering in order to spatially partition the image into regions of interest. A data mining process follows to preserve only the matched points that belong to high-activity clusters as they are proven to be regions that bear significant information for the object identification. The proposed identifier relies on the idea that a cluster on the complex scene encloses a part of the prototype object if it corresponds to a neighborhood on the prototype shape, with high density and if the boundary parts included in the cluster follow the same figural continuity as their corresponding parts on the prototype's contour. Otherwise, the included points are mismatched points. The robustness of the proposed method to image transformations, such as rotation, translation and scaling and to ambient conditions, such as partial occlusion and distortion, is verified by the presented experimental results.

Acknowledgements. The investigations which are the subject of this paper were initiated by Dstl under the auspices of the United Kingdom Ministry of Defence Systems Engineering for Autonomous Systems Defence Technology Centre.

References

1. Belongie, S., Malik, J., Puzicha, J.: Shape matching and object recognition using shape contexts. IEEE Trans. Pattern Anal. Machine Intell. 24(4), 509–522 (2002)
2. Bezdek, J.C., Pal, M.R., Keller, J., Krisnapuram, R.: Fuzzy models and algorithms for pattern recognition and image processing. Kluwer Academic Publishers (1999)
3. Canny, J.F.: A computational approach to edge detection. IEEE Trans. Pattern Anal. Machine Intell. 8(6), 679–698 (1986)
4. Carter, J.R.: Boundary tracing method and system. European Patent Applcation EP341819-A3 (1989)
5. Chellappa, R., Bagdazian, R.: Fourier coding of image boundaries. IEEE Trans. Pattern Anal. Machine Intell. 6(1), 102–105 (1984)
6. Chiu, S.: Fuzzy model identification based on cluster estimation. Journal of Intelligent and Fuzzy Systems 2(3) (1994)
7. Davies, E.R.: Machine Vision: Theory, Algorithms, Practicalities. Academic Press (1997)
8. Deriche, R., Cocquerez, J.P., Almouzny, G.: An efficient method to build early image description. In: 9th International Conference on Pattern Recognition, Rome (November 1988)
9. Haralick, R., Shapiro, L.G.: Computer and Robot Vision, vol. I, pp. 28–48. Addison-Wesley (1992)
10. Harris, C.J.: Tracking with rigid models. In: Blake, A., Yuille, A. (eds.) Active Vision, ch. 4, pp. 59–73. MIT Press, Cambridge (1992)
11. Hu, M.K.: Visual pattern recognition by moment invariants. IRE Transactions on Information Theory IT (8), 179–187 (1962)

12. Jonker, R., Volgenant, A.: A shortest augmenting path algorithm for dense and sparse linear assignment problems. Computing 38, 325–340 (1987)
13. Kauppinen, H., Seppanen, T., Pietikainen, M.: An experimental comparison of autoregressive and fourier-based descriptors in 2D shape classification. IEEE Trans. Pattern Anal. Machine Intell. 17(2), 201–207 (1995)
14. Kopf, S., Haenselmann, T., Effelsberg, W.: Enhancing curvature scale space features for robust shape classification. In: Proceedings of IEEE International Conference on Multimedia and Expo. ICME 2005 (July 2005)
15. Latecki, L.J., Lakamper, R.: Shape similarity measure based on correspondence of visual parts. IEEE Trans. Pattern Anal. Machine Intell. 22(10), 1185–1190 (2000)
16. Loncaric, S.: A survey of shape analysis techniques. Pattern Recognition 31(8), 983–1001 (1998)
17. Lowe, D.: Integrated treatment of matching and measurement errors for robust model-based motion tracking. In: Proceedings IEEE 3rd International Conference on Computer Vision, pp. 436–440 (1990)
18. Lowe, D.: Distinctive image features from scale-invariant key-points. International Journal of Computer Vision 60(2), 91–110 (2004)
19. Mikolajczyk, K., Schmid, C.: A performance evaluation of local descriptors. IEEE Trans. Pattern Anal. Machine Intell. 24(10), 1615–1630 (2005)
20. Mokhtarian, F., Mackworth, A.: Scale-based description and recognition of planar curves and two-dimensional shapes. IEEE Trans. Pattern Anal. Machine Intell. 8(1), 34–43 (1986)
21. Van Otterloo, P.J.: A contour-oriented approach to shape analysis. Prentice Hall International (UK) Ltd. (1991)
22. Rattarangsi, A., Chin, R.T.: Scale-based detection of corners of planar curves. IEEE Trans. Pattern Anal. Machine Intell. 14(4), 430–449 (1992)
23. Ross, T.J.: Fuzzy logic with engineering applications. Publishing House of Electronics Industry, Beijing (2004)
24. Di Ruberto, C., Morgera, A.: A comparison of 2-D moment-based description techniques. In: International Conference on Image Analysis and Processing, pp. 212–219 (2005)
25. Taubin, G., Cooper, D.B.: Geometric invariance in computer vision. MIT Press (1992)
26. Teague, M.R.: Image analysis via the general theory of moments. Journal of the Optical Society of America 70(8), 920–930 (1980)
27. Terzopoulos, D., Witkin, A., Kass, M.: Constraints on deformable models: recovering 3d shape and nongrid motion. Artificial Intelligence 36(1), 91–123 (1988)
28. Veltkamp, R.C., Hagedoorn, M.: State of the art in shape matching. Principles of Visual Information Retrieval, 87–119 (2001)
29. Yager, R., Filev, D.: Generation of fuzzy rules by mountain clustering. Journal of Intelligent and Fuzzy Systems 2(3), 209–219 (1994)
30. Zhang, D., Lu, G.: Enhanced generic fourier descriptors for object-based image retrieval. In: IEEE International Conference on Acoustics, Speech, and Signal Processing, vol. 4, pp. 3668–3671 (2002)
31. Zhang, D., Lu, G.: Shape based image retrieval using generic fourier descriptors. Signal Processing: Image Communication 17(10), 825–848 (2002)
32. Zhang, D., Lu, G.: Review of shape representation and description techniques. Pattern Recognition 37(1), 1–19 (2004)
33. Zhong, Y., Jain, A.K., Dubuisson-Jolly, M.P.: Object tracking using deformable templates. IEEE Trans. Pattern Anal. Machine Intell. 22(5), 544–549 (2000)

Chapter 10
Historical Consistent Neural Networks: New Perspectives on Market Modeling, Forecasting and Risk Analysis

Hans-Georg Zimmermann, Christoph Tietz, and Ralph Grothmann

Abstract. From a mathematical point of view, neural networks allow the construction of models, which are able to handle high-dimensional problems along with a high degree of nonlinearity. In this chapter we deal with a special type of time-delay recurrent neural networks. In these models we understand a part of the world as a large recursive system which is only partially observable. We model and forecast all observables, avoiding the problem in open systems that we do not know the external drivers from present time on. This framework goes far beyond the paradigms of standard regression theory and allows us to forecast financial markets and perform a new way of risk analysis.

10.1 Introduction

Many business management disciplines require precize forecasts in order to enhance the quality of planning throughout the value chain. The complex planning and decision-making scenarios generally require the forecast to take account of a wide range of external factors with largely non-linear cause and effect relationships with one another and with the parameters. Furthermore, allowance must be made for the uncertainty in forecasting, which can result for example from hidden factors.

Modern neuroinformatics processes offer significant benefits for dealing with the typical challenges associated with forecasting. With their universal approximation properties, neural networks (NN) make it possible to describe non-linear relationships between a large number of external factors and parameters [6]. In contrast, conventional econometrics generally uses linear models (e.g. autoregressive models (AR), multivariate linear regression) which, for all that they facilitate efficient calculation of links, provide only inadequate models for non-linear dynamics. Other

Hans-Georg Zimmermann · Christoph Tietz · Ralph Grothmann
Siemens AG, Corporate Technology, Intelligent Systems and Control, Munich

P. Georgieva et al. (Eds.): Adv. in Intelligent Signal Processing and Data Mining, SCI 410, pp. 259–274.
springerlink.com

conventional time series analysis procedures (such as ARMA, ARIMA, ARMAX) remain confined to linear systems [14].

The widest imaginable range of models is discussed within the class of neural networks [10, 11]. For example, in terms of the data flow in the model, it is possible to draw a distinction between "feedforward" and (time) recurrent neural networks. In a feedforward neural network (FNN), data from an upstream layer is propagated to downstream layers only. Furthermore, there is no provision for any links between the neurons in a single layer. In contrast, (time-delay) recurrent neural networks (RNN) include links which transfer data from a downstream layer (time-delayed) to upstream layers [1].

It is noteworthy that any equation for a neural network can be portrayed in graphic form by means of an architecture which represents the individual layers of the network in the form of nodes and the links between the layers in the form of borders. This relationship will be described hereinafter as the correspondence principle between equations, architectures and the local algorithms associated with them. For example, the "error back propagation algorithm" needs only locally available data from the forward and reverse flow of the network in order to calculate the partial formulations of the error function of the neural network according to the weights of a given layer during training [15, 19]. The use of local algorithms here provides an elegant basis for the expansion of the neural network with a view to the modeling of high-complexity systems. Used in combination with an appropriate (random) learning rule, it is possible to use the gradients as a basis for the identification of robust minima for network error function [19].

In this article, we present a new type of recurrent neural network, known as historical consistent neural networks (HCNN). The special feature of this network architecture is that it facilitates modeling not just of individual dynamics, but also of high-complexity systems made up of a number of sub-dynamics interacting. The models are symmetrical in their input and output variables, i.e. the system description does not draw any distinction between input, output and internal state variables. In fact, its models are produced using a large-scale state vector, only part of which is observable. The expected values are calculated from these observables and then adjusted by the values observed. An open system thus becomes a closed system. Since there is no longer any separation between the input and output variables, we refer to the consistent modeling of the observables for a large dynamic system.

The remainder of the article is broken down into three sections. Section 10.2 is dedicated to the theoretical origins of HCNN. Section 10.3 reports on the use of HCNN to solve industrial forecasting and planning problems. Section 10.4, the final section, summarizes (in brief) the primary findings of the current article once again, whereby particular attention is paid to new risk concepts which may derive from the HCNN presented.

10.2 Historical Consistent Neural Networks (HCNN)

In order to examine the origins of HCNN, we consider first the model of a simple RNN for modeling of a single dynamic system. The model of the HCNN is then developed from the design weakness of this approach to modeling.

10.2.1 Modeling Open Dynamic Systems with RNN

We start from the assumption that a time series y_τ is created by an open dynamic, which can be described in discrete time τ using a state transition and output equation, thus a discrete state space model [4]:

$$\begin{aligned} &\text{State transition equation:} \quad s_\tau = f(s_{\tau-1}, u_\tau) \\ &\text{Output equation:} \quad y_\tau = g(s_\tau) \end{aligned} \tag{10.1}$$

The hidden time-delay recurrent state transition equation $s_\tau = f(s_{\tau-1}, u_\tau)$ describes the current state s_τ by means of a function of the historical system state $s_{\tau-1}$ and of the external factors u_τ. The system formulated in the state transition equation can therefore be interpreted as a partially autonomous and partially externally driven dynamic. In this context, we refer to an open dynamic system.

The output equation y_τ uses the non-observable system state s_τ in every time phase τ to calculate the observable output of the dynamic system y_τ. The data-driven system identification is conducted by means of the selection of appropriate (parameterized) functions f() and g() by minimization of an error function:

$$E = \frac{1}{T} \sum_{\tau=1}^{T} (y_\tau - y_\tau^d)^2 \rightarrow \min_{f,g}! \tag{10.2}$$

The two f() and g() functions are calculated using the quadratic error function (Equation 10.2) in such a way that the average interval between the observed data y_τ^d and the system outputs y_τ over a period of observation ($\tau = 1, \ldots, T$) is minimal.

Thus far, we have given a general description of the state transition and the output equation for the dynamic system. Hereinafter, we define functions f() and g() by means of RNN with weighting matrices A, B and C [17]:

$$\begin{aligned} &\text{State transition equation:} \quad s_\tau = \tanh(As_{\tau-1} + Bu_\tau) \\ &\text{Output equation:} \quad y_\tau = Cs_\tau \end{aligned} \tag{10.3}$$

Equation 10.3 uses an RNN with weighting matrices A, B and C to model the open dynamic system. The RNN is designed as a non-linear state-space model, which is able to approximate any functions f() and g() [13]. In principle, any non-linearity can be applied to the neurons of the hidden network layer s_τ as long as it is continuous and differentiable [4, 19]. Most common activation functions are sigmoidal, logistic or radial basis functions [4]. We prefer to use hyperbolic tangent non-linearities tanh() as squashing functions for the hidden network layer s_τ, because the hyperbolic tangent is a smooth step function which is symmetric around the origin [19]. For small weights information flows pass the hyperbolic tangent nearly unchanged, which allows to model linear relationships [19].

The output equation is specified as a linear function. The time recurrence in the state transition equations requires intertemporal dependencies: in an RNN, a state s_τ can in theory be explained by means of superimposition of all of the preceding states and of the corresponding external factors observed [4].

We use the technique of finite unfolding in time [12] as a solution for data-driven system identification, i.e. for the selection of appropriate parameter matrices A, B and C to minimize the error function (Equation 10.2). The underlying idea here is that any RNN can be reformulated to form an equivalent feedforward neural network, if matrices A, B and C are identical in the individual time steps ("shared weights" [12]). The actual training on the network can then be conducted using the "error-back propagation-through-time", an extension of the well-known error-back propagation algorithm, in order to calculate the partial definitions of the error function according to weights and to select an appropriate learning rule for the weight update [12, 19]. For algorithmic solution methods, the reader is referred to the overview article by B. Pearlmatter [9]. The (limited) physical development of the RNN (see Equation 10.3) is illustrated in Figure 10.1.

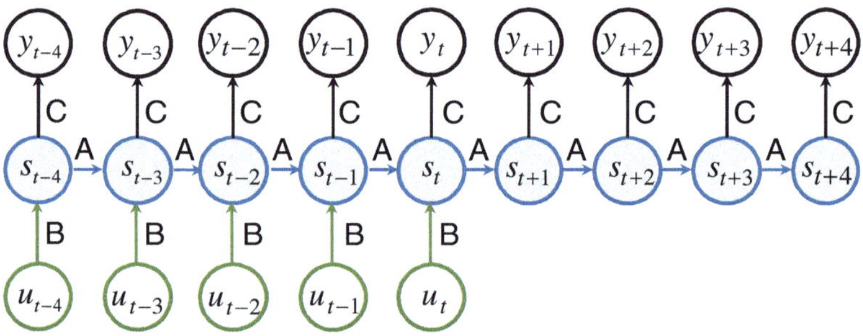

Fig. 10.1. Recurrent neural network (RNN) using finite unfolding in time [17]

The neural network architecture shown in Figure 10.1 is a physical representation of the state transition and output equation (see Equation 10.3). The network forecast is generated by superimposition of two components: (i.) the autonomous system dynamic (coded in state transition matrix A), the data from the individual time steps accumulated in a memory, and (ii.) the influence of external factors (coded in Matrix B). One of the benefits of the shared weights technique used here is the moderate number of free parameters (weights) in the model. In comparison with FNN, expansion of the time horizon does not lead to an increase in the number of network weights, since only shared weights A, B and C are reused. The risk of overgeneralization (known as "overfitting") therefore has only a secondary role to play [20].

It is apparent in practical applications that RNNs focus largely upon the latest time steps of development (and here upon the development of the external factors in particular) in order to explain dynamics. In order to strike a balance between the autonomous and externally driven elements of the dynamic (and thus to improve the model's forecasting ability as well), we suggest that the autonomous section of the system dynamic should be projected into the future across a number of time steps (here t +2, t +3 and t +4) [20]. In order to describe the development of the dynamic in one of the (additional) future time steps adequately, the state transition matrix A must be structured in training in such a way that data from preceding time steps (from time steps t at least) can be used. The shared weights principle means that this capacity of Matrix A can be used to interpret the system characteristic in every time steps. The outcome of this is that consistent multi-step forecasts can be produced [20]. However, it is assumed here that there will not be any significant changes in the external factors u_τ in the future ($u \approx 0$), whereby the changes ($u_\tau = x_\tau - x_{\tau-1}$) are selected as the transformation for the raw data x. The effectiveness of this mechanism increases in relation to the proportion of the autonomous element in the dynamic at large.

10.2.2 Modeling of Closed Dynamic Systems with HCNN

The RNN discussed in the preceding paragraph is used to model and forecast an open dynamic system. The time characteristic for the system is described partially by means of an autonomous (sub-)dynamic and partially by means of external variables. This modeling framework is comparable with a (non-linear) regression approach. However, a critical assessment highlights the fundamental design weakness of the RNN clearly: the model describes the dynamic of a parameter against many input variables. The interaction of various sub-dynamics within a large system is shown indirectly by means of allowance for external variables only. Matters are further complicated by the assumption, in forecasts of future system states, that there will not be any significant change in the environment ($u_\tau \approx 0 \mid \tau > 0$).

Many real technical and economic applications must however be seen in the context of major systems in which various (non-linear) dynamics interact with one another (in time). For example, the international financial flows which are part and parcel of globalization have contributed to the ever increasing interlinking of different markets. Changes in expected returns on stock markets, for example, can trigger transmission mechanisms which bring about changes in the bond, currency and commodities markets. An isolated assessment of an individual market or of a sub-dynamic of the system at large would not fit the bill here. This example also demonstrates that an individual variable or sub-dynamic here has no particular significance in terms of drawing a distinction between external factors and parameters, in the same way as conventional regression models. The individual variables of a high-complexity system should in fact be regarded as symmetrical. Projected on to a model, this means that all variables must be influencing factors and parameters at once in the model. In order to emphasize this symmetry in the modeling, we refer hereinafter only to the observables Y_τ in a large system. The term "observables" embraces the input and output variables for conventional modeling approaches (i.e. $Y_\tau := \{y_\tau, u_\tau\}$). This also makes the problem with consistency in the RNN even more obvious: on the output side, the RNN in the example (Fig. 10.1) provides forecasts of the observables Y_τ up to time steps τ+4, whereas the input side assumes that the observables Y_τ will not vary from their current value. This lack of consistency represents a clear contradiction within the model framework. If, on the other hand, we are able to implement a model framework in which common descriptions and forecasts can be used for all of the observables, we will be in a position to close the open system – in other words, we will model a closed large dynamic system.

In addition to the consistency problem characterized above and the RNN's exclusive focus upon a sub-dynamic, the use of different matrices (A, B and C in this case) in the modeling should be regarded as critical. Since all of the observables can have a reciprocal effect upon one another, viewing the externally driven element of the system (coded in Matrix B, Fig. 10.1) in isolation is a questionable approach. This also applies to the derivation of observables from the system state (see Matrix C, Fig. 10.1). For modeling of large systems, therefore, we suggest that only a (large) Matrix A should be used, which characterizes the system's transition from one (time) state to the other and through which all of the sub-dynamics can interact with one another. The model does not include any further free parameters apart from this matrix. The development of the overall system in the time steps of the past ($\tau = t - m$), of the present ($\tau = t$) and of the future ($\tau = t + n$) is therefore always described using the same state transition matrix A; there is not therefore any special distinction in the treatment of the individual time steps. This is further evidence of the symmetry of the modeling. Restriction to one matrix does not impose any limitation upon the universal approximation properties [13].

Motivated by the design principles for modeling of large dynamic systems and the design weaknesses of the RNN outlined above (see Fig. 10.1), we develop the equations for HCNN as follows [21]:

$$\begin{array}{ll} \text{State transition equation:} & s_\tau = \tanh(As_{\tau-1}) \\ \text{Output equation:} & Y_\tau = [Id, 0]s_\tau \\ \text{System identification:} & E = \sum_{\tau=t-m}^{t} (Y_\tau - Y_\tau^d)^2 \to \min_A! \end{array} \quad (10.4)$$

In Equation 10.4, the HCNN is formulated as a (non-linear) state state-space model with a (hidden) state transition and an output equation. The development in the partial dynamics for all observables is characterized by the sequence of states s_τ. In each of the time steps, the observables ($i = 1, \ldots, N$) are arranged on the respective first N neurons of a state s_τ, whereas the non-observable (internal) variables are represented by means of the subsequent neurons. This format does not amount to any limitation in general, because it represents a simple rearrangement of a general image matrix. The connector [Id, 0] in the output equation is a fixed matrix of appropriate size which makes available the first N components of the internal states which are affected by the observables ($i = 1, \ldots, N$). The first block of the connector therefore equates to an identity matrix Id, all of the elements in the second block of the matrix equal zero. Since the matrix [Id, 0] is fixed, the state transition matrix A contains the only free parameters in the network. The network layers Y_τ supply the expected values (forecasts) for observables Y_τ in every time steps. In order to identify the overall dynamic (see system identification, Equation 10.4), the HCNN is unfolded across the entire time range. Upon the basis of Equation 10.4, we can use the correspondence principle to formulate the following network architecture for the HCNN (see Section 10.1):

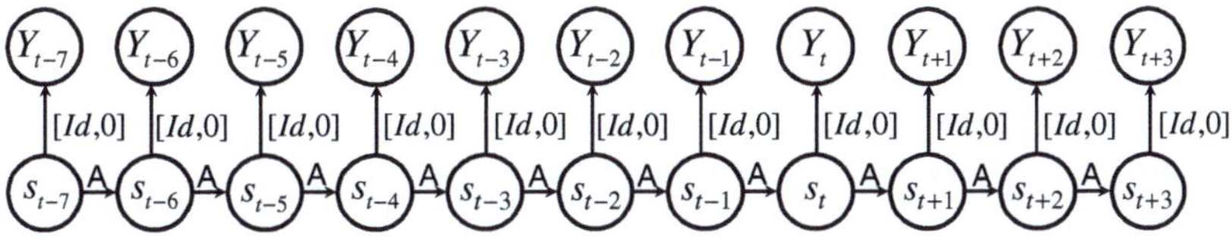

Fig. 10.2. Architecture of the Historical Consistent Neural Network (HCNN) [21]

In the historical consistent network architecture for the HCNN (see Fig. 10.2), the internal states are designed technically as hidden layers in the network and supplied with a tanh activation function. The forecasts for the observables are supplied by the output layers Y_τ. The HCNN learns the system dynamics as a whole from a single historical consistent sequence of observations, which equates to the entire observed history of the system. The output layers Y_τ are therefore provided with target values for the observables and an error function up until the present day ($\tau = t$) only. There are no observed values available in training for the future development time steps. However, the expected values $Y_{\tau > t}$ for the future can be determined by recursive projection of the network.

In practice, we use a technique along the lines of error correction [17] or "Teacher Forcing" [8, 16] in order to make the best possible use of the data from the observables made available and to accelerate training on the HCNN. To this end, we modify the fundamental architecture of the HCNN (Fig. 10.2) in such a way that the expected values for the observables in the present and all past time steps are replaced by the corresponding observed values. The resulting network architecture for the HCNN is shown in Figure 10.3 below.

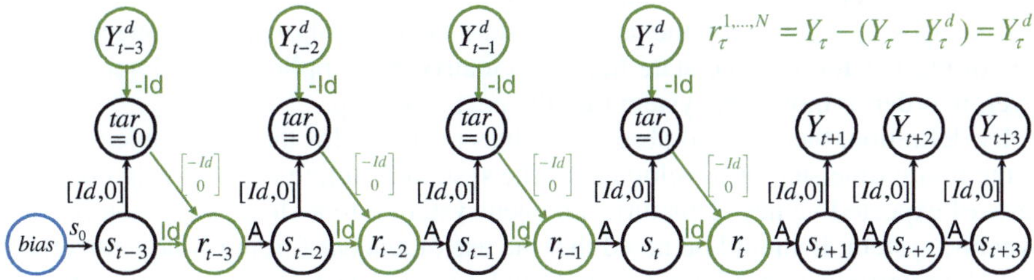

Fig. 10.3. Architecture for the identification of closed dynamic systems with a Historical Consistent Neural Network (HCNN) along the lines of "Teacher Forcing" [16, 8]

The HCNN modified for training purposes (see Fig. 10.3) includes the Teacher Forcing mechanism in a new architectonic formulation. In every step of development up to the present day ($\tau = t$), the expected values for all observables, which are generated by the network for the first components of the internal state space, are replaced by the actual observations.

To this end, we introduce a hidden intermediate layer r_τ, which has the identity of the activation function and has the same dimensions as the state vector s_τ, in every time step of development up to the present time step. In order to explain the way in which this intermediate layer works, we consider first the output layers of the network up to the present day ($\tau = t$). Up to this time step ($\tau = t$), the output layers are given fixed target values of zero (tar = 0) and an error function. The observed values Y_τ^d for the observables are written, preceded by a minus symbol, on the fixed target value of zero specified. The observed values Y_τ^d for the observables are made available to the network by means of additional input layers which have the same dimensions as the output layers. The (fixed) connector -Id, designed as a diagonal matrix with values of -1, links the input layers Y_τ^d with the output layers and thus ensures that the observed values are written with a minus symbol in the output layer. This causes the network to learn in training to create the expected values Y_τ to compensate for the negative observed values - Y_τ^d already applied to the output layer, in order to achieve the fixed target value of zero specified.

The content of the output layer, i.e. $Y_\tau - Y_\tau^d$, is now transferred to the first N neurons of the hidden intermediate layer r_τ on a component-by-component basis with a minus symbol in every time step. The $[-Id, 0]^T$ connector (also fixed) is used for this purpose. Furthermore, the content of the state vector s_τ which includes the expected values for the observables Y_τ on its first N components is transferred to every hidden intermediate layer r_τ with an identity matrix Id on a component-by-component basis. In the first N components of the intermediate layer r_τ, therefore, the expected values for the observables Y_τ on the first N components of the state vector s_τ are replaced by the corresponding observed values Y_τ^d (i.e. $r_\tau(1, \ldots, N) = Y_\tau - (Y_\tau - Y_\tau^d) = Y_\tau^d$, see Fig. 10.3). None of the other components of the state vector s_τ are modified in the intermediate layer r_τ. The upshot of this mechanism therefore is that, in the intermediate layer r_τ, the expected values of the observables Y_τ are replaced by their actual observations Y_τ^d, without any changes to other components of the state vector. Following the intermediate step of teacher forcing, the state transition matrix A is applied to the intermediate layer, in order to move the system into the next time step.

By definition, there are no observed values available for the observables for any of the future time step of development. The system is iterated exclusively upon the basis of expected values. This turns an open system into a closed high-complexity dynamic system. It should be noted that the use of teacher forcing is not concerned with the reintroduction of input/output modeling, but in fact with the principle of replacement of expected values with actual observations, in order to enhance the system's learning characteristic. For sufficiently high-complexity HCNNs, which describe the observables perfectly, the architecture converges towards the fundamental HCNN architecture shown in Figure 10.2, if the forecasting error converges to zero. One may argue that teacher forcing disturbs the symmetry between past and future time steps of the HCNN modeling. Our experiments show that this is not the case at least for an ensemble of HCNN forecasts.

Since all of the additional connectors for the mechanism introduced are fixed and are used only to transfer data in the network, teacher forcing does not lead to a larger number of free network parameters. State transition matrix A still contains the only free parameters in the model. In order to learn the first state s(0) of the closed dynamic system in the first step of time development, we suggest that an additional bias vector should be introduced. This bias vector represents a starting value s(0), compatible with the rest of the dynamic characteristic.

Note, that it is also possible to switch the position of the tanh() activation function from the internal states s_τ to the hidden intermediate layers r_τ. Proceeding this way the outputs of the HCNN are not limited to the domain [-1, 1] of the activation function. If the hidden intermediate layers r_τ are equipped with the non-linear activation function, we also have to include entities of r_τ at all future time steps of the HCNN ($\tau > t$).

If additional external inputs u_τ are available, which should not be treated as observables within the HCNN, we introduce additional input layers u_τ and connect them to the internal state s_τ of the HCNN at all past and future time steps. Examples for external inputs u_τ are calendar information, weather forecasts or problem specific offsets.

In accordance with the correspondence principle (see Sec. 10.1), the following equations can be formulated for the HCNN with teacher forcing (Fig. 10.3) [21]:

$$\begin{array}{lll} \text{State transition equation:} & \forall \tau \le t & s_\tau = \tanh\left(A\left(s_{\tau-1} - \begin{bmatrix} Id \\ 0 \end{bmatrix}(Y_{\tau-1} - Y_{\tau-1}^d)\right)\right) \\ & \forall \tau > t & s_\tau = \tanh(A s_{\tau-1}) \\ \text{Output equation:} & \forall \tau \in t & Y_\tau = [Id, 0] s_\tau \\ \text{System identification:} & & E = \sum_{\tau=t-m}^{t} (Y_\tau - Y_\tau^d)^2 \rightarrow \min_A! \end{array} \quad (5)$$

Equation 10.5 shows the state transition and output equation derived from the architecture of the HCNN. It is to be noted that the teacher forcing principle is applied to the time steps up to the present time step ($\tau = t$) only. In the time steps up to the present step ($\tau = t$), the system is readjusted in each case using the actual observations Y_τ^d for the observables in the intermediate layers r_τ. For all future time steps, the system uses the expected values for the observables Y_τ only. The principle of using expected and observed values means that the HCNN's dynamic remains consistent in all time steps. If no observed values are available, the model uses the actual expected values for the observables. All time steps, from the past, via the present day to the future, are treated symmetrically. The state transition is coded in a single matrix. Since the model is developed over time across the entire data volume, the HCNN learns the historically consistent behavior of the system dynamic throughout the time line. The system output equation is identical in all time steps. It should be noted that, due to the design of the HCNN, the system outputs are restricted to the domain of the activation function used for the hidden system state s_τ (in this case: tanh()). There is no further scaling of the system outputs, since the outputs across the fixed connectors [Id, 0] only are derived from the hidden system state. Preprocessing of the observables Y_τ should therefore be selected by means of the selection of appropriate pretransformations in such a way that the transformed data is appropriate for the domain of the activation function for the hidden layer s_τ.

The HCNN models the dynamics for all of the observables and their interaction in parallel. The dimension of the state transition matrix A for a high-complexity system must be adjusted as appropriate for this purpose. A high-dimension state transition matrix A is required to model all of the observables in an HCNN. Because of the way in which the error-back propagation algorithm or its expansion for recurrent neural networks (error-back propagation-through-time) operates, a fully interconnected high-complexity (i.e. large-scale) state transition matrix A can, however, lead to an information overload or explosion, resulting in back propagation of the error, which not only has a detrimental effect upon the HCNN's learning, but also destroys the entire network. Our suggested solution to this problem is to select a thin interconnection for state transition matrix A ("sparse matrix"), since the small number of matrix elements not equal to zero limit the flow of information. A thin interconnection (sparsity) can therefore be regarded as an essential requirement for modeling of high-complexity systems.

Two dimensions must be configured for the selection of an appropriate sparse interconnection: the dimension of the state space dim(s) and the degree of sparsity. We intend to show that these variables are linked to two lower level metaparameters: connectivity (con) and memory length (mem), where connectivity (con) is the number of items of data required in one time step which must be superimposed in order to calculate the follow-up state. The memory length (mem) describes the accumulation of data over time in order to arrive at a system state which includes all of the relevant data for a best possible forecast ("Markovian state") [18]. The connectivity (con) and memory length (mem) metaparameters cannot therefore be considered as independent of one another. For example, a very sparse interconnection promotes the development of a long memory, whereas a dense interconnection favors the superimposition of data (and therefore connectivity) in the network [18]. We suggest the following (pragmatic) guides to balance connectivity (con), on the one hand and memory length (mem), on the other hand, in an appropriate way and to infer an appropriate parameterization for the variables of the dimension (dim(s)) and the density of interconnection (sparsity) [18]:

$$\begin{aligned} &\text{Dimension of Matrix}A \quad && \dim(s) = \text{mem} \cdot \text{con} \\ &\text{Sparsity of Matrix}A \quad && \text{Sparsity} = random\left(\frac{\text{con}}{\text{mem} \cdot \text{con}}\right) = random\left(\frac{1}{\text{mem}}\right) \end{aligned} \tag{10.6}$$

In Equation 10.6, a link is established between the functional property memory (mem) and the structural property of sparsity for the optimum size of network. Zimmermann et al. [18] provides more extensive research into this matter. This link has an important consequence for the random initialization of the interconnection in an HCNN. The random factor means that the interconnection of parts of Matrix A will be particularly sparse whereas for others it will be above average. Equation 10.6 implies the fluctuation in the degree of interconnection for different memory lengths in parts of the model. Models of this nature are therefore able to integrate the characteristics of dynamic systems into different time scales.

If system identification is calculated repeatedly, an ensemble of solutions will be produced, which all have an imaging error of zero in the past, but which differ from one another in the future. The form of the ensemble is governed by differences in the reconstruction of the hidden system variables from the observables observed: for every finite volume of observations there is an infinite number of explanation models which describe the data perfectly, but differ in their forecasts, since the observations make it possible to reconstruct the hidden variables in different forms in model training. Underdefined, high-complexity HCNNs represent just such a model class.

Since every model gives a perfect description of the observed data, we use the simple arithmetical average of the individual forecasts as the expected value, provided the ensemble histogram is unimodal in every time step (see Sec. 10.3.2., Fig. 10.4). In other words, the diversification of the ensemble, whose individual members describe past observables virtually without residual errors, produces an expected value in the form of the balanced average of the ensemble. The uncertainty in the modeling can be balanced by this averaging. It is to be noted that multimodal distributions can lead to inferences of a number of main scenarios which should be considered separately from one another.

In addition to the expected value, we also wish to consider the bandwidth of the ensemble channel, i.e. its distribution. We interpret distribution of the ensemble as the forecast risk, whose risk density is defined by data on the one hand and by the modeling framework on the other hand. The forecasting risk is at least as large as the distribution of the ensemble, since all individual forecasts explain the historical observations perfectly and we cannot therefore know which model represents the genuine dynamic. The fact that the genuine dynamic is a part of the ensemble is due to the universal approximation characteristics of the HCNN modeling framework. The results of a series of experiments and our findings from different projects (see Sec. 10.3) lead us to believe that the ensemble density form is consistent with the characteristics of financial market series if interactive (non-linear) decision-making models, such as the HCNN, are used to describe the dynamics.

The expected value and the breadth of the ensemble channel can be used directly in the design of systems to aid with decision making, such as trading indicators, whereby for example the forecast of a market price development (the arithmetical average of the ensemble) is set against the forecasting risk (distribution of the ensemble in the time step of the forecast). In Section 10.3.2, we describe an approach to the measurement of risk upon the basis of ensemble forecasting and distribution.

10.3 Applications in Financial Markets

This Section sets out to present current project applications for those forecasting models whose theory has already been described.

10.3.1 Price Forecasts for Procurement

Many of the raw materials and supplies which Siemens AG requires, such as copper for the construction of generators or energy (oil, gas and electricity), can be purchased on the commodities exchange on a daily basis by means of standardized forward transactions ("futures"). In the context of sharp increases in commodity prices and increased price volatility in particular, optimum timing of purchasing decisions assumes a great deal of importance for the costing and hedging of the project business.

Since 2004, we have provided Siemens Real Estate and the central Procurement function with appropriate price forecasts to assist them with the procurement of electricity, gas and oil. Last December, in response to the convincing results, the forecast models were extended to the copper market, at the behest of the energy sector. The neural networks forecast the price trend in the respective markets for the next 20 days (in one-day steps). Since the energy, raw materials, interest, stock and currency markets are, in the age of international financial flows, very closely interlinked, the approach selected was that of coherent market modeling upon the basis of the HCNNs presented, whereby every individual market is interpreted as an interactive sub-dynamic of an integrated dynamic.

Table 10.1 below provides information on the HCNN's performance for the 20-day forecast for the price trend for copper contracts traded on the spot market at the London Metal Exchange (LME). The analysis is based upon the average ensemble forecast of approx. 250 individual HCNN forecasts in each case.

Table 10.1. Quality of the 20-day forecast for the LME copper spot price using HCNN

Forecast range	Saving in comparison with average price	Hit rate
Dec 8 – Jan 2 09	€183.23	70.0%
Dec 15 – Jan 9 09	€158.15	70.0%
Jan 12 – Feb 6 09	€36.33	60.0%
Jan 19 – Feb 13 09	€-90.45	50.0%
Feb 2 – Feb 27 09	€-75.28	50.0%
Feb 16 – Mar 13 09	€216.43	55.0%
Mar 2 – Mar 27 09	€399.38	50.0%
Mar 16 – Apr 10 09	€352.45	65.0%
Mar 30 – Apr 24 09	€74.10	65.0%
Apr 13 – May 8 09	€71.38	70.0%
Apr 27 – May 22 09	€314.95	50.0%
May 11 – June 5 09	€122.78	50.0%
May 25 – June 19 09	€395.25	70.0%
June 1 – June 26 09	€58.68	55.0%
June 15 – July 10 09	€-88.55	50.0%
June 29 – July 24 09	€27.68	55.0%
July 13 – Aug 7 09	€556.46	75.0%
July 27 – Aug 21 09	€57.52	70.0%
Aug 10 – Sep 4 09	€203.85	75.0%
Aug 24 – Sep 18 09	€14.38	75.0%
Average:	**€149.43**	**61.5%**

The first column of Table 10.1 shows the 20-day forecast range considered in the respective instance. The forecasts were produced on a Friday for the next 20 trading days in each case and as a rule renewed on a rolling 10-day basis. The second column of Table 10.1 includes a performance characteristic, which includes the purchasing performance of the forecast. The motivation for this characteristic is that a purchasing decision for a given volume of copper (in this case: 1 metric ton), which is made upon the basis of the HCNN forecast in the respective forecast range, should be more advantageous than the average LME copper spot price for the volume of copper in the period under review. The purchasing decision is determined by the minimum price forecast by the HCNN in the respective forecast range. Negative values in the characteristic are an indication that the purchasing decision in comparison with the average price was less advantageous. The third column of Table 10.1 includes another performance characteristic, the "hit rate". In order to calculate the hit rate for a forecast range, we compare the forecast price changes with those actually observed. The greater the hit rate, the better equipped the model is to record the direction of market price changes during the period under review.

As Table 10.1 shows, during the entire period under review, from Dec. 2008 to Sept. 2009, the HCNN generated a significant added value in comparison with a purchasing strategy without timing based upon the average market price. Over the period under review, the average saving made in comparison with the average price amounted to €149.43 per metric ton of copper. The HCNN's average hit rate is 61.5%. The results must be considered in the context of the difficult market conditions in particular, which were significantly affected by the effects of the financial crisis in the year of 2009 as well.

10.3.2 Risk Management

The experience gained during the latest financial crisis has triggered a far-reaching discussion on the limitations of quantitative forecasting models and made investors very conscious of risk [3].

Risk management frequently considers the probability distribution of market prices/returns [5]. In order to understand risk distributions, traditional risk management uses historical simulations which require strong model assumptions. Risk is understood as a random walk, in which the diffusion process is calibrated by the observed past error of the underlying model [7].

For our approach this concept fails, because the (past) residual error of the HCNN is zero. Our risk concept is based on the partial observability of the world, leading to different reconstructions of the hidden variables and thus, different future scenarios. Since all scenarios are perfectly consistent with the history, we do not know which scenario describes the future trend best and risk emerges [21].

Our approach directly addresses the model risk. For HCNN modeling we claim that the model risk is equal to the forecast risk. The reasons can be summarized as follows: First, HCNNs are universal approximators, which are therefore able to describe every future market scenario. Second, the form of the ensemble distribution is caused by underlying dynamical equations, which interpret the market dynamics as the result of interacting decisions [19]. Third, in experiments we have shown that the ensemble distribution is independent from the details of the model configuration, iff we use large models and large ensembles.

In contrast to Monte Carlo Simulations [7] in which we have one model creating a diversity of forecasts based on input distributions, our approach is based on a single input history. The diversity of the forecast ensemble emerges from the non-unique reconstruction of hidden variables from the observed data.

In the application presented in Section 10.3.1. above, we focused our attention upon the ensemble forecast for the HCNN, in order to forecast a price dynamic. However, an ensemble based upon an HCNN also provides important insights into complex risk relationships: The diagram below (Fig. 10.4, left) shows the approach described, using as an example a forecast of the Dow Jones Industrial Index (DJX) in weekly steps over a range of three months. For the ensemble, a HCNN was used to generate 250 individual forecasts for the DJX.

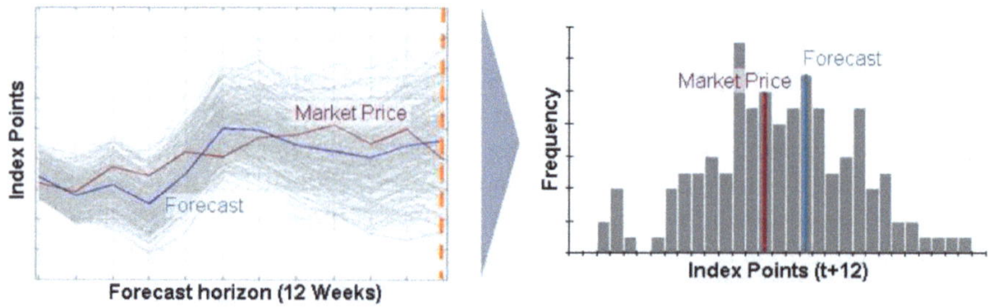

Fig. 10.4. HCNN ensemble forecast for the Dow Jones Index (12-week forecast range), left, and associated index point distribution for the ensemble in forecast time step t+12 weeks, right.

For every forecast date, all of the individual forecasts for the ensemble represent the empirical density function, i.e. a probability distribution over many possible exchange rates at a single point in time (see Fig. 10.4, right). The density function is produced by the assessment of the ensemble distribution at a single point in time (see Fig. 10.4, left). It is noticeable that the actual trend in the DJX does not at any point in time move outside the ensemble channel (see grey lines, Fig. 10.4, left). The expected value for the forecast distribution is also an adequate point forecast for the DJX (see Fig. 10.4, right).

It is our intention to use the distribution of index points derived from the HCNN ensemble in a future project for the evaluation of (plain vanilla) call and put options. We abide by the idea of a simulation-based option price evaluation [5]: firstly, the individual ensemble forecasts for the expiration date of the call or put option to be evaluated are evaluated using the strike price. If the option in respect of an individual forecast within the ensemble expires without any value, the respective individual forecast is no longer included in considerations. The ensemble forecast for which the option under consideration exhibits a value as at the expiration date are discounted to today's date, added up and divided by the total number of ensemble members. It is possible to infer trading recommendations from a comparison of the option evaluation calculated in this way with the actual price for the option observed on the market.

10.4 Summary and Outlook

Recurrent neural networks form dynamic systems in the form of high-complexity and non-linear state models. The reciprocal effect of model-internal (hidden) and observed variables in large recurrent neural networks offer new prospects for risk management. The ensemble approach based upon HCNN offers an alternative approach to forecasting of future probability distributions, which differs fundamentally from traditional risk management procedures in terms of the use of historical data and the form of model dependency. The HCNN can provide a perfect description of the dynamic of the observables in the past. However, the historical data makes it possible to reconstruct the hidden variables in various forms in model training. All of the models are consistent with the observed data, but plot different development paths for the observables in the future. Since the genuine development path for the dynamic is not known and all paths have the same probability of occurrence, the average of the ensemble may be regarded as the best possible forecast, whereas the bandwidth (spread) of the distribution describes the market risk.

Work currently in progress concerns the analysis of the properties of the ensemble and the implementation of these concepts in practical risk management and financial market applications.

References

[1] Calvert, D., Kremer, S.: Networks with Adaptive State Transitions. In: Kolen, J.F., Kremer, S. (ed.) A Field Guide to Dynamical Recurrent Networks, pp. 15–25. IEEE Press (2001)
[2] Elton, E.J., Gruber, M.J., Brown, J., Goetzmann, W.N.: Modern Portfolio Theory and Investment Analysis, 7th edn. John Wiley & Sons (2007)
[3] Föllmer, H.: Alles richtig und trotzdem falsch? Anmerkungen zur Finanzkrise und Finanzmathematik. In: MDMV, vol. 17, pp. 148–154 (2009)
[4] Haykin, S.: Neural Networks. A Comprehensive Foundation, 2nd edn. Macmillan College Publishing, New York (1998)
[5] Hull, J.: Options, Futures and Other Derivative Securities. Prentice-Hall, Englewood Cliffs (2001)
[6] Hornik, K., Stinchcombe, M., White, H.: Multilayer Feedforward Networks are Universal Approximators. Neural Networks 2, 359–366 (1989)
[7] McNeil, A., Frey, R., Embrechts, P.: Quantitative Risk Management: Concepts. Princeton University Press, Princeton (2005)
[8] Pearlmatter, B.: Gradient Calculations for Dynamic Recurrent Neural Networks. In: Kolen, J.F., Kremer, S. (eds.) A Field Guide to Dynamical Recurrent Networks, pp. 179–206. IEEE Press (2001)

[9] Pearlmatter, B.: Gradient Calculations for Dynamic Recurrent Neural Networks: A survey. IEEE Transactions on Neural Networks 6(5), 1212–1228 (1995)
[10] Poddig, T., Huber, C.: Renditeprognose mit Neuronalen Netzen. In: Kleeberg, J.M., Rehkugler, H. (eds.) Handbuch Portfoliomanagement, pp. 349–484. Bad Soden/Ts (1998)
[11] Poddig, T., Sidorovitch, S.: Künstliche Neuronale Netze: Überblick, Einsatzmöglichkeiten und Anwendungsprobleme. In: Hippner, H., Küsters, U., Meyer, M., Wilde, K. (eds.) Handbuch Data Mining im Marketing, pp. 363–402 (2001)
[12] Rumelhart, D.E., Hinton, G.E., Williams, R.J.: Learning Internal Representations by Error Propagation. In: Rumelhart, D.E., McClelland, J.L., et al. (eds.) Parallel Distributed Processing, Foundations. vol. 1. MIT Press, Cambridge (1986)
[13] Schäfer, A.M., Zimmermann, H.-G.: Recurrent Neural Networks are Universal Approximators. In: Kollias, S.D., Stafylopatis, A., Duch, W., Oja, E. (eds.) ICANN 2006. LNCS, vol. 4131, pp. 632–640. Springer, Heidelberg (2006)
[14] Wei, W.S.: Time Series Analysis: Univariate and Multivariate Methods. Addison-Wesley Publishing Company, N.Y. (1990)
[15] Werbos, P.J.: Beyond Regression: New Tools for Prediction and Analysis in the Behavioral Sciences. PhD. Thesis, Harvard University (1974)
[16] Williams, R.J., Zipser, D.: A Learning Algorithm for continually running fully recurrent neural networks. Neural Computation 1(2), 270–280 (1989)
[17] Zimmermann, H.G., Grothmann, R., Neuneier, R.: Modeling of Dynamical Systems by Error Correction Neural Networks. In: Soofi, A., Cao, L. (eds.) Modeling and Forecasting Financial Data, Techniques of Nonlinear Dynamics. Kluwer (2002)
[18] Zimmermann, H.G., Grothmann, R., Schäfer, A., Tietz, C.: Modeling Large Dynamical Systems with Dynamical Consistent Neural Networks. In: Haykin, S., Principe, J.C., Sejnowski, T.J., McWhirter, J. (eds.) New Directions in Statistical Signal Processing: From Systems to Brain. MIT Press, Cambridge (2006)
[19] Zimmermann, H.G.: Neuronale Netze als Entscheidungskalkül. In: Rehkugler, H., Zimmermann, H.G. (eds.) Neuronale Netze in der Ökonomie, Grundlagen und wissenschaftliche Anwendungen. Vahlen, Munich (1994)
[20] Zimmermann, H.G., Neuneier, R.: Neural Network Architectures for the Modeling of Dynamical Systems. In: Kolen, J.F., Kremer (eds.) A Field Guide to Dynamical Recurrent Networks, pp. 311–350. IEEE Press (2001)
[21] Zimmermann, H.G., von Jouanne-Diedrich, H., Grothmann, R., Tietz, C.: Market Modeling, Forecasting and Risk Analysis with Historical Consistent Neural Networks. In: Hu, B., et al. (eds.) Operations Research Proceedings 2010, Selected Papers of the Annual International Conference of the German Operations Research Society (GOR), pp. 531–536. Springer, Heidelberg (2011)

Chapter 11
Reinforcement Learning with Neural Networks: Tricks of the Trade

Christopher J. Gatti and Mark J. Embrechts

Abstract. Reinforcement learning enables the learning of optimal behavior in tasks that require the selection of sequential actions. This method of learning is based on interactions between an agent and its environment. Through repeated interactions with the environment, and the receipt of rewards, the agent learns which actions are associated with the greatest cumulative reward.

This work describes the computational implementation of reinforcement learning. Specifically, we present reinforcement learning using a neural network to represent the valuation function of the agent, as well as the temporal difference algorithm, which is used to train the neural network. The purpose of this work is to present the bare essentials in terms of what is necessary for one to understand how to apply reinforcement learning using a neural network. Additionally, we describe two example implementations of reinforcement learning using the board games of Tic-Tac-Toe and Chung Toi, a challenging extension to Tic-Tac-Toe.

11.1 Introduction

Reinforcement learning is a machine learning method that enables the learning of optimal behavior by an agent through the repeated interaction with an environment. Optimal behavior in this case can be defined as the set of sequential decisions that result the best possible outcome. This learning process can essentially be regarded as a process of trial and error, which is coupled with the receipt of feedback that guides the learning of the best actions to pursue during the decision process.

Board games are prime examples of reinforcement learning. Players take turns assessing the state, or the configuration, of the game, and then select the action that

Christopher J. Gatti · Mark J. Embrechts
Rensselaer Polytechnic Institute, Troy, NY
e-mail: gattic@rpi.edu, embrem@rpi.edu

P. Georgieva et al. (Eds.): Adv. in Intelligent Signal Processing and Data Mining, SCI 410, pp. 275–310.
springerlink.com

they believe will most likely result in an outcome that is in their best interest. Game play therefore consists of the making of sequential decisions in order to achieve a goal, which is most often to win the game. During the game however, the true utility and benefit of each action is not known; only at the end of the game is there certain evidence of the utility of all of a players' actions played during the game.

Consequently, board games have been a prominent domain for the development of computational implementations of reinforcement learning. This is largely due to the fact that board games have well-defined and completely understood rules and actions. Real-world domains, in comparison, may have complex underlying processes that are not fully understood, thus complicating the process of learning how to interact within such domains. Reinforcement learning has been applied to many board games, including checkers [12], chess [10], and Othello [2]. The most notable and successful implementation is the application of reinforcement learning to the game of backgammon by Tesauro [16, 17, 18]. In this work, reinforcement learning was used to train a neural network to play backgammon to such a high level that it could challenge world-class human opponents.

Board games are also useful for explaining reinforcement learning in terms of both the methodology and the practical implementation details. This is also due to the well-defined domains of board games and because the elements of reinforcement learning and board games are quite analogous to each other. The purpose of this work is to present reinforcement learning at a basic level, including only what is essential to understanding and implementing this learning method. More specifically, this work explains the application of reinforcement learning using a neural network to the game of Tic-Tac-Toe, which is then extended and adapted to the game of Chung Toi, a challenging extension to Tic-Tac-Toe. Additionally, we present various tricks and techniques that have the potential to improve the efficiency of training the neural network for reinforcement learning. This work aims to provide the essential background knowledge that is required to implement reinforcement learning to any game-like learning scenario which relies on sequential decisions.

This work begins with both a general overview and a formal definition of reinforcement learning in Section 11.2. Section 11.3 focuses on the implementation of reinforcement learning with a neural network where we describe all of the necessary components to apply reinforcement learning and discuss parameter and algorithm settings as well as modifications to the basic implementation. We then describe in detail two example applications of reinforcement learning in Section 11.4, that of Tic-Tac-Toe and Chung Toi. Finally, we conclude and summarize this chapter in Section 11.5. Throughout this chapter some material is occasionally repeated using similar wording, and this is purposely done in order to reinforce important concepts of this learning paradigm or to present the material in a slightly different manner.

11.2 Overview of Reinforcement Learning

Reinforcement learning is based on an *agent* that repeatedly interacts with an *environment* (Fig. 11.1). Interactions between the agent and the environment proceed by the agent observing the state of the environment, selecting an action which it believes is likely to be beneficial, and then receiving feedback, or a *reward*, from the environment that provides an indication of the utility of the action (i.e., whether the action was good or bad). More specifically, the agent selects actions based on the its perception of the value of the subsequent state. The feedback provided to the agent is used to improve its estimation of the value of being in each state. In the general reinforcement learning paradigm, rewards may be provided after each and every action. After repeated interaction with the environment, the agent's estimation of the true state values slowly improves, which enables the selection of more optimal moves in future interactions.

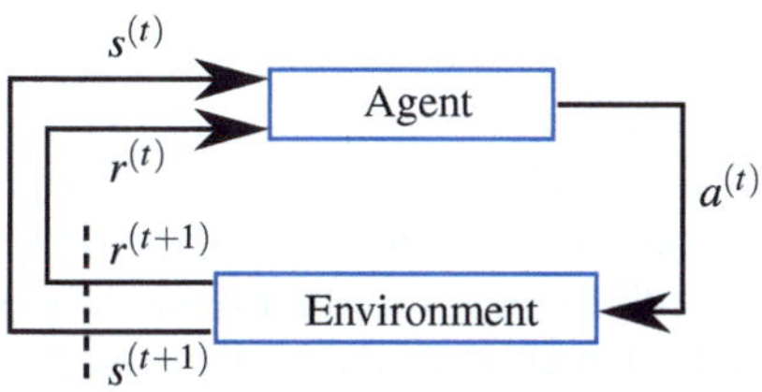

Fig. 11.1. The reinforcement learning paradigm consists of an agent interacting with an environment. The agent observes the state of the environment $s^{(t)}$ and pursues an action $a^{(t)}$ at time t in order to transition to state $s^{(t+1)}$ at time $t+1$, and the environment issues a reward $r^{(t+1)}$ to the agent. Over time, the agent learns to pursue actions that lead to the greatest cumulative reward. Figure adapted from [15].

This method was largely developed as a machine learning technique by Sutton and Barto [14, 15]. Many formulations of reinforcement learning have been developed in order to either accommodate various types of environments or to utilize slightly different learning mechanisms. One of the fundamental reinforcement learning methods is the temporal difference algorithm, which is the main focus of this work. This algorithm uses information from future states, and effectively propagates this information back through time, in order to improve the agent's valuation of each state that was visited during an episode. Consequently, the improvements to the estimations of the state values has a direct impact on the action selection processes as well. This algorithm is particularly useful in scenarios that do not necessarily follow the schema shown in Fig. 11.1 where rewards are provided following every action. In some scenarios, with board games being a prime example, feedback is often only provided at the end of the game in the form of a win, loss, or draw. The temporal difference algorithm can be used to determine the utility of actions played early in the game based on the outcome of the game, which results in a refined action selection procedure.

11.2.1 Sequential Decision Processes

Reinforcement learning is more formally described as a method to determine optimal action selection policies in sequential decision making processes. The general framework is based on sequential interactions between an *agent* and its *environment*, where the environment is characterized by a set of *states* S, and the agent can pursue *actions* a from the set of possible actions A. The agent interacts with the environment and transitions from state $s_t = s$ to state $s_{t+1} = s'$ by selecting an action $a_t = a$. Transitions between states are typically based on some defined probability $\mathcal{P}^a_{ss'}$. Note that the subscript t indicates the particular time step at which states are visited or actions are pursued, and thus the sequences of states and actions can then be thought of as a progression through time.

The sequential decision making process therefore consists of a sequence of states $s = \{s_0,\ s_1,\ ...,\ s_T\}$ and a sequence of actions $a = \{a_0,\ a_1,\ ...,\ a_T\}$ for time steps $t = 0, 1, ..., T$, where state s_0 is considered to be the initial state. Feedback may be provided to the agent at each time step t in the form of a *reward* r_t based on the particular action pursued. This feedback may be either rewarding (positive) for pursuing beneficial actions that lead to better outcomes, or they may be aversive (negative) for pursuing actions that lead to worse outcomes. Processes that provide feedback at every time step t have an associated reward sequence $r = \{r_1,\ r_2,\ ...,\ r_T\}$. A complete process of agent-environment interaction from $t = 0, 1, ..., T$ is referred to as an *episode* consisting of the set of states visited, actions pursued, and rewards received. The total reward received for a single episode is the sum of the rewards received during the entire decision making process, $\mathcal{R} = \sum_{t=1}^{T} r_t$.

For the general sequential decision making process define above, the transition probabilities between states $s_t = s$ and $s_{t+1} = s'$ may be dependent on the complete sequences of previous states, actions, and rewards:

$$Pr\{s_{t+1} = s', r_{t+1} = r \mid s_t,\ a_t,\ r_t,\ s_{t-1},\ a_{t-1},\ r_{t-1},\ ...,\ s_1,\ a_1,\ r_1,\ s_0,\ a_0,\} \quad (11.1)$$

Processes in which the state transition probabilities depend only the most recent state information (s_t, a_t) are said to satisfy the *Markovian assumption*. Under this assumption, the state transition probabilities can be expressed as:

$$Pr\{s_{t+1} = s', r_{t+1} = r \mid s_t,\ a_t\} \quad (11.2)$$

which is equivalent to the expression in (11.1) because all information from $t = 0, 1, ..., t-1$ does not affect the state transition at time t.

If a process can be assumed to be Markovian and can be posed as a problem following the general formulation defined above, this process is amenable to both modeling and analysis. The Markov decision process (MDP) is the specific process which entails sequential decisions and consists of a set of states S, a set of permissible actions A, and a set of rewards R. Each state $s \in S$ has an associated true state value $V^*(s)$. The decision process consists of repeated agent-environment interactions in which the agent attempts to learn the true value of each state. The true state values are unobservable to the agent however, and thus the agents' estimates of the state values $V(s)$ are merely approximations to the true state values.

The transition between states $s_t = s$ and $s_{t+1} = s'$ when pursuing action a may be represented by a transition probability $\mathcal{P}^a_{ss'} = Pr\{s_{t+1} = s \mid s_t = s,\ a_t = a\}$. For problems in which the transition probabilities between all states are known, the transition probabilities can be used to formulate an explicit model of the agent-environment process. A policy π is defined to be the particular sequence of actions $a = \{a_0,\ a_1,\ ...,\ a_T\}$ taken throughout the decision making process. Similarly, the expected reward for transitioning from state s to state s' when pursuing action a may be represented by $\mathcal{R}^a_{ss'} = E\{r_{t+1} \mid s_t = s,\ a_t = a,\ s_{t+1} = s'\}$. An optimal policy π^* is considered to be the set of actions pursued during the process maximizes the total cumulative reward R received.

The optimal policy can be determined if the true state values $V^*(s)$ are known for every state. The true state values can be expressed using the Bellman optimality equation, which states that the value of being in state s and pursuing action a is a function of both the expected return and the optimal policy for all subsequent states:

$$\begin{aligned} V^*(s) &= \max_{a \in A} E\left\{r_{t+1} + \gamma V^*(S_{t+1}) \mid s_t = s,\ a_t = a\right\} \\ &= \max_{a \in A} \sum_{s^+} \mathcal{P}^a_{ss'} \left[\mathcal{R}^a_{ss'} + \gamma V^*(s')\right] \end{aligned}$$

The Bellman equation provides a conceptual solution to the sequential decision making problem in that the value of being in a state is essentially a function of the value of all future states. The solution to reinforcement learning problems is often regarded as a policy π, or an action selection procedure, that leads to an optimal outcome. When implementing reinforcement learning however, the Bellman equation does not have to be explicitly solved. Rather, the identification of optimal policies in reinforcement learning problems is merely based on the notion that future information is relevant to the valuation of current states, and that accurate estimations of state values can be used to determine the optimal policy.

As previously mentioned, if the transition probabilities $\mathcal{P}_{ss'}$ between states are explicitly known, these probabilities can be used to formulate a model of the system. If however, these transition probabilities are not known, transitions between states must be made in a different manner. In this case, we can use the structure of the environment together with the allowable actions to essentially constrain and generate the transitions between states. Using the game of Tic-Tac-Toe as an example, the environment (including the board layout, players, and pieces), the rules of the game, and the possible actions are clearly defined. When state transition probabilities are not known, as in these cases, the agent-environment system is considered to be *model free*. The transition probabilities are essentially learned however, through the repeated interaction between the agent and environment.

11.2.2 Reinforcement Learning with a Neural Network

The overarching concept of reinforcement learning is that the state values are learned through repeated agent-environment interaction, and that the behavior, or the action policy, of the agent emerges from the knowledge of the state values. Initially, the agents' state value estimations are erroneous because it has no knowledge of the environment. Agent-environment interactions improve the accuracy of the agents' state value estimates, such that these estimates become closer to the true, unknown state values.

A computational implementation that models the agent-environment interaction therefore requires some method of keeping track of all state values as well as the ability to update the state values based on additional knowledge gained through interactions with the environment. A basic method to track and update state values is to explicitly store each state value in a look-up table. This method however, is only practical for problems that have a relatively small state spaces. Additionally, when using a reinforcement learning algorithm that is based on state-action pair values, as opposed to just state values, this table must track the values of all state-action pair combinations, and can thus be quite large even for relatively small problems.

Another method that can be used to track all state values and that is not limited by the size of the problem is the use a function approximator that learns to approximate the true state values. This is accomplished by associating an input state vector $\mathbf{x}$ with a corresponding state value $V(\mathbf{x})$ as an output. In this work, we focus on using a neural network as the function approximator, and thus the neural network represents the agent which learns the state values. This approach is advantageous over look-up tables because it requires a relatively small number of parameters and because neural networks can approximate nonlinear state value functions. Other types of function approximators may be used as long as there are free parameters which can be updated and adjusted in order to produce better approximations of the true state values. In the case of a neural network, the connection weights between the nodes are adjusted such that the network output becomes a better approximation of the true state values. This weight updating procedure is driven by the agent-environment interactions and the feedback received by the agent, and this process is represents the training of the neural network.

The neural network interacts with the environment in two ways: 1) pursuing actions, 2) and learning. More specifically, the action selection procedure proceeds by the neural network observing the current state of the environment and using the state-value estimates of the potential subsequent states to select which action to pursue. When the agent selects actions that always correspond to next-state values with the greatest value, the agent is said to follow a *greedy* action selection policy. The learning procedure uses the rewards received, which are a direct result of the actions pursued, to adjust the network weights, which thus improves the state value estimates. In other words, the rewards received are what guides the learning of the state values.

11.3 Implementing Reinforcement Learning

Applying reinforcement learning to a problem formulated as a sequential decision making process requires the representation of a few key elements. The main components are: 1) the *environment* and 2) the *agent*. The environment is characterized by a set of *states* S. The agent transitions between states $s \in S$ by selecting *actions* a from a set of possible actions A based on the *values* of potential subsequent states $V(s')$. The agents' state value estimates (i.e., its knowledge about the environment) is refined through interacting with the environment and by the receipt of *rewards*, such that positive rewards are received for beneficial actions and negative rewards are received for unfavorable actions. Additionally, the learned knowledge, or the expertise, of the agent can be evaluated using some form of a knowledge evaluation method to determine how well the agent is learning.

The section that follows describes various aspects of each of these elements with the aim of explaining how each element can be applied in practice. An implementation of reinforcement learning also requires the user to specify numerous settings and parameters, and many different combinations may yield successful results. This section also provides some insight appropriate parameter and algorithm settings.

The description of the implementation of reinforcement learning is facilitated with the aid of a tangible example of a decision making process. The well-known game of Tic-Tac-Toe will be used in the following section to make some concepts clear. For those unacquainted with Tic-Tac-Toe, a brief overview follows. Tic-Tac-Toe is a board game between 2 players that is played on a board consisting of 9 positions configured in a 3×3 square. One player is assigned pieces represented by $\bigcirc$ and the other is assigned pieces represented by $\times$. Players alternate taking turns and place their pieces on open board positions with the goal of trying to get 3 of their own pieces in a row either along a horizontal, a vertical, or a diagonal.

11.3.1 Environment Representation

The construction of the environment for use within a reinforcement learning framework is specific to the task at hand. As the actions (i.e., state transitions) of the agent are essentially governed by the environment, it is important to develop an in silico environment that accurately reflects that of the real environment.

State Encoding Scheme

The state of the environment is numerically represented by a state vector $\mathbf{x}$, which serves as the input to the neural network. The construction of this vector and what this vector represents can influence the ability of the agent to learn. All information that is relevant to the environment and that may influence the behavior of the agent must be fully contained within this state vector by the encoding of specific features about the environment or domain. The particular features that are used to describe each state are not limited by any particular encoding scheme, and thus there may be multiple state encoding schemes that work well for the same environment.

In the game of Tic-Tac-Toe for example, the most apparent state encoding scheme is a *raw* encoding scheme. This scheme represents the state of the environment in terms of the configuration of the pieces on the board. This encoding scheme and state vector may also include information regarding which player is to select the next move, as this information may also affect which actions are selected. The configuration of the pieces on the board may be represented by a 9×1 vector $\mathbf{p}$, with each element of the vector corresponding to one particular position on the Tic-Tac-Toe board. The presence of a piece on the board can be indicated by a 1 for $\bigcirc$, a -1 for $\times$, and a 0 for an open board position. A binary 2×1 vector $\mathbf{t}$ may also be used to indicate the player that is to select the next move, such that $[1,\ 0]^T$ corresponds to $\bigcirc$ selecting the next move and $[0,\ 1]^T$ corresponds to $\times$ selecting the next move. These two vectors could then be concatenated to form a single 11×1 state vector $\mathbf{x} = \left[\mathbf{p}^T\ \mathbf{t}^T\right]^T$.

A simple raw encoding scheme however, may not be best state representation in terms enabling the neural network to learn the state values. When a raw encoding scheme is not used, the idea is to extract features from the environment that are thought to be relevant to the valuation of each state, and thus also relevant to the goal of the learning process. Examples of feature-based encoding schemes for the game of Tic-Tac-Toe are presented by [8]. These encoding schemes use additional features beyond the raw encoding, which include singlets (lines with exactly one piece), doublets (lines with exactly two of the same pieces), diversity (number of different singlet directions), and crosspoints (an empty position for which two singlets of the same piece intersect). The utility of hand-crafted features has shown to be considerably advantageous in the game of backgammon [17]. In this work, the encoding scheme used both the locations of each piece on the board (raw encoding) as well as specific features that aimed to provide significant expert insight to the state valuations. These additional features proved to be very effective, enabling the neural network to be able to challenge world-class human opponents.

Training Methodology

The methodology used to train the agent to learn a domain can also have a significant effect on the efficiency at which the agent learns the game and on the ultimate performance of the agent to interact with the environment. In the game of Tic-Tac-Toe, the opponent may also be considered a component of the environment that interacts with the agent. In this case, a simple approach to training the agent is to use an opponent which simply selects moves at random, thus introducing a stochastic element to the environment. This strategy is likely to result in an agent which has little expertise in the game however, because very little skill and knowledge is required to match the random actions selected by the opponent.

Another strategy is to use an opponent that has equivalent skill to that of the agent. This can be accomplished simply by using the agent as both the agent as well as the opponent, and this strategy is known as *self-play* training. The output of the neural network can be thought of as the confidence in which the episode will result in a successful outcome for the agent. If, in the game of Tic-Tac-Toe for example, the agent is regarded as ○, the output of the neural network is therefore the confidence in which ○ will win the game. Furthermore, if a reward of 1 is received by the agent for a win, and a reward of -1 is received for a loss, output values from the neural network that are closer to 1 indicate greater confidence that ○ will win the game and output values closer to -1 indicate greater confidence that ○ will lose the game. The confidence in which ○ will win the game can equivalently be thought of as the confidence in which × will lose the game. Conversely, the confidence in which ○ will lose the game is equivalent to the confidence in which × will win the game.

This relationship between ○ winning and × losing (and vice versa) can be used to train a single neural network that learns from all games, regardless of the outcome. As previously described, the action selection procedure for ○ may follow a greedy policy in which actions resulting in subsequent states with the greatest positive value are selected. This policy can be used because a positive reward corresponds to ○ winning the game. The same neural network may also be used to select actions for × using a similar policy, except that actions resulting in subsequent states with the smallest value (which can be negative) are selected. This policy can be used because a negative reward corresponds to × winning the game. Although the action selection policies for each player are converses of each other, the weights of the neural network are updated uses the same procedure as outlined in Sect. 11.3.2.1 regardless of which player selects the action.

The self-play training scheme is just one method that allows for the use of a training opponent that has some expertise. Another method that can provide a challenging training opponent includes having the agent play against a program that is known to have considerable expertise such as those trained using other methods (e.g., minimax, alpha-beta pruning, etc.) [21]. Two additional training procedures rely on merely presenting game scenarios to the agent or having it learn from observations. One of these methods uses database games, which consists of complete sets of state vectors from a large number of games that have been stored in a database. The learning strategy when database games are used is similar to that described above however, in this case the agent does not select moves, but rather it is merely presented with sequential state vectors. A second methods similar to that of database games is to present game scenarios to the agent which have been selected by expert human players. This method attempts to embed specific expert knowledge into the agent by using specific game scenarios. The use of methodologies that are not based on self-play may only allow the agent to gain expertise that nearly equals that of the training opponent or training method. The self-play training method on the other hand, seems to be able to continually increase its playing expertise as it continues to play training games, as is exemplified by [17, 18]. Self-play also training allows the agent to potentially learn novel strategies that have not been conceived or widely used by human players [18].

Evaluation of Agent Expertise

The success of a reinforcement learning implementation to a board games could be defined in various ways. As the ultimate goal of such implementations is to develop a machine learner that rivals, and potentially exceeds, the knowledge and expertise of humans, the playing ability of the agent versus opponents with different levels of expertise provides a measure of success. Note that the opponent that is used to evaluate the expertise of the agent is separate from that used to train the agent using the self-play training method defined above. The agent's learned expertise may first be evaluated using the simplest opponent, one that selects all moves at random. One would expect that the agent should be able to win consistently against this opponent. A next step may be to evaluate the agent using an opponent that has some expertise. This could be done by allowing the opponent to take greedy end-game moves. Such moves allow a player to select moves such that the resulting move allows the player to win the game (greedy win) or to block the win of the opposing player (greedy block), regardless of the estimated state value of the particular move. Another method to create a more challenging opponent is the allow the opponent to use the knowledge gained by the agent, but for only a small percentage of its actions. More extensive evaluations could consist of playing humans with varying levels of expertise or by having expert human players evaluate the quality of the actions selected by the agent [17]. Additionally, the expertise of the agent may also be evaluated under different conditions of the environment which may provide an advantage to one player. In Tic-Tac-Toe for example, this could be done by allowing ○ always to play the first move, thus giving ○ and advantage.

Reward Values

Rewards play an important role in temporal difference learning as these ultimately allow for the agent to learn by guiding the state value estimates toward their true values. The particular reward values used should be based on the possible outcomes for an episode of agent-environment interaction, and the reward values should reflect the desirability of the outcomes. For the game of Tic-Tac-Toe, the possible outcomes include a win, a loss, or a draw, and these correspond to outcomes that may be regarded as good, bad, and indifferent, respectively. Thus, a possible set of rewards may be 1 for a win, -1 for a loss, and 0 for a draw. Modifying the rewards from these values, such as using a negative reward for a draw, will change the state values and ultimately change the behavior of the agent.

The example reward values described above are one potential option for scenarios in which rewards are only provided at the end of an episode. For scenarios in which feedback can be provided during an episode that reflects the quality of actions with some certainty, rewards can be provided at these time points as well. Example of this include when there are clear sub-goals that must be achieved while achieving the overall goal [1], or for continuous domains in which there are continuous measures of performance [3]. This chapter focuses only on domains where feedback is provided at the end of an episode.

11.3.2 Agent Representation

As previously mentioned, the agent can be represented by a neural network that acts as a function approximator. The neural network attempts to learn the true values of each state by interacting with the environment and refining the weights so that the outputs of the neural network become better approximations of the true state values. This section discusses exactly how the the weights of the neural network are updated using the temporal difference algorithm as well as various settings and parameters that can be used with this weight refinement algorithm.

11.3.2.1 Temporal Difference Algorithm

The temporal difference algorithm, also referred to as TD(λ), can be thought of as an extension to the back-propagation algorithm [13, 19]. The back-propagation algorithm is used to train neural networks, for example, for classification or function approximation by solving the credit-assignment problem. In other words, the network weights are individually adjusted based on how much each weight contributed to the error of the network output. TD(λ) on the other hand, solves the *temporal* credit-assignment problem, which is also referred to as back-propagation through time. This method uses information (i.e., network errors) from past time steps to update network weights at current time steps. As there is considerable similarity between the back-propagation algorithm and the temporal difference algorithm, the back-propagation algorithm will first be reviewed and will then be extended to the temporal difference algorithm. Both of these algorithms will be based on the 3-layer neural network configuration shown in Fig. 11.2. The following descriptions assume the reader has a basic understanding of neural networks and of the back-propagation algorithm.

The weight update equation of the back-propagation algorithm for a weight w_{jh} of a neural network (a weight from a node in layer h to a node in layer j) takes the general form:

$$\begin{pmatrix} \textit{weight} \\ \textit{correction} \\ \Delta w_{jh} \end{pmatrix} = \begin{pmatrix} \textit{learning} \\ \textit{parameter} \\ \alpha \end{pmatrix} \times \begin{pmatrix} \textit{prediction} \\ \textit{error} \\ E \end{pmatrix} \times \begin{pmatrix} \textit{local} \\ \textit{gradient} \\ \delta_j \end{pmatrix} \times \begin{pmatrix} \textit{input of} \\ \textit{neuron } j \\ y_h \end{pmatrix} \quad (11.3)$$

where the learning parameter α modulates the magnitude of the weight adjustment, E is the prediction error, δ_j is the local gradient that is based on the derivative of the transfer function evaluated at the node in layer j, and y_h is the output of hidden node h (which is also the input to output node j) and is computed as $y_h = f(v_h)$ where the induced local field is $v_h = \sum_i w_{hi} y_i$ and $f(\cdot)$ is a transfer function. The prediction

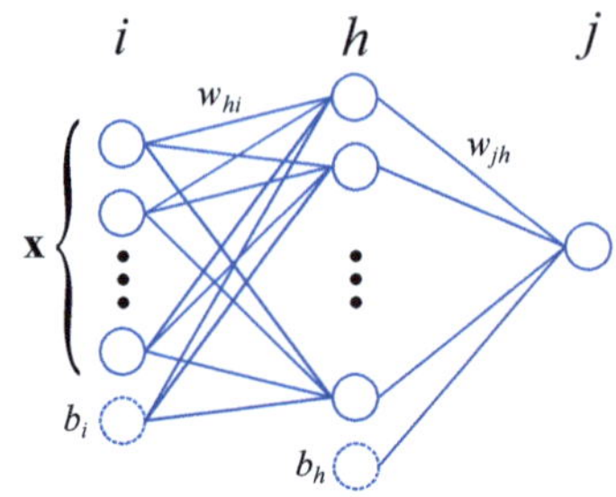

Fig. 11.2. Neural network with input layer i, hidden layer h, and output layer j. State vector $\mathbf{x}$ is input to the network and outputs a value from the single node in layer j. Weights are defined such that w_{hi} represents the weight from a node in layer i to a node in layer h, and that w_{jh} represents the weight from a node in layer h to a node in layer j. Bias nodes b_i and b_h are also included in the input and hidden layers, respectively, and can be implemented by increasing the input and hidden layers by one node each and using a constant input of 1 to both bias nodes.

error from this network is stated as $E = (y_j^* - y_j)$ where y_j is the value of output node j and y_j^* is the corresponding target output value. The expression for Δw_{jh} can be written more explicitly using the partial derivative of the network error E with respect to the network weights:

$$\begin{aligned}\Delta w_{jh} &= -\alpha \frac{\partial E}{\partial w_{jh}} \\ &= -\alpha \frac{\partial E}{\partial y_j} \frac{\partial y_j}{\partial v_j} \frac{\partial v_j}{\partial w_{jh}} \\ &= \alpha \, (y_j^* - y_j) \, f'(v_j) \, y_h\end{aligned}$$

where $f'(v_j)$ is the derivative of the transfer function evaluated for the induced local field v_j. This weight adjustment expression can be extended for updating the weights Δw_{hi} connecting nodes from input layer i to nodes in hidden layer h and can be expressed as:

$$\begin{aligned}\Delta w_{hi} &= -\alpha \frac{\partial E}{\partial w_{hi}} \\ &= -\alpha \frac{\partial E}{\partial y_h} \frac{\partial y_h}{\partial v_h} \frac{\partial v_h}{\partial w_{hi}} \\ &= \alpha \, (y_j^* - y_j) \, f'(v_j) \, w_{jh} \, f'(v_h) \, x_i\end{aligned}$$

where v_h is the induced local field at hidden node h and x_i is the output of input node i (which is also the input to input node i). Note that the neural network used in this work has a single output node (Fig. 11.2), and thus the network error E is computed from only this output node. This is a slight simplification of the general form of the neural network with multiple output nodes, in which case the back-propagation algorithm propagates all errors at the output nodes in layer j back to the hidden layer h.

The back-propagation algorithm can then be extended to the temporal difference algorithm with a simple modification of the weight update equations. Using the example of a game play scenario, the basic implementation of this algorithm adjusts network weights following every play during the game (i.e., iterative updates). This algorithm works by discounting state value prediction errors by a temporal difference factor λ. The general form of the TD(λ) algorithm can be stated as:

$$\begin{pmatrix} weight \\ correction \\ \Delta w_{jh} \end{pmatrix}^{(t)} = \begin{pmatrix} learning \\ parameter \\ \alpha \end{pmatrix} \times \begin{pmatrix} prediction \\ error \\ E \end{pmatrix}^{(t)} \times \sum_{k=0}^{t} \begin{pmatrix} temporal \\ discount \\ \lambda \end{pmatrix}^{t-k} \times \begin{pmatrix} local \\ gradient \\ \delta_j \end{pmatrix}^{(k)} \times \begin{pmatrix} input\ of \\ neuron\ j \\ y_h \end{pmatrix}^{(k)} \quad (11.4)$$

which adds a temporal discount factor λ and a summation from the initial time step $t = 0$ up to the current time step t. Note that from here on, the convention of using a superscript enclosed in parentheses (e.g., $y_j^{(t)}$) will be used to indicate the particular time step t of the corresponding variable. The weight updates occur at every time step during an episode up to and including the terminal time step $t = T$. The superscript-parentheses convention is used to distinguish the values of variables at a particular time step from exponents and subscripts that also may be associated with variables. Thus superscripts enclosed in parentheses are *not* exponents but merely refer to the value or a variable at a particular time step; however, superscripts that are not enclosed in parentheses are exponents, as in the case of λ^{t-k} for example.

In Eq. (11.4), the prediction error E is also modified from that for the back-propagation algorithm. This error, also known as the temporal difference error, is instead based on the difference between the predicted state values at $t+1$ and t, as well as the reward received:

$$E^{(t)} = \left(r^{(t+1)} + \gamma V^{(t+1)} - V^{(t)}\right) \quad (11.5)$$

where $V^{(t)}$ and $V^{(t+1)}$ are the predicted state values at the current and subsequent time steps, respectively, and $r^{(t+1)}$ is the reward provided for transitioning to state $s^{(t+1)}$. The values of $V^{(t)}$ and $V^{(t+1)}$ are determined by evaluating the respective state values ($\mathbf{x}^{(t)}$ and $\mathbf{x}^{(t+1)}$) through the neural network using forward propagation, where the next-state vector $\mathbf{x}^{(t+1)}$ is determined based on an action selection procedure, which is explained later in this chapter. This expression for the temporal difference error also discounts the subsequent state value $V^{(t+1)}$ by a factor γ, which serves to attenuate the value that the network is attempting to learn. In many applications, the discount factor γ is set to 1 however, there is some support for the notion that learning improves when γ is set between 0.5 and 0.7 [6]. Note that this error expression is from where the temporal difference algorithm gets its name, as it is based on the difference in the predicted state values at two different time steps.

The general form of the TD(λ) algorithm can be more explicitly written for updating the network weights such that $w \leftarrow w + \Delta w$. The weight updates between nodes in the output layer j and nodes in the hidden layer h ($\Delta w_{jh}^{(t)}$) at time step t can be stated as:

$$\Delta w_{jh}^{(t)} = \alpha \left(r^{(t+1)} + \gamma V^{(t+1)} - V^{(t)} \right) \sum_{k=0}^{t} \lambda^{t-k} f'(v_j^{(k)}) y_h^{(k)} \tag{11.6}$$

where $f'(v_j^{(k)})$ is the derivative of the transfer function at node j evaluated at the induced local field $v_j^{(k)}$ at time step k. Equation (11.6) can then be extended to updating the weights between nodes in the hidden layer h and nodes in the input layer i ($\Delta w_{hi}^{(t)}$) at time step t as:

$$\Delta w_{hi}^{(t)} = \alpha \left(r^{(t+1)} + \gamma V^{(t+1)} - V^{(t)} \right) \sum_{k=0}^{t} \lambda^{t-k} f'(v_j^{(k)}) w_{jh}^{(t)} f'(v_h^{(k)}) x_i^{(k)} \tag{11.7}$$

A basic implementation of the TD(λ) algorithm requires only the use of Eqs. (11.6) and (11.7). Extending these equations using some relatively simple techniques however, can significantly reduce the computational cost in terms of both time and space, and can improve the efficiency of the learning algorithm.

The use of a momentum term with coefficient η can be added to Eqs. (11.6) and (11.7) in order to incorporate a portion of the weight update from the previous time step $t-1$ into that for the current time step t. This has the effect of smoothing out the network weight changes between time steps and is often most effective when training the network in a batch manner. Batch training is where weight updates are computed during every time step, but the updates are only applied after every n time steps where $n > 1$ and could extend across multiple episodes. After adding the momentum term, Eqs. (11.6) and (11.7) become, respectively:

$$\Delta w_{jh}^{(t)} = \eta \Delta w_{jh}^{(t-1)} + \alpha \left(r^{(t+1)} + \gamma V^{(t+1)} - V^{(t)} \right) \sum_{k=0}^{t} \lambda^{t-k} f'(v_j^{(k)}) y_h^{(k)} \tag{11.8}$$

$$\begin{aligned} \Delta w_{hi}^{(t)} &= \eta \Delta w_{hi}^{(t-1)} \\ &\quad + \alpha \left(r^{(t+1)} + \gamma V^{(t+1)} - V^{(t)} \right) \sum_{k=0}^{t} \lambda^{t-k} f'(v_j^{(k)}) w_{jh}^{(t)} f'(v_h^{(k)}) x_i^{(k)} \end{aligned} \tag{11.9}$$

There are a few important notes about the above two equations. The error term, the difference between subsequent state values $V^{(t+1)}$ and $V^{(t)}$ (neglecting the reward term for now), is essentially the information (i.e., feedback) that is used to update network weights. Drawing from terminology of the back-propagation algorithm, $V^{(t+1)}$ can be considered the target output value y_j^*, and $V^{(t)}$ can be considered the predicted output value y_j. The network error is therefore based on the next state value $V^{(t+1)}$, in spite of the fact that this value is merely an estimate and may actually be quite different than the true state value.

The general form of a sequential decision making processes proceeds over $t = 0, 1, ..., T$. For all intermediate (i.e., non-terminal, $t \neq T$) time steps, the next-state values $V^{(t+1)}$ are available based on their predicted value after pursuing an action; an associated reward $r^{(t+1)}$ may also be provided at these time steps as well. The incorporation of rewards at intermediate time points is dependent on the specific problem. For example, some problems have well-defined subgoals that must be achieved en route to an ultimate goal. In such cases, non-zero reward values may be provided when these subgoals are achieved. In other problems such as board games, there are often no such subgoals, and thus reward values at all intermediate time steps $t = 0, 1, ..., T-1$ are 0. The temporal difference error is then only based on the difference between subsequent state values (i.e., $\gamma V^{(t+1)} - V^{(t)}$).

At the terminal time step $t = T$, there is no next state value $V^{(t+1)}$, and this value is set to 0. The reward at this time step $r^{(t+1)}$ is non-zero however, and the temporal difference error is then based on the previous state value and the reward value (i.e., $r^{(t+1)} - V^{(t)}$). For problems in which there are no intermediate rewards, it is important to note that the reward provided at time step T is the only information which is known with complete certainty, and thus this is the only true information from which the neural network can learn the values of all previously visited states.

The specific values used as rewards are often problem-dependent and have an influence on the ability of the neural network to learn the state values. In many board games, the use of rewards consisting of a 1 for a win and a -1 for a loss (and possibly a 0 for a draw) often work well. For problems that are based on a more continuous performance metric, reward values may take on values over a range (e.g., $r = [-1, 1]$) depending on the performance of the agent.

Equations (11.8) and (11.9) require that information (terms within the summation) from all previous states $s^{(0)}$, ..., $s^{(t-1)}$ be used to determine the appropriate weight adjustment at state $s^{(t)}$. At time step t, information from all previous states is merely discounted by λ. This can be exploited to reduce the number of computations required for each weight update, which can be significant for episodes with many time steps. Weight updates at time t can be made based only on information from the current time step t as well as a cumulative aggregation of the information from previous time steps up to $t-1$.

Let the summation in Eqs. (11.8) and (11.9) be called g. For each network weight there is a corresponding value of g, which is initially set to 0 at the beginning of each game. The value of $g^{(t)}$ at the current time step t can be computed by discounting its previous value $g^{(t-1)}$ by λ and adding it to the current state's weight adjustment.

The values of g for the hidden-output weights ($g_{jh}^{(t)}$) and the input-hidden weights ($g_{hi}^{(t)}$) become, respectively:

$$g_{jh}^{(t)} = f'(v_j^{(t)})y_h^{(t)} + \lambda g_{jh}^{(t-1)} \tag{11.10}$$

$$g_{hi}^{(t)} = f'(v_j^{(t)})w_{jh}^{(t)} f'(v_h^{(t)})x_i^{(t)} + \lambda g_{hi}^{(t-1)} \tag{11.11}$$

Replacing the summations in Eqs. (11.8) and (11.9) with g as in Eqs. (11.10) and (11.11), respectively, yields the following weight update equations:

$$\Delta w_{jh}^{(t)} = \eta \Delta w_{jh}^{(t-1)} + \alpha \left(r^{(t+1)} + \gamma V^{(t+1)} - V^{(t)} \right) g_{jh}^{(t)} \tag{11.12}$$

$$\Delta w_{hi}^{(t)} = \eta \Delta w_{hi}^{(t-1)} + \alpha \left(r^{(t+1)} + \gamma V^{(t+1)} - V^{(t)} \right) g_{hi}^{(t)} \tag{11.13}$$

11.3.2.2 Additional Insight into the Temporal Difference Equation

There are two ways to think about the reinforcement learning algorithm, as presented by Sutton and Barto [15]: the forward view and the backward view. The forward view focuses on the values and rewards of future states which the agent is attempting to accurately estimate. In other words, the agent is always looking at future information in order to improve its present state value estimates. In Eqs. (11.8) and (11.9), this forward-looking concept is evident in the error term that is comprised of the current state value, next-state value, and the next-state reward.

In the case of board games where all non-terminal rewards are 0, the agent adjusts its value predictions based on the error between the future state value $V^{(t+1)}$ and $V^{(t)}$ (for $t \neq T$), and thus $V^{(t)}$ evolves toward $V^{(t+1)}$. Note however, that $V^{(t+1)}$ may be erroneous due to the random initialization of the neural network weights, and thus the weight updates may also be erroneous, especially in the early stages of training. When $t = T$, the agent adjusts its value predictions based on the reward value $r^{(t+1)}$ and the current state value $V^{(t)}$, and thus $V^{(t)}$ evolves toward $r^{(t+1)}$. In this case, the value of $r^{(t+1)}$ is known with complete certainty, and thus the final weight updates are in the correct direction. When many episodes are performed, the state values at latter time points evolve and become better approximations of their true state values because of their temporal proximity to the reward value. With even more episodes, state values at early time points in the episode can better approximate their true values. The ability of the agent to approximate the true state values is dependent on many things, such as the parameters in the temporal difference equations above, and this will be shown later in this chapter.

The backward view focuses on the mechanics of the temporal differencing. More specifically, this view considers how past states at time steps $0, 1, \ldots, t-1$ influence learning at the current state t. This concept is evident the summation term in Eqs. (11.8) and (11.9) in that the gradient, or the potential for influencing weight updates, from past states is discounted. Thus, past states essentially have less influence on how the state value error affects the weight updates. To see this more clearly, suppose that at time step $t = 2$, the summation in Eq. (11.8) is written explicitly as (while neglecting the momentum term for simplicity):

$$\Delta w_{jh}^{(2)} = \alpha \left(r^{(3)} + \gamma V^{(3)} - V^{(2)} \right) \left[\lambda^2 f'(v_j^{(0)}) y_h^{(0)} + \lambda^1 f'(v_j^{(1)}) y_h^{(1)} + \lambda^0 f'(v_j^{(2)}) y_h^{(2)} \right]$$

It can be seen that the effect of the prediction error $(r^{(3)} + \gamma V^{(3)} - V^{(2)})$ on the weight change $\Delta w_{jh}^{(2)}$ due to the gradient at time step $t = 0$, is discounted by λ^2, thus reducing the effect that the gradient at this time point has on the weight update.

Using the Temporal Difference Algorithm

The reinforcement learning framework, and the TD(λ) algorithm described above proceeds in a simulation-like fashion with the agent repeatedly interacting with the environment for a specified number of episodes N. This process is outlined in Algorithm 1. Recall that the state values of the current state $V(s)$ and the subsequent state $V(s')$, where $s = \mathbf{x}^{(t)}$ and $s' = \mathbf{x}^{(t+1)}$, are determined by evaluating the respective state vectors through the neural network using forward propagation.

```
Initialize neural network (initialize all weights w);
for N episodes do
    Initialize s;
    repeat for each step of episode:
        a ← action given by policy π for s;
        Take action a, observe reward r and next state s';
        E ← r + γV(s') − V(s);
        w ← w + Δw where Δw = f(E);
        s ← s';
    until s is terminal state;
end
```

Algorithm 1. The TD(λ) algorithm using an iterative weight updating scheme. Weight changes Δw are performed according to Eqs. (11.8) and (11.9).

11.3.2.3 Neural Network Settings

The neural network lies at the heart of the TD(λ) algorithm and thus the settings and parameters of the neural network can influence the ability of the network to learn accurate state value estimates. There has been a fair bit of work concerning the setting of parameters for neural networks. Some of the most effective methods are known as Efficient BackProp training methods, which are discussed in [9] and [7]. As these techniques have been found to be useful for efficiently training neural networks with the back-propagation algorithm in classification and function approximation problems, these methods also have the potential to be useful for training using the TD(λ) algorithm. This section discusses some of the important points regarding neural network settings as they relate to the TD(λ) algorithm based on results from the literature and from empirical results.

Neural Network Architecture

As with all uses of neural networks, the architecture of the network, in terms of the number of layers and number of nodes within each layer, is a primary decision. In most applications of reinforcement learning with neural networks, a three-layer neural network is used, consisting of an input layer, a single hidden layer, and an output layer. While the number of hidden layers can be increased, it seems that this does not necessarily improve the learning of the neural network. This may be due to the fact that a single hidden layer is all that is necessary to approximate nonlinear functions with a neural network. Furthermore, while the state value function is often nonlinear, it may not be overly complex, and a single hidden layer may therefore be sufficient. Another reason for not using multiple hidden layers is due to the increased computational expense required for training the additional weights, which has not been justified by a significant increase in performance.

For the use of the TD(λ) algorithm described herein, the number of input nodes is determined by the number of elements in the state vector $\mathbf{x}$. As discussed above, there is not a unique state encoding scheme for each problem or environment, and thus the size of the state vector may vary depending the encoding scheme, which then dictates the number of nodes in the input layer. The number of output nodes is often just 1, the value of which represents the estimated state value of the corresponding input state vector. The number of hidden nodes however, can have a significant effect on the ability of the neural network to learn. In general, hidden layers with more nodes seem to perform better, and hidden layers with 40–80 hidden nodes have been used to with great success [17]. This could be due to the fact that the larger number of hidden nodes is required to accurately approximate the nonlinear state value function. Note again though, that using more hidden nodes increases the total number of weights in the network, which increases the computational expense of using such a network. A very large number of hidden nodes will also likely not improve the performance of the network significantly, as many of the weights may not contributed significantly to the state value estimates. Finally, bias nodes are often used in the input and hidden layers, which only connect to the immediate latter layer (as in Fig. 11.2).

Transfer Functions

There are a number of choices for the type of transfer functions used at the nodes of the neural network. The most common transfer function is the basic sigmoid transfer function of the form $f(v) = \frac{1}{1+e^{-v}}$, where v corresponds to the induced local field at a particular node, as previously described. This transfer function was used for both hidden and output nodes in the network created by [17]. However, there are other transfer functions that can be used that have had similar success, either in reinforcement learning or in neural networks in general. For the hidden layers, these include the hyperbolic tangent function $f(v) = \tanh(v)$ and a modified modified hyperbolic tangent function of the form $f(v) = 1.7159\tanh(\frac{2}{3}v)$ [9]. The purpose behind the modified hyperbolic tangent function lies in the notion that the values of v stay away from the saturation region of the transfer function, and thus the derivative of the transfer function at these values is non-zero and more effectively contributes to adjusting network weights. For the output layer, the basic linear function $f(v) = v$ has been used as an alternative to the sigmoid transfer function.

There does not seem to be any single transfer function that has been found to be generally applicable across applications of reinforcement learning. One thing that is important is the pairing of the output transfer function and the reward values provided to the agent (i.e., the values which the output node is ultimately driven toward). The selection of the output transfer function should be guided by ensuring that the range over which this transfer function operates corresponds to the reward values. For example, if reward values consist of a 1 for a win and a 0 for a loss, the output transfer function should be the sigmoid transfer function which operates over $[0,1]$. If the reward values consist of a 1 for a win and a -1 for a loss, an output transfer function should be selected that has the ability to operate over $[-1,1]$, such as the linear transfer function or the hyperbolic transfer function.

Weight Initialization

The weights of the neural network can be initialized in a number of ways. The simplest method is to sample the network weights from a uniform distribution which extends over a small range, such as [-0.2, 0.2] as used in [20]. Sampling from a uniform distribution over a small range that is centered at zero allows for the transfer function to operate near its linear region. When this is the case, the gradients of the transfer function are large, enabling efficient learning. When the transfer function operates in the saturation regions however, the gradients are small, and this results in small weight changes [9].

LeCun et al. [9] suggest another method to initialize network weights, which is said to further increase the learning efficiency. In this method, the weights are initialized based on the number of incoming weights to each node. More specifically, for each node, weights are initialized by randomly sampling from a distribution with a mean of zero and a standard deviation of $\sigma_w = m^{-1/2}$ where m is the number of weights leading into node w. LeCun et al. also notes that the success of this method also relies on normalizing the data that are input to the network and using a modified version of the hyperbolic tangent transfer function (previously described).

Learning Rates

The learning rate parameter α can have a substantial impact on the learning and the performance of the neural network. When neural networks are used for problems such as function approximation or classification, high values of α may not allow the weights of the network to converge, whereas low values of α are more likely to allow for weight convergence, although training may be very slow. Thus, a moderate learning rate is optimal.

Another important point is that each layer should have its own learning rate. This is due to the fact that latter layers (layers closer to the network output) tend to have larger gradients than earlier layers (layers closer to the input layer). When a single value of α is used for all layers, the weights of the latter layers may change considerably while those of the earlier layers may change very little, leading to unbalanced weight changes between layers and inefficient training.

An effective method to overcome the differences in the size of the layer gradients is described in [4], and the method presented here follows a similar method. Setting the learning rates consists of a number of steps. Initially, the learning rates for all layers are set to 1. The learning rates of all layers are then increased by $n_l\sqrt{2}$ where n_l is the number of previous layers from the output layer. For example, in a 3 layer neural network, $\alpha = 1$ for the hidden-output layer and $\alpha = \sqrt{2}$ for the input-hidden layer. The final step is to scale all learning rates proportionally such that the largest learning rate is equal to $\frac{1}{n_p}$ where n_p is the number of training patterns.

For reinforcement learning, this method may not be directly applicable as n_p is dependent on the weight updating scheme, such as iterative weight updates (following every move) or batch (epoch) weight updates (following sets of moves or episodes). Additionally, if a batch size of 1 episode is used, n_p is dependent on the number of time steps within the episode. The most important detail about setting the learning rates however, is the that the ratio of the learning rates between layers must be set appropriately. Determining the largest learning rate of all layers may be determined by starting with the simple case of iterative weight updates and using trial and error while monitoring the weight changes over the course of training. When batch weight updates are used instead, the learning rates should be scaled by the number of episodes per batch update. Batch training has been found to work well in other applications of neural networks however, and this may also be beneficial for reinforcement learning provided parameters are adjusted accordingly.

11.3.2.4 TD(λ) Settings

In addition to the settings and parameters for the neural network, there are also a few settings that are specific to the TD(λ) algorithm that can also influence ability of the neural network to learn state values. The primary settings that relate to the TD(λ) algorithm include the temporal discount factor λ and the action selection policy of the agent.

Temporal Discount Factor λ

The main parameter of the TD(λ) algorithm, λ, determines how much influence previous time steps have on weight updates in current time steps. This parameter can be set over the range [0, 1]. When $\lambda = 0$, only the most recent time step $t-1$ influences the weight update at time step t, and thus information from time step t is only passed back 1 time step. When $\lambda \to 0$ but $\lambda \neq 0$, time steps prior to $t-1$ have some, though very little, influence on weight updates at time step t. Thus information from time step t is propagated backward in time very slowly. As $\lambda \to 1$ but $\lambda \neq 1$, time steps prior to $t-1$ have increasingly more influence on the weight updates at time step t, and information at time step t is propagated backward in time more quickly. Finally when $\lambda = 1$, all time steps have equal influence on weight updates at time step t, and this is considered to follow Monte Carlo behavior.

The effects of λ can be seen by considering the reward at the terminal time step T. Suppose that $\lambda \approx 0$, but $\lambda \neq 0$. If a single reward value (either positive or negative) is received at the terminal time step T, this information will not propagate backward in time much beyond time step $T-1$. In order for this reward to reach states at time steps close to the initial time step, many episode of a similar process must be followed, with the reward slowly being propagated backward through time steps with each episode. Consider a second case in which $\lambda \approx 1$, but $\lambda \neq 1$. In this case, again with a single reward provided at the terminal time step T, all time steps have nearly equal influence on future time steps, and thus information is propagation backward in time extending over many time steps. Note that in the case of board games, the only true information about the environment is provided at the end of the episode, and the information propagated backward through time at all other time steps is based on the agents' state value estimates. If the agents' state value estimates are erroneous for the entire episode, erroneous information is being propagated backward in time. This then conflicts with the true information provided as a reward at the end of the episode.

The two scenarios above lead to the notion that λ should be set somewhere between 0 and 1. Doing so attempts to balance the speed of information propagation and the amount of certainty in the information being propagated. Numerous implementations of TD(λ) to board games have suggested that using $\lambda \approx 0.7$ seems to work well (see, for example, [5, 17, 20]).

Exploration Versus Exploitation

In order for the agent to form estimations of all state values it must be able to visit each state. For the agent to learn *accurate* estimations of all state values, it must visit each state numerous times. The type of learning method for the TD(λ) algorithm described in this chapter is referred to as an *on-policy* learning method because the agent only learns about state values for those states which it explicitly visits. (Other temporal difference methods allow for the learning of state values for states which are not explicitly visited by the agent, and these are referred to as *off-policy* learning methods.)

Learning is therefore dependent on the actions that are pursued by the agent. While the notion of action selection is absent from the weight update Eqs. (11.8) and (11.9), the action selection policy π is central to reinforcement learning and is a key step during agent-environment interaction (as in the action selection step in Algorithm 1).

A naïve action selection policy, also called a *greedy* policy, is as follows. All possible actions at time step t are evaluated by the agent by evaluating all possible next-state vectors $\mathbf{x}^{(t+1)}$ through the neural network. These next-state vectors are evaluated in order to obtain estimated state values for each possible subsequent state. The action that corresponds to the greatest next-state value is then selected and pursued. This approach is also called a pure exploitive action selection policy such that the agent always *exploits* its learned knowledge.

The greedy action selection policy is not often the most efficient policy for learning accurate estimates of all state values. This is because the agents' state value estimates are not necessarily equivalent to the true state values. The difference between the estimated and true state values results from the fact that the agents' state value estimates are dependent on the particular states visited and the rewards received (i.e., the state trajectories of the agent). Additionally, in early interactions between the agent and environment, the agent has no knowledge about the environment. In other words, its estimated state values are quite erroneous. Thus, if a greedy action selection policy is pursued all of the time, the agent will often be following a suboptimal policy because of the erroneous state value estimates. Following such an exploitative policy is likely to be a very inefficient learning strategy because it may take many episodes for the agent to realize relatively accurate state value estimates. Relating this to a neural network implementation, the agents' state value estimates are erroneous because of the randomly initialized weights between the nodes. As the agent begins to interact with the environment however, the agent gains knowledge about the environment, and this is reflected in the adjustments of the network weights leading to improved state value estimates.

Another action selection policy that could be followed is one in which the agent *explores* actions that are regarded as suboptimal based on its state value estimates. By following this policy, the agent would again evaluate all possible next-states, but it would select an action at random from the set of actions which does not include that with the greatest next-state value. This policy is referred to as an explorative action selection policy. Similar to the deficiency of following a purely exploitative action selection policy, following a purely explorative action selection policy is also likely an inefficient learning strategy because the agent never gets a chance to test its learned knowledge.

A relatively efficient learning strategy can be produced by combining the policies of action exploitation and action exploration. In such a policy, the agent largely relies on exploiting its knowledge, and does so by selecting actions that have the greatest next-state value $\varepsilon\%$ of the time. The other $(1-\varepsilon)\%$ of the time the agent uses an explorative policy by selecting its action at random from the set of available actions. This action selection policy is referred to as an ε-*greedy* policy. This policy allows for the agent to both test its learned knowledge about the environment while also exploring actions that may be optimal despite the fact that they may be regarded as suboptimal by the agent's current knowledge.

11.4 Examples of Reinforcement Learning

Reinforcement learning can be applied to many different types of problems that can be formulated as a sequential decision making process. Many different benchmark problems have been developed in order to test the applicability and extensibility of various reinforcement learning methods. Some of these problems include gridworld, the pole-balancing problem, and the mountain-car problem. Gridworld consists of an environment represented by a 2-dimensional grid with a specific starting location, a goal location, and potential hazard locations. The goal of the agent is to navigate through the gridworld from the start to the goal while avoiding the hazards. The pole-balancing problem consists of a cart, which can only move horizontally, and that has a vertical pole attached by a hinge. The goal of this problem is to keep the pole oriented nearly vertically by moving the cart to the left or to the right. The mountain-car problem consists of a car that is attempting to climb out of a valley; however, the car does not have enough power to drive out of the valley [11]. In order to drive out of the valley, the car must drive up the opposite side of the valley to gain sufficient inertia in order achieve its goal. This problem is challenging because the car must first move away from the goal in order for it to ultimately achieve its goal.

This section however, will demonstrate how reinforcement learning performs on two board games, Tic-Tac-Toe and Chung Toi, which is a challenging variation of Tic-Tac-Toe. The implementations shown in this section will apply some of the environment, neural network, and TD(λ) settings that were previously discussed, and will explore some of the effects of these settings.

11.4.1 Tic-Tac-Toe

A base scenario for the implementation of Tic-Tac-Toe is described below, which includes all of the settings used in this scenario. The parameters and settings used in this base scenario are also provided more concisely in Table (11.1); note that parameters α, η, ε, and λ were kept constant through training. Following this base scenario, the parameter ε was changed to different values to determine its effect on the ability of the agent to learn the game.

Environment Settings

The game of Tic-Tac-Toe was implemented by first constructing the environment, which consisted of coding the structure of the environment including the board and the rules of the game, defining a state encoding scheme, and developing an evaluation opponent. The environment consisted of a 3×3 matrix with each element corresponding to one location of the Tic-Tac-Toe board. The rules of the largely game consisted of defining the allowable actions that either player could take; i.e., players could only place pieces on open locations of the board. Additionally, players would alternate taking turns by placing pieces on the board. Note that this form of alternating play pairs well with the self-play training scheme, and self-play training was just what was used in this case. Following the placement of each piece, the board was evaluated to determine if either player had won by having three of the same piece in a row. If either player had won, the game (i.e., episode) was terminated and the corresponding reward was provided. The agent, represented by $\bigcirc$, received a reward of 1 if it had won the game, -1 if $\times$ had won the game, or 0 if the game resulted in a draw.

The state encoding scheme was the same as that previously described in Sect. 11.3.1. Briefly to review, $\bigcirc$ was encoded by a 1 and $\times$ was encoding by a -1. The 3×3 board was represented by a 9×1 vector, with one element in this vector corresponding to each location on the board. This vector represented the presence or absence of pieces at each location on the board such that a 1 or a -1 was placed in an element of this vector if a $\bigcirc$ or a $\times$ occupied the corresponding board position, respectively, or a 0 was placed if the board position was open. A 2×1 turn-encoding vector was used to indicate which player was to play the next move. If $\bigcirc$ was to play the next move, this vector was $[1,\ 0]^T$; if $\times$ was to play the next move, this vector was $[0,\ 1]^T$. The complete state vector $\mathbf{x}$ was created by concatenating the board and turn-encoding vectors, resulting in an 11×1 state vectors.

As previously mentioned, the agent was trained using the self-play training scheme. Greedy end-game moves, including greedy wins and greedy blocks, were allowed during the training games. During evaluation games, the agent was not allowed to take greedy end-game moves but instead had to rely only on its learned knowledge. The playing performance of the agent was evaluated by playing 500 evaluation games against an opponent upon initialization (0 training games) and after every 500 training games; 500 evaluation games were found to be sufficient for stable performance estimates at each evaluation session. The evaluation opponent consisted of a random opponent, such that all moves for this player were randomly selected. During both training and evaluation games, $\bigcirc$ always played the first move of the game.

Neural Network Settings

The agent was represented by a 3-layer neural network with 11 input nodes (corresponding to the 11×1 state vector), 40 hidden nodes, and 1 output node. Bias nodes were used on the input and hidden layers, which had a constant input of 1. Nodes in the hidden layer used the hyperbolic tangent transfer function and a linear transfer function was used at the output node. The weights of the network were initialized using the method by [9], previously described. The learning rates α for each layer were set individually, with $\alpha = 0.01$ for the input-hidden layer weights and $\alpha = 0.007$ for the hidden-output layer weights. Network weights were updated using an iterative method such that they were updated after every play.

TD(λ) Settings

The temporal discount factor λ was set to 0.7, similar to that found to be successful in other implementations of reinforcement learning [17, 20]. During training, the agent pursued an exploitative action selection policy 90% of the time ($\varepsilon = 0.9$), and thus pursued what it perceived to be suboptimal actions 10% of the time.

Table 11.1. Settings for the base scenario of Tic-Tac-Toe.

General settings	
Number of games	2500
Starting player during training	○
Starting player during evaluation	○
Number of evaluation games	500
Frequency of evaluation games	500
Neural network settings	
Layers	3
Nodes per layer [i, h, j]	[11, 40, 1]
Transfer functions [h, j]	[tanh, linear]
Weight initialization method	as per [9]
Learning rate [α_{hi}, α_{jh}]	[0.0100, 0.0071]
Momentum coefficient (η)	0.0
Weight update method	iterative
TD(λ) settings	
Temporal discount factor (λ)	0.7
P[exploitation] (ε)	0.9
Environment settings	
Rewards [win, loss, draw]	[1, -1, 0]
Training methodology	self-play
Evaluation opponent	random player

The results of all scenarios are presented using performance curves obtained by playing the agent against the evaluation player over the course of training. Performance curves consisted of the % of wins, losses, and draws at each evaluation session. The curves that are shown are slightly smoothed such that the performance values at any point during training was the average of that particular value and the two adjacent values. This was done to more clearly show the average performance of the agent, which can fluctuate to some degree due to the random nature of the training process.

Results for Tic-Tac-Toe

Using only the basic settings outlined in the base scenario above, the agent was able to learn to play the game of Tic-Tac-Toe against a random opponent to an acceptable level with relatively little training (Fig. 11.3). The game was learned sufficiently within about 500 games to win almost 95% of the evaluation games. Although this particular example used a trivial opponent, it serves as a good basis to ensure that the implementation is working correctly.

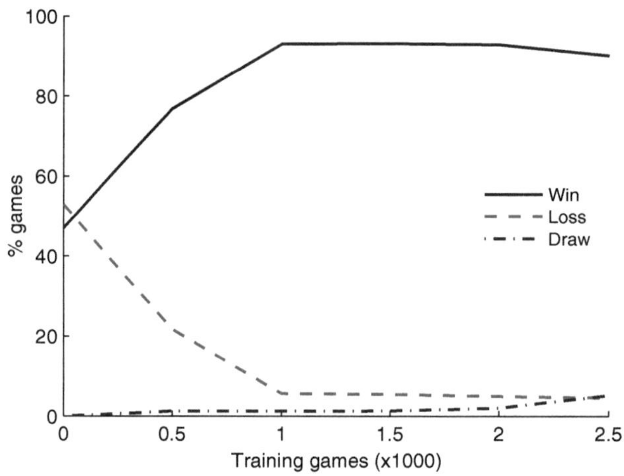

Fig. 11.3. Performance of the agent playing Tic-Tac-Toe in terms of % of wins, losses, and draws against an opponent that selects random actions.

Effect of Action Exploitation/Exploration

The effect of the agent's action selection policy was evaluated by changing the parameter ε. Recall that this parameter affects how often the agent exploits its learned knowledge versus how often it explores other actions. Four scenarios were tested with ε set to 0.1, 0.4, 0.7, and 1.0. These experiments used the same scenarios as described above such that ○ always selected the first move, self-play training was used, and the evaluation opponent selected random moves.

Figure (11.4) shows that the agent is able to learn the game relatively quickly and almost all evaluation games for ε = 0.1, 0.4, and 0.7. For ε = 1.0 however, the agent learns very little, which is due to it never exploring what it perceives to be suboptimal actions. An additional interesting note about these scenarios is the similarities between the cases with ε = 0.1, 0.4, and 0.7. It seems that in this type of environment with a simple game against a completely random opponent, the agent can learn very well regardless of how often it exploits its knowledge, so long as it can explore other options.

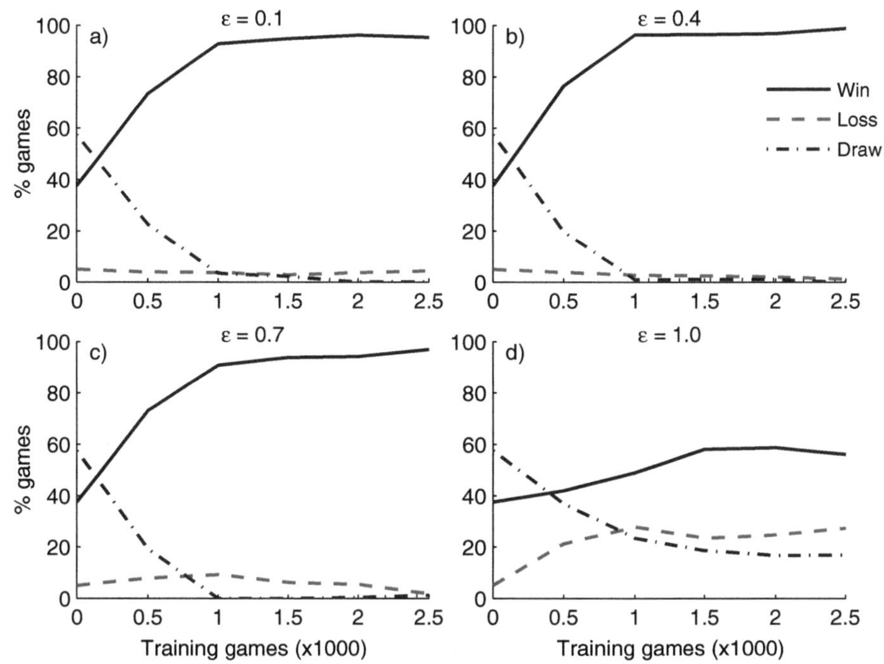

Fig. 11.4. Performance of the agent playing Tic-Tac-Toe in terms of % of wins, losses, and draws against the random player for different values of ε: **a)** $\varepsilon = 0.1$, **b)** $\varepsilon = 0.4$, **c)** $\varepsilon = 0.7$, **d)** $\varepsilon = 1.0$.

11.4.2 Chung Toi

Reinforcement learning was next applied to the game of Chung Toi, which is an extension to the game of Tic-Tac-Toe. Due to the similarities between the games, implementing Chung Toi required relatively minor modifications to the components developed for Tic-Tac-Toe. The following describes the implementation of Chung Toi, and all parameters used in the test scenarios are presented in Table (11.2). Parameters α, η, ε, and λ were also kept constant through training as with Tic-Tac-Toe.

The game of Chung Toi is played on the same 3×3 board that is used for Tic-Tac-Toe. Also similarly to Tic-Tac-Toe, the goal of the game is to get three of ones' pieces in a row along a horizontal, vertical, or diagonal of the board. One difference between Chung Toi and Tic-Tac-Toe however, is that in Chung Toi each player has only three pieces, with one player playing white pieces and the other playing red pieces. These pieces are octagonally-shaped and are labeled with arrows, and these arrows play an important role in the game: the orientation of the arrows restricts the allowable moves of the pieces.

The game of Chung Toi proceeds in two phases. The first phase consists of players placing their pieces on open board positions (alternating turns). Pieces can be placed such that their arrows are oriented either cardinally (parallel to the board axes) or diagonally. If, after each player has placed their 3 pieces on the board, neither player has three of their own pieces in a row, the second phase of play begins. In this phase, players are allowed to translate and/or rotate their pieces. Rotations can be applied to any piece on the board however, pieces may only be translated in directions which correspond to the direction in which their arrows are pointed. For example, if a piece is oriented diagonally, it may only move to an open board position that is diagonal to its current location. Additionally, if a combined move is selected that both translates and rotates a piece on move, the translation occurs before the rotation. Thus, because each player has only 3 pieces and because players can move pieces around the board, the game of Chung Toi is open-ended and there is no upper limit on the number of moves in each game (as in Tic-Tac-Toe).

Environment Settings

The environment for the game of Chung Toi was based on that of Tic-Tac-Toe and required small adaptations and extensions, the first of which was extending the state encoding scheme. As the orientation of the pieces plays an important role in Chung Toi, this information also needed to be included in the state encoding. The state vector for Chung Toi consisted of three parts: a 9×1 vector $\mathbf{p}$ indicating the presence (or absence) of the pieces; a 9×1 vector $\mathbf{r}$ indicating the orientation of the pieces; and a 2×1 vector $\mathbf{t}$ indicating which player was to play the next move. The vector $\mathbf{p}$ was similar to the position vector used in Tic-Tac-Toe where a 1 indicated the presence of a white piece (which represented the agent), a -1 indicated the presence of a red piece, and a 0 indicated an open board position. The vector $\mathbf{t}$ was the same as for as for Tic-Tac-Toe where $[1,\ 0]^T$ indicated white was to play the next move, and $[0,\ 1]^T$ indicated red was to play the next move. The vector $\mathbf{r}$ was used to encode the orientations of each piece where a 1 indicated that the corresponding piece in $\mathbf{p}$ was oriented cardinally, a -1 indicated that the corresponding piece in $\mathbf{p}$ was oriented diagonally, and a 0 indicated an open board position. The complete state vector $\mathbf{x}$ consisted of the concatenation of vectors $\mathbf{p}$, $\mathbf{r}$, and $\mathbf{t}$ into a single 20×1 state vector.

The other elements of the environment including the reward scheme and the performance evaluation method were quite similar to those of Tic-Tac-Toe. The agent (white) received a reward of 1 if it won a game and received a -1 if red won a game. In reality, the game of Chung Toi cannot end in a draw, as players keep translating and/or rotating pieces around the board until one player wins. In order to constrain the open-ended nature of the game however, the maximum number of moves allowed was 100, and games that exceed this number were considered draws, in which case the agent received a reward of 0.

Similar to Tic-Tac-Toe as well, the training scheme consisted of self-play with greedy end-game moves only allowed during training and not allowed during evaluation. The performance of the agent was evaluated against a *'smart'* random opponent this time however. This opponent largely played randomly selected moves, but whenever possible, it could take greedy end-game moves. Thus, this type of opponent is more challenging than the pure random opponent. The agent's performance was evaluated using 500 evaluation games that were played upon initialization and following every 1000 training games. For both training and evaluation games, the starting player of each game was the agent (white).

Neural Network Settings

The neural network for the game of Chung Toi used slightly different settings than that for Tic-Tac-Toe. This was done to explore the utility of changing such settings and these changes do not necessarily indicate that these settings are superior. The input layer of the neural network used 20 nodes, corresponding to the number of elements in the state vector. The learning rates α for each layer were scaled downward compared to Tic-Tac-Toe, resulting in learning rates of $\alpha = 0.001$ for the input-hidden layer weights and $\alpha = 0.0007$ for the hidden-output layer. The magnitude of the learning rates was determined by monitoring the magnitude of the weight changes during training. The transfer function used in the hidden layer was the modified hyperbolic tangent function suggested by [9]. Neural network weights were initialized using the method in [9]. These two features were previously discussed in Sect. (11.3.2.3).

Results for Chung Toi

Figure (11.5) shows the performance of the agent against the *'smart'* random opponent. In spite of the more challenging game and environment, the agent is able to learn the game and perform very well, winning nearly 90% of the evaluation games. A *'smart'* opponent was also tested with the game of Tic-Tac-Toe although the results were much worse; the agent won approximately 55% of the games and the remainder of games largely resulted in draws. The difference in performance between the two games with the same settings may therefore be due to the differences in the environment. In the case of Tic-Tac-Toe, the environment could be considered constrained such that there is a limited number of moves per game, whereas in Chung Toi, the games are open-ended.

Effect of Temporal Discount Factor λ

The effect of the temporal discount factor λ was explored by using values of λ = 0.0, 0.4, 0.7, and 1.0, and the results under these different conditions are shown in Fig. (11.6). These results seem to indicate that, from the values evaluated, $\lambda = 0.7$ allows the agent to learn and perform best whereas very little is learned with other values of λ. This agrees with other works using the temporal difference algorithm that have also suggested that $\lambda = 0.7$ enables efficient learning [17, 20]. Thus, this particular setting for λ may be more universally applicable to different domains

Table 11.2. Settings for the base scenario of Chung Toi.

General settings	
Number of games	5000
Starting player during training	White
Starting player during evaluation	White
Number of evaluation games	500
Frequency of evaluation games	1000
Neural network settings	
Layers	3
Nodes per layer [i, h, j]	[20, 40, 1]
Transfer functions [h, j]	[modified tanh, linear]
Weight initialization method	as per [9]
Learning rate [α_{hi}, α_{jh}]	[0.0010, 0.0007]
Momentum coefficient (η)	0.5
Weight update method	iterative
TD(λ) settings	
Temporal discount factor (λ)	0.7
P[exploitation] (ε)	0.9
Environment settings	
Rewards [win, loss, draw]	[1, -1, 0]
Training methodology	self-play
Evaluation opponent	*'smart'* random player

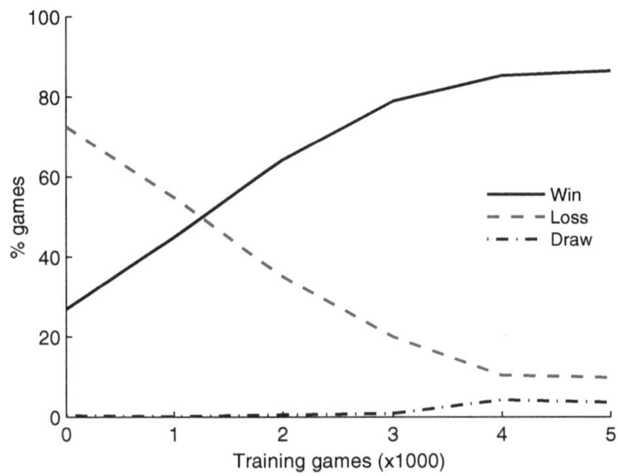

Fig. 11.5. Performance of the agent playing Chung Toi in terms of % of wins, losses, and draws against the *'smart'* random player.

and environments than other parameters. Additionally, in these scenarios evaluating λ, it was observed that the magnitude of the weight changes was proportional to the different values of λ. Intuitively, this makes sense based on the weight update equations (11.8) and (11.9). Considering this and the fact that the magnitude of weight updates are also largely based on α, it is possible that an optimal setting for α may be dependent on λ.

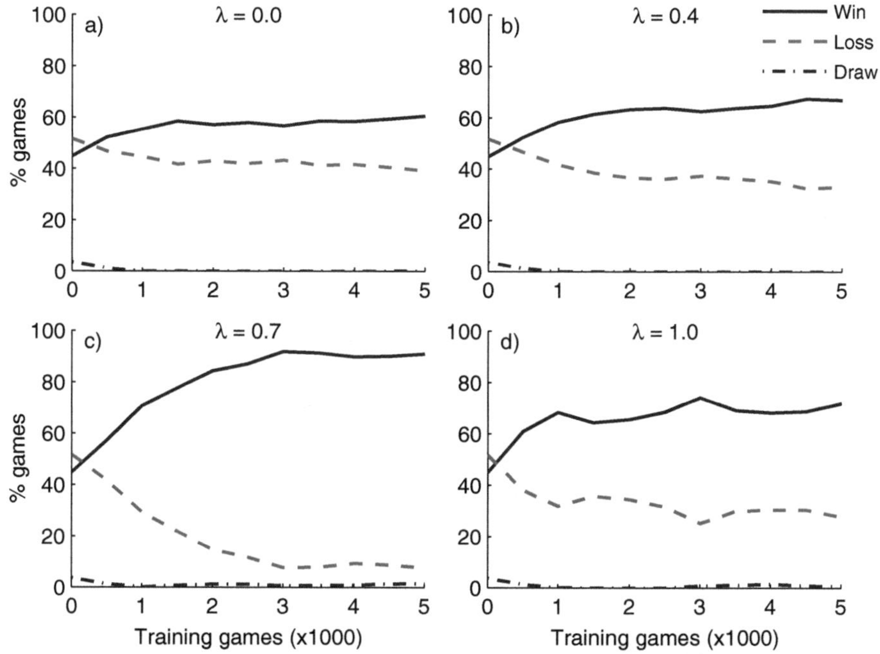

Fig. 11.6. Performance of the agent playing Chung Toi for different values of λ: **a)** $\lambda = 0.0$, **b)** $\lambda = 0.4$, **c)** $\lambda = 0.7$, **d)** $\lambda = 1.0$.

11.4.3 Applying/Extending to Other Games/Scenarios

From the above examples it should be clear that the implementation of reinforcement learning is based on a framework that entails two main entities: an agent and an environment. The agent, represented by a neural network and trained using TD(λ), is not specific to the application, but can be used in any reinforcement learning application where state value function approximation is required. Once this neural network and its learning algorithm are created, it can easily be used in many domains. The environment, on the other hand, is specific to the particular application and requires more extensive development. The major elements of the environment are listed below, along with possible adaptations that could be made when extending reinforcement learning to new domains.

State encoding: The state encoding scheme is central to any reinforcement learning implementation and it is from this that the agent (neural network) must learn. Although raw board encoding schemes were used in the example implementations in this work, the importance of a quality state encoding scheme should not be understated. Raw encoding schemes are useful for easy human interpretation however, they may not be optimal for enabling efficient learning by a neural network. Encoding schemes that include special, expert-crafted features that are relevant to the ultimate learning goal may be very useful to the neural network. A carefully constructed encoding scheme may also reduce the state space of environment, making accurate function approximation easier as well. For example, encoding Tic-Tac-Toe with features such as singlets, doublets, diversity, and crosspoints [8] results in a state vector that is invariant to the orientation of the board. In other words, symmetric board configurations (mirror images) would result in the same state vector, whereas with a raw encoding the state vectors would be unique for each board configuration and orientation. A reduced state space may then allow for the use of a smaller neural network that can be trained faster.

Action selection: The action selection procedure first determines the set of all possible actions $a_{ss'} \in A$ that can take the agent from the current state s to the next state s'. Each of these potential actions are then evaluated by passing the corresponding next-state vectors $\mathbf{x}_{s'}$ through the neural network. Finally, a particular action $a_{ss'}$ is selected based on some action selection policy.

State evaluation: The state of the domain or scenario must be continually assessed in order to determine when an episode terminates. In the games of Tic-Tac-Toe and Chung Toi, the boards were assessed after every action to determine if there were three of the same piece in a row.

Reward distribution: Rewards must be distributed based on the state of the environment. While the development of this component is not as involved as the those listed above, it is very important to the reinforcement learning framework. In the games of Tic-Tac-Toe and Chung Toi, rewards were only distributed at the terminal time step. However, there are many domains in which rewards, or some form of feedback, can be provided *during* an episode in addition to at the end of an episode. The equations for updating the weights of the neural network do not change; the reward term r in Eqs. (11.8) and (11.9) can simply be assigned a non-zero value at intermediate time steps $t = 1, 2, ..., T-1$.

Agent evaluation: In order to determine if the agent is learning, its knowledge must be tested in some manner either during or following the agent-environment interaction process. With respect to board games, and as was used in Tic-Tac-Toe and Chung Toi, opponents can be developed with different skill levels. Agent evaluation may also be performed using other methods, including other computer programs or using human experts.

Table 11.3. Variables used in text.

Variable	Description
t	Current time step
T	Terminal time step
S	Set of all states
A	Set of all actions
R	Set of all rewards
$s^{(t)}$ and s	State s at time t
s'	State s at time $t+1$
$a^{(t)}$	Action a pursued at time t
$r^{(t)}$	Reward r received at time t
$\mathcal{P}^a_{ss'}$	Probability of transitioning from state s to state s' when pursuing action a
$\mathcal{R}$	Cumulative reward received
$\mathcal{R}^a_{ss'}$	Probability of receiving a reward when transitioning from state s to state s' while pursuing action a
$V^*(s)$	True state value of state s
$V(s)$	(Perceived) state value of state s
π	Action policy
π^*	Optimal action policy
γ	Next-state discount factor
$\mathbf{x}$	State vector
$V()$	Value of state $\mathbf{x}$
i	Input layer to neural network
h	Hidden layer of neural network
j	Output layer of neural network
w_{hi}	Weight to a node in layer h, from a node in layer i
w_{jh}	Weight to a node in layer j, from a node in layer h
Δw_{hi}	Change to weight w_{hi}
Δw_{jh}	Change to weight w_{jh}
b_i	Bias node in the input layer i
b_h	Bias node in the hidden layer j
α	Learning rate parameter
E	Error term
δ_h	Local gradient at a node in layer h
y_i	Output of node in layer i; also the input to a node in layer h
y_h	Output of node in layer h; also the input to a node in layer j
y_j	Output of node in layer j
y^*_j	Target output of node in layer j
v_h	Induced local field at a node in layer h
v_j	Induced local field at a node in layer j
$f(\cdot)$	Transfer function at a node in neural network
$f'(\cdot)$	Derivative of transfer function at a node in neural network
λ	Temporal discount factor
η	Momentum parameter
g_{jh}	Summation of temporal gradient for weights connecting nodes in layer h to nodes in layer j
g_{hi}	Summation of temporal gradient for weights connecting nodes in layer i to nodes in layer h

Table 11.3. *(continued).*

Variable	Description
σ_w	Standard deviation of distribution for sampling weights for initialization
m	Number of weights leading into a node
n_l	Number of layers previous to the output layer of the neural network
n_p	Number of training patterns
ε	Proportion of actions for which knowledge is exploited

The implementations presented in this work were based on the basic TD(λ) algorithm with no significant modifications. There are many settings, parameters, and extensions to this method that could be modified or added, and these changes have the potential to improve learning both in terms of efficiency and performance. It is also very likely that there are interactions between multiple settings, thus complicating the behavior of the algorithm. Parameter settings in reinforcement learning applications often seem to be domain-specific and concrete rules for specifying parameters remain to be developed.

One easily implementable extension to the basic TD(λ) algorithm is to use parameters that change dynamically over the course of training, either by simple annealing or based on the learning of the agent. One example of this could be to integrate the agent's performance into the learning process. The rationale for doing this could be that parameters of the TD(λ) algorithm, or of the neural network, may be better set based on the performance and knowledge of the agent, and thus these settings and parameters would dynamically change during the learning process. As a simple example, ε, which dictates how often the agent exploits its knowledge versus how often it explores other actions, could be set equal to the agent's performance. When the agent's performance is high, indicating that it has gained substantial knowledge, ε would similarly be high, and the agent would rely on its learned knowledge more frequently and only explore suboptimal actions occasionally. Conversely, when the agent's performance is low, indicating it has little knowledge, ε would also be low, and the agent would rely less on its knowledge and instead explore its action space more frequently with the intention of increasing its knowledge.

11.5 Summary

In summary, reinforcement learning is a powerful learning method that is ideally suited for situations which can be posed as sequential decision making problems. The use of a neural network allows for the learning of behavioral policies in environments where the state space is large, thus prohibiting the use of conventional methods such as dynamic programming. This learning method also does not rely on state transition probabilities to be known, which are often required for more explicit sequential state models such as Hidden Markov Models. The use of the temporal difference algorithm enables the learning of associations between an agent's actions and rewards received despite the fact that rewards may be received at different points in time. The temporal difference algorithm is based on solving the temporal credit-assignment problem, which extends the credit-assignment problem solved with the back-propagation algorithm.

This work aimed to provide an introduction to the implementation of reinforcement learning using a neural network. The work is not comprehensive however, and there are many extensions and possible modifications that may enable improved learning. There are also other learning algorithms, beyond TD(λ), that can be used for reinforcement learning. Some of these algorithms include Q-learning, Sarsa, and actor-critic methods [15], all of which exploit slightly different learning strategies.

The games of Tic-Tac-Toe and Chung Toi were used to show just how reinforcement learning could be applied to a sequential decision making task. These implementations relied on the basic form of the TD(λ) algorithm with settings that have proved to be useful in other applications with neural networks. Even with this simple implementation the neural network was able to perform well in these different domains.

The use of a neural network for learning the state values for reinforcement learning may be considered a trick in itself. In addition, the Efficient BackProp techniques for training neural networks, based on the approaches of LeCun et al. [7, 9], may also be advantageous. These techniques concern the network architecture, transfer functions, weight initialization method, setting of the learning rates, as well as the use of batch (epoch) training, and these methods may allow for more efficient training in some applications. Additionally, the setting of parameters related to the temporal difference algorithm, such as the temporal discount factor λ and the action exploitation parameter ε, can also affect the training efficiency, and there are likely interactions between neural network and temporal difference algorithm parameters as well.

Future work may extend and explore the utility of the techniques discussed in this work. One potential approach that could lead to more efficient training is an epoch-based training scheme which a sliding memory stores a large number of the recently observed states (e.g., 5000), and then epoch training randomly selects a batch of states from this memory. This approach has similarities to that of database games [21] however, the use of this modified epoch training method may enable more efficient training.

Acknowledgements. This work was partially supported by the Natural Sciences and Engineering Research Council of Canada.

References

1. Bakker, B., Schmidhuber, J.: Hierarchical reinforcement learning based on subgoal discovery and subpolicy specialization. In: Groen, F., Amato, N., Bonarini, A., Yoshida, E., Kröse, B. (eds.) Proc. of the 8th Conf. on Intell., Amsterdam, The Netherlands, pp. 438–445 (2004)

2. Binkley, K.J., Seehart, K., Hagiwara, M.: A study of artificial neural network architectures for Othello evaluation functions. Trans. Jpn. Soc. Artif. Intell. 22(5), 461–471 (2007)
3. Doya, K.: Reinforcement learning in continuous time and space. Neural Comput. 12(1), 219–245 (1999)
4. Embrechts, M.J., Hargis, B.J., Linton, J.D.: An augmented efficient backpropagation training strategy for deep autoassociative neural networks. In: Proc. of the 15th European Symposium on Artificial Neural Networks (ESANN), Bruges, Belgium, April 28-30, pp. 141–146 (2010)
5. Gatti, C.J., Linton, J.D., Embrechts, M.J.: A brief tutorial on reinforcement learning: The game of Chung Toi. In: Proc. of the 19th European Symposium on Artificial Neural Networks (ESANN), Bruges, Belgium, April 27-29 (2011)
6. Ghory, I.: Reinforcement Learning in Board Games. Technical Report CSTR-04-004, Department of Computer Science. University of Bristol (2004)
7. Haykin, S.: Neural Networks and Learning Machines, 3rd edn. Prentice-Hall, New York (2008)
8. Konen, W., Bartz–Beielstein, T.: Reinforcement Learning: Insights from Interesting Failures in Parameter Selection. In: Rudolph, G., Jansen, T., Lucas, S., Poloni, C., Beume, N. (eds.) PPSN 2008. LNCS, vol. 5199, pp. 478–487. Springer, Heidelberg (2008)
9. LeCun, Y.A., Bottou, L., Orr, G.B., Müller, K.-R.: Efficient Backprop. In: Orr, G.B., Müller, K.-R. (eds.) NIPS-WS 1996. LNCS, vol. 1524, p. 9. Springer, Heidelberg (1998)
10. Mannen, H., Wiering, M.: Learning to play chess using TD(λ)–learning with database games. In: Benelearn 2004: Proc. of the 13th Belgian-Dutch Conference on Machine Learning, pp. 72–79 (2004)
11. Moore, A.: Efficient memory-based learning for robot control. PhD Thesis. University of Cambridge (1990)
12. Patist, J.P., Wiering, M.: Learning to play draughts using temporal difference learning with neural networks and databases. In: Benelearn 2004: Proc. of the 13th Belgian-Dutch Conference on Machine Learning, pp. 87–94 (2004)
13. Rumelhart, D.E., Hinton, G.E., Williams, R.J.: Learning internal representations by error propagation. In: Rumelhart, D.E., McClelland, J.L. (eds.) Parallel Distributed Processing, vol. 1. MIT Press, Cambridge (1986)
14. Sutton, R.S.: Learning to predict by the method of temporal difference. Mach. Learn. 3, 9–44 (1988)
15. Sutton, R.S., Barto, A.G.: Reinforcement Learning: An Introduction. MIT Press, Cambridge (1988)
16. Tesauro, G.: Neurogammon: A neural network backgammon program. In: Proc. of the International Joint Conference on Neural Networks., vol. 3, pp. 33–40 (1990)
17. Tesauro, G.: Practical issues in temporal difference learning. Mach. Learn. 8, 257–277 (1992)
18. Tesauro, G.: Temporal difference learning and TD-Gammon. Communications of the ACM 8(3), 58–68 (1995)
19. Werbos, P.: Beyond regression: New tools for prediction and analysis in the behavioral sciences. Ph.D. Dissertation. Harvard University, Cambridge, MA (1974)
20. Wiering, M.A.: TD learning of game evaluation functions with hierarchical neural architectures. Master's Thesis. University of Amsterdam (1995)
21. Wiering, M.A.: Self-play and using an expert to learn to play backgammon with temporal different learning. J. Intell. Learn. Syst. & Appl. 2, 57–68 (2010)

Chapter 12
Sliding Empirical Mode Decomposition-Brain Status Data Analysis and Modeling

A. Zeiler, R. Faltermeier, A.M. Tomé, I.R. Keck, C. Puntonet,
A. Brawanski, and E.W. Lang

Abstract. Biomedical signals are in general non-linear and non-stationary. *Empirical Mode Decomposition* in conjunction with *Hilbert-Huang Transform* provides a fully adaptive and data-driven technique to extract *Intrinsic Mode Functions* (IMFs). The latter represent a complete set of locally orthogonal basis functions to represent non-linear and non-stationary time series. Large scale biomedical time series necessitate an online analysis which is presented in this contribution. It shortly reviews the technique of EMD and related algorithms, discusses the newly proposed *SEMD* algorithm and presents some applications to biomedical time series recorded during neuromonitoring.

12.1 Introduction

12.1.1 Empirical Mode Decomposition

Recently an empirical nonlinear analysis tool for complex, non-stationary time series has been pioneered by N. E. Huang et al. [1]. It is commonly referred to as *Empirical Mode Decomposition* (EMD), and if combined with Hilbert spectral analysis

A. Zeiler · I.R. Keck · E.W. Lang
CIML Group, Biophysics, University of Regensburg, D-93040 Regensburg, Germany
e-mail: elmar.lang@biologie.uni-regensburg.de

R. Faltermeier · A. Brawanski
Neurosurgery, University Regensburg Medical Center, D-93040 Regensburg, Germany
e-mail: rupert.faltermeier@klinik.uni-regensburg.de

A.M. Tomé
DETI/IEETA, Universidade de Aveiro, 3810-193 Aveiro, Portugal
e-mail: ana@ieeta.pt

C. Puntonet
DATC, ETSIIT, Universidad de Granada, 38170 Granada, Spain
e-mail: carlos@atc.ugr.es

P. Georgieva et al. (Eds.): Adv. in Intelligent Signal Processing and Data Mining, SCI 410, pp. 311–349.
springerlink.com

it is called *Hilbert - Huang Transform* (HHT). It adaptively and locally decomposes any non-stationary time series in a sum of *Intrinsic Mode Functions* (IMF) which represent zero-mean amplitude and frequency modulated oscillatory components. The EMD represents a fully data-driven, unsupervised signal decomposition technique and does not need any *a priori* defined basis system. The empirical nature of EMD offers the advantage over other empirical signal decomposition techniques like *exploratory matrix factorization* (EMF) of not being constrained by conditions which often only apply approximately in case of biomedical signals. Especially with the latter one often has only a rough idea about the underlying modes and frequently their number is unknown. Furthermore, large scale biomedical time series recorded over days necessitate an on-line analysis while EMD can analyze data only globally so far. This contribution will review the technique of empirical mode decomposition and its recent extension, called Ensemble Empirical Mode Decomposition (EEMD). The latter represents a noise assisted EMD which alleviates certain problems which inevitably accompany any EMD application. This contribution also proposes a new on-line EMD variant, called *weighted Sliding EMD* (wSEMD), and discusses its application to biomedical time series recorded during neuromonitoring of brain injured patients in intensive care units.

12.1.2 Neuromonitoring

The continuous and simultaneous monitoring of different physiological brain status parameters like Intra-Cranial Pressure (ICP), brain Tissue Partial Oxygen pressure (TiPO2) and the cerebral Blood Flow Velocity (BFV) in distinct cerebral vessels, is an increasingly employed technique during *neuromonitoring* in neurosurgical intensive care units [2], [3]. Such complex monitoring practice is used for early detection of neurological worsening of patients suffering from a brain injury. A major therapeutic goal is maintaining a sufficient supply of cerebral matter with blood and oxygen to avoid ischemic insults. But the human brain exhibits several control mechanisms, like the cerebral autoregulation or the cerebrospinal fluid circuit, which guaranty sustained cerebral perfusion. A severe brain trauma may suspend such control systems temporarily, partially or even globally, hence alter the response of cerebral perfusion to a neurosurgical treatment. Failure of the autoregulation often turns out to be fatal.

Knowledge of cerebral regulatory mechanisms in advance of any frightening event is thus mandatory to adapt patient management to an acute change in the physiological status of the patient. But the interpretation of recorded brain status data is difficult. Cerebral mechanisms lead to complex interdependencies between individual compartments depending on their actual functionality. Recently several mathematical methods have been developed based on correlation analysis simplifying the interpretation of brain status data [4]. However, up till now there is no widely accepted comprehensive mathematical model of the cerebral physiology that reliably interprets the interrelation of the different data under different states of cerebral regulatory mechanisms, although some serious attempts have been undertaken [5], [6],

[7], [8]. Based on the characteristics of the data, a method for their analysis should be used which can reliably separate different oscillatory components belonging to different physiological mechanisms and simultaneously identify noise components.

12.1.3 Dynamic Cerebral Autoregulation

The assessment of the *dynamic cerebral autoregulation* (DCA) provides a major challenge in neuromonitoring in intensive care units. DCA can be described as the ability of the cerebral microvasculature to modulate cerebral perfusion in response to fast blood pressure fluctuations or an altered oxygen demand. Transcranial doppler ultrasound (TCD) measurements of the cerebral blood flow velocity (BFV) enables monitoring DCA during externally induced blood pressure variations (Vasalva maneuver). Usually the interdependencies of ABP and BFV are analyzed by means of coherence and transfer function analysis which are based on linear methods. But nonlinear interactions between non-stationary signals cannot be quantified based on theories of stationary and linearly superimposed signals. They cannot reliably assess nonlinear interactions in physiological systems.

A possible solution to this dilemma is discussed in [9]. A new technique called *multi-modal pressure flow* (MMPF) method is proposed, which utilizes the Hilbert-Huang transform to quantify interactions between non-stationary cerebral blood flow velocity (BFV) and blood pressure (BP) in specific frequency ranges as for example the respiratory component of the blood pressure (0.1-0.4 Hz). Using EEMD of both signals, BP and BFV are decomposed into their IMFs. Afterwards the IMFs containing the respiratory signal are identified in both signals by analysis of their instantaneous frequency. Subsequently the phase shift of the instantaneous phases is used to quantify the offset of both signals. Using this technique, the authors showed that the *fast dynamic* cerebral autoregulation can be characterized by specific phase delays between the decomposed blood pressure and blood flow velocity oscillations. They demonstrated that the phase shifts are significantly reduced in hypertensive, diabetics and stroke subjects with impaired cerebral autoregulation. Additionally, the new technique can reliably assess cerebral autoregulation using both induced blood pressure/blood flow velocity oscillations during clinical tests and spontaneous blood pressure/blood flow velocity fluctuations during resting conditions.

12.1.4 Modeling of Cerebral Circulation and Oxygen Supply

Recently [7] we presented a mathematical seven compartment model of cerebral circulation and oxygen supply including an autoregulation mechanism and a pressure dependent production and absorption of cerebrospinal fluid (CSF). The model is designed to interpret the slow temporal dynamics of neurophysiological parameters, recorded at a neurosurgical intensive care unit (ICU), and their interrelationships. In particular, we focused on the arterial blood pressure (ABP), the intracranial pressure (ICP), the partial oxygen pressure in brain tissue (TiPO2) and the relative oxygen

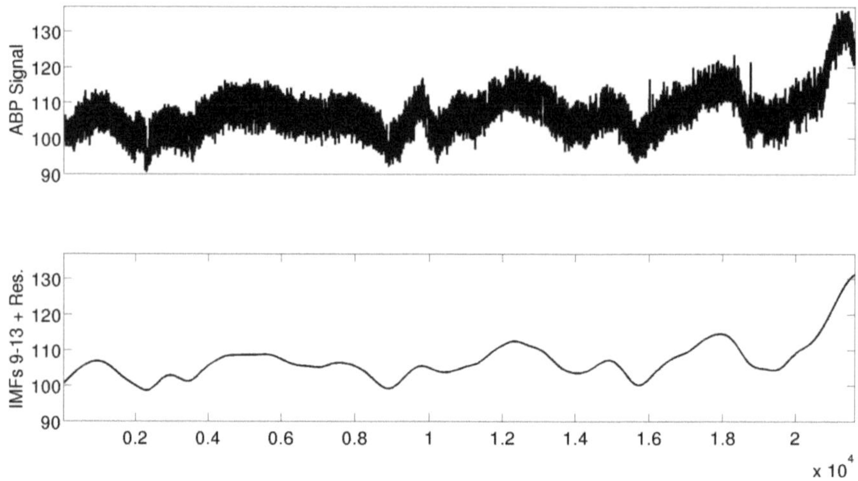

Fig. 12.1. The upper trace shows the original time series of an ABP. Below the low frequency part of the signal is depicted which consists of the sum of IMFs 9-13 and the residuum.

saturation of hemoglobin in mixed arterial/venous cerebral blood (SatHbO2). To minimize the number of model parameters, we focused on a simple model which resembles the work of Ursino and Lodi [5], [10], which only incorporates the basic processes of cerebral circulation and oxygenation. The circulatory part of the model is built up of seven compartments including arteries, capillaries, veins, brain tissue, cerebrospinal fluid, the sagittal sinus and an artificial compartment for the simulation of brain swelling. Currently we combine an oxygen diffusion model based on the theory of the Krogh cylinder with our hydrodynamic model. The combination of the hydrodynamic model with an oxygen diffusion model enables us to simulate the behavior of the partial oxygen pressure in brain tissue (TiPO2), a parameter which can be monitored continuously at an ICU. Using this hybrid model approach, we can reproduce the experimentally well known autoregulation curve [11] which describes a constant cortical blood flow over a wide range of ABP if the autoregulation mechanism is intact. In case of an impaired autoregulation, a linear relationship between pressure and flow emerges. Furthermore, similar interdependencies could be shown for ABP and TiPO2. Although the simulations show a clear nonlinear behavior in case of an impaired autoregulation, in the physiologically meaningful range of the arterial blood pressure a nearly linear correlation exists. Using the built-in artificial compartment for the simulation of brain swelling, we can show that the model reproduces exactly the behavior of experimentally determined pressure-volume curves [12], [13]. Such measurements are used to determine the cerebral compliance which describes alterations in ICP due to volume changes. A disturbed compliance may lead to an exponentially rising ICP for little variations in cerebral volume which in turn has a drastic impact on cerebral perfusion. Also recently

[14], [8], we showed that different combinations of impaired or well functioning autoregulation and compliance lead to significantly different correlations of the above mentioned cerebral status parameters. From a medical point of view this can be used to determine which cerebral regulatory mechanisms are disturbed if we can identify simulated correlations with correlations appearing in time series data recorded from impaired subjects. Despite some promising results the rather large signal-to-noise ratio (SNR) provides a major challenge for the identifications of these correlations. Additionally, the recorded data are composed of different dynamical modes which need to be separated reliably and un-distortedly. A favorable aspect lies in the time scale separation of the various superimposed and intermingled modes, though. The non-stationary part of the ABP signal needs special attention, as it serves as input for the model. Hence, EMD can be applied to separate the non-stationary components of ABP, ICP and TiPO2 from the recorded signals. As an example, Fig. 12.1 shows the extracted non-stationary part of an ABP signal which can be used as input to corresponding model calculations. The latter show promising preliminary results regarding the functionality of the proposed physiological regulatory mechanisms. These modeling efforts will thus be combined in future work with an EMD analysis to extract physiologically meaningful modes which can be directly compared to corresponding model predictions. Both aspects, signal processing and modeling are needed to extract the full biomedical information buried in brain status data recorded during neuromonitoring.

12.2 Empirical Mode Decomposition

The EMD method was developed from the assumption that any non-stationary and non-linear time series consists of different simple intrinsic modes of oscillation. The essence of the method is to empirically identify these intrinsic oscillatory modes by their characteristic time scales in the data, and then decompose the data accordingly. Through a process called *sifting*, most of the *riding waves*, i.e. oscillations with no zero crossing between extremes, can be eliminated. The EMD algorithm thus considers signal oscillations at a very local level and separates the data into *locally* non-overlapping time scale components. It breaks down a signal $x(t)$ into its component IMFs obeying two properties:

1. An IMF has only one extremum between zero crossings, i.e. the number of local minima and maxima differs at most by one.
2. An IMF has a mean value of zero.

Note that the second condition implies that an IMF is *stationary* which simplifies its analysis. But an IMF may have amplitude modulation and also changing frequency.

12.2.1 The Standard EMD Algorithm

The sifting process can be summarized in the following algorithm. Decompose a time series $x(t)$ into IMFs $x_n(t)$ and a residuum $r(t)$ such that the signal can be represented as

$$x(t) = \sum_n x_n(t) + r(t) \tag{12.1}$$

Note that a residuum is a monotonous, i. e. non-oscillatory signal which does not fulfill the conditions for an IMF. Sifting then means the following steps:

- *Step 0*: Initialize: $n := 1, r_0(t) = x(t)$
- *Step 1*: Extract the n-th IMF as follows:

  a) Set $h_0(t) := r_{n-1}(t)$ and $k := 1$
  b) Identify all local maxima and minima of $h_{k-1}(t)$
  c) Construct, by cubic splines interpolation, for $h_{k-1}(t)$ the envelope $U_{k-1}(t)$ defined by the maxima, and the envelope $L_{k-1}(t)$ defined by the minima
  d) Determine the mean $m_{k-1}(t) = \frac{1}{2}(U_{k-1}(t) - L_{k-1}(t))$ of both envelopes of $h_{k-1}(t)$. This running mean is called the low frequency local trend. The corresponding high-frequency local detail is determined via a process called *sifting*.
  e) Form the $(k) - th$ component $h_k(t) := h_{k-1}(t) - m_{k-1}(t)$
     1) if $h_k(t)$ is not in accord with all IMF criteria as given above, increase $k \to k+1$ and repeat the sifting process starting at step $b)$
     2) if $h_k(t)$ satisfies the IMF criteria, then set $x_n(t) := h_k(t)$ and $r_n(t) := r_{n-1}(t) - x_n(t)$

- *Step 3*: If $r_n(t)$ represents a residuum, stop the sifting process; if not, increase $n \to n+1$ and start at Step 1 again.

The sifting process separates the non-stationary time series data into locally non-overlapping intrinsic mode functions (IMFs) which are *locally orthogonal*. Global orthogonality is not guaranteed as neighboring IMFs might have identical frequencies at different time points (typically in $< 1\%$ of the cases).

Fig. 12.2 provides a simple example of an EMD decomposition of a toy signal composed from a superposition of a chirp signal ($|a(t)| = 1[a.u.], 25[Hz] \leq \omega/(2\pi) \leq 100[Hz]$), a sinusoid ($|a(t)| = 1, \omega/(2\pi) = 10[Hz]$) and a straight line given by $x(t) = 0.5 \cdot t$. Fig. 12.2 also presents the related first four IMFs with highest energy extracted by the algorithm. Note that the number of IMFs extracted depends on the stopping criterion employed. In the given example seven IMFs were extracted, the last IMF representing the monotonous trend.

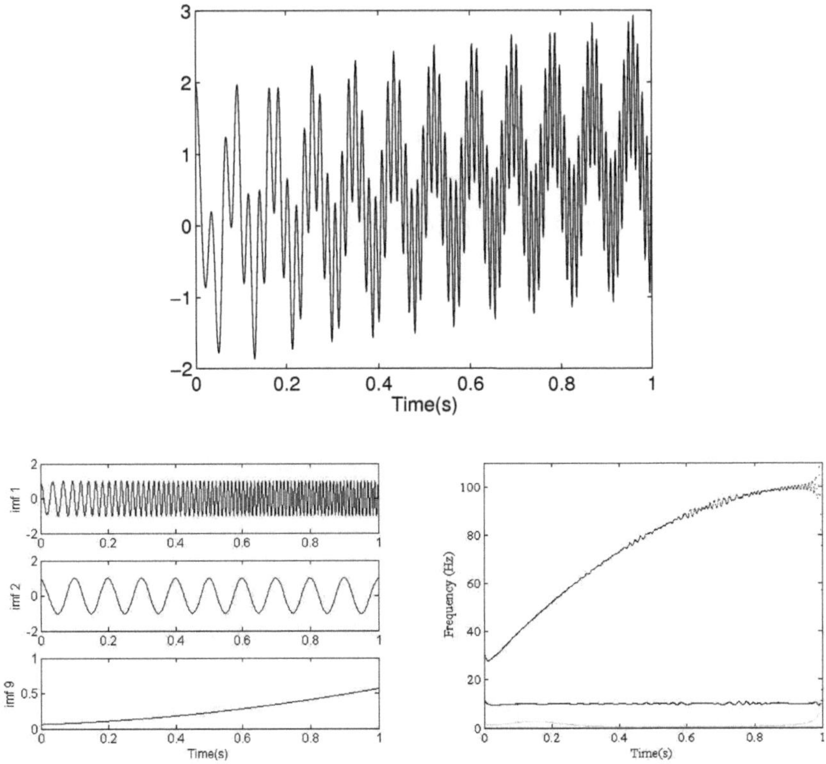

Fig. 12.2. *Top:* The toy signal consisting of a superposition of a chirp, a sinusoid and a monotonous trend as detailed in the text. *Bottom Left:* The three IMFs corresponding to the toy signal components as obtained with plain EMD. *Bottom Right:* Hilbert - Huang spectrum of the first two IMFs of the toy signal.

12.2.2 The Hilbert - Huang Transform

After having extracted all IMFs, they can be analyzed further by applying the Hilbert Transform (HT) or processing them in any other suitable way [15], [16]. The combination of EMD decomposition of a signal into its IMFs plus residue and a subsequent Hilbert spectral analysis to extract instantaneous frequencies and phases is called *Hilbert Huang Transform* (HHT) [17], [18]. The Hilbert transform calculates the conjugate pair of $x_n(t)$ via

$$H\{x_n(t)\} = \frac{1}{\pi} P \left\{ \int_{-\infty}^{\infty} \frac{x_n(\tau)}{(t-\tau)} d\tau \right\} \tag{12.2}$$

where P indicates the *Cauchy principal value*. This way an analytical signal $z_n(t)$ can be defined via

$$z_n(t) = x_n(t) + iH\{x_n(t)\} = a_n(t)\exp(i\theta_n(t)) \tag{12.3}$$

with amplitude $a_n(t)$ and instantaneous phase $\theta_n(t)$ given by

$$a_n(t) = \sqrt{x_n^2(t) + H\{x_n(t)\}^2} \tag{12.4}$$

$$\theta_n(t) = \arctan\left(\frac{H\{x_n(t)\}}{x_n(t)}\right) \tag{12.5}$$

Each IMF can now be expressed as

$$x_n(t) = Re\left[a_n(t)\exp\left(i\int \omega_n(t)dt\right)\right] \tag{12.6}$$

The signal can then be expressed as

$$x(t) = Re\left\{\sum_{n=1}^{N} a_n(t)\exp\left(i\int \omega_n(t)dt\right)\right\} + r(t) \tag{12.7}$$

An IMF expansion thus provides a *generalized* Fourier expansion. Note that because

$$\theta_n(t) = arg(z_n(t)) = \int_{-\infty}^{t} \omega_n(\tau)d\tau = \int_0^t \omega_n(\tau)d\tau + \theta_n(0) \tag{12.8}$$

an instantaneous frequency $\nu_n(t)$ can be obtained as

$$\nu_n(t) = \frac{\omega_n(t)}{2\pi} = \frac{1}{2\pi}\frac{d}{dt}\theta_n(t) \tag{12.9}$$

Fig. 12.2 also shows the Hilbert - Huang spectrum of the toy signal considered.

12.2.3 Ensemble Empirical Mode Decomposition

Plain EMD faces problems resulting from boundary effects which lead to mode splitting and over-sifting. To fight these disturbances, a noise assisted variant, called *Ensemble Empirical Mode Decomposition* (EEMD) has been proposed by [19], [20] which assists and considerably improves the sifting process. In practice EEMD works as follows:

- Add white noise to the data set
- Decompose the noisy data into IMFs
- Iterate these steps and at each iteration add white noise
- Calculate an ensemble average of the respective IMFs to yield the final result.

An illustrative example of the performance of EEMD vs EMD is given in Fig. 12.3. Two signals ($x_1(t) = 0.1 \cdot \sin(20t), x_2(t) = \sin(t)$) are superimposed whereby signal $x_1(t)$ is present only during certain time spans to simulate a situation which often happens with biomedical signals. Clearly, standard EMD shows strong mode-mixing in this case while EEMD (see Fig. 12.3), using an ensemble of 15 different noisy signals, copes quite well with this complicated signal.

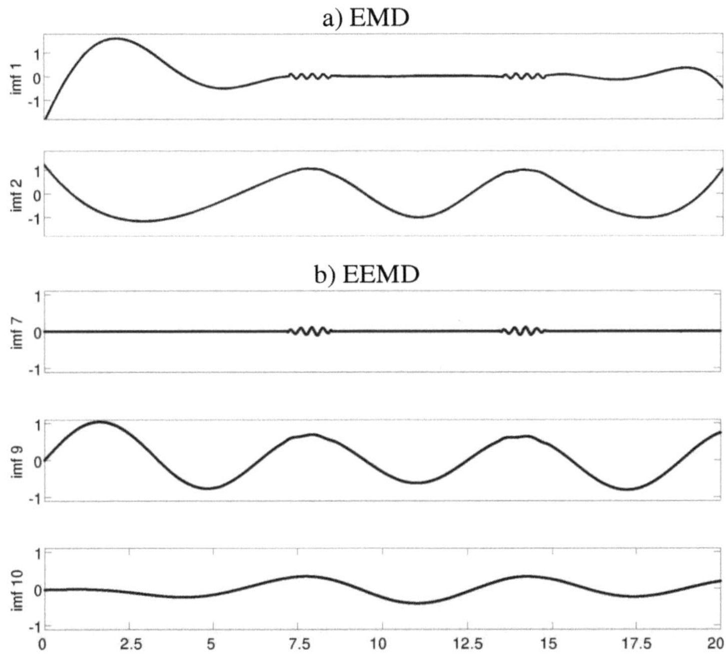

Fig. 12.3. a) The first two IMFs obtained with an EMD analysis. Mode mixing is clearly visible due to the partial absence of mode $x_1(t)$. b) IMFs obtained with an EEMD analysis. The component signals underlying the original signal are extracted almost perfectly.

12.3 Sliding Empirical Mode Decomposition

Plain EMD is applied to the *full length* signal which in view of limited resources like computer memory also limits the length of the time series to be dealt with. This is an especially serious problem with biomedical time series data, which often are recorded over very long time spans. Biomedical data, for example acquired during neuromonitoring accumulate to huge amounts of data when monitoring patients in intensive care units (ICU) is extended over days with an appropriate (up to 1000 Hz) sampling rate. Analyzing such large data sets is hardly possible because of the computational load involved when using conventional EMD. Even more important, however, is the fact that data analysis has to wait until monitoring is finished. But an immediate *on-line* analysis of such time series data is of utmost importance in ICUs. As no proper EMD algorithm was available yet to achieve this goal, we recently proposed a robust and easy to implement version of EMD, called *Sliding Empirical Mode Decomposition* or SEMD [21]. This algorithm decomposes time series of arbitrary length into a residuum and a particular number of IMFs. In the subsequent paragraphs we will discuss the operating mode of SEMD and some characteristics of its IMFs and the residuum. We will focus on the impact of well known boundary problems of EMD on the reconstruction quality of SEMD. We will use toy data to analyze the reconstruction quality of the residuum an the IMFs for different parameter settings. We also present a new extension of SEMD, called weighted SEMD which accounts for unwanted boundary effects and tries to suppress them efficiently. An application of wSEMD to brain status data is presented also.

Note that SEMD combines ideas from *localEMD* as well as *onlineEMD* [22].

- *Local EMD* [22] pursues the idea to iterate the sifting process only in regions where the mean is still finite to finally meet the stopping criterion everywhere. Localization can be implemented via a weighting function $w(t)$ which is $w(t) = 1$ in regions where sifting is still necessary and decays to zero at the boundaries. This can be easily integrated into the EMD algorithm via

$$h_k(t) = h_{k-1}(t) - w_{k-1}(t)m_{k-1}(t) \tag{12.10}$$

  This procedure essentially improves the sifting process and tries to avoid *oversifting*.
- *On-line EMD*: Recently, a blockwise processing, called *on-line EMD*, has been proposed [22]. The method is still in its infancy and yields unsatisfactory results so far. It still needs to be developed to a robust and efficient *on-line* technique as the one proposed in this contribution.

12.3.1 The Principle of SEMD

In a first step, the recorded time series $x(t_n), n = 1, \ldots, N$ encompassing N samples, is split into M segments $s_m, m = 1, \ldots, M$, each encompassing $\delta_m = t_{n+\delta_m} - t_n$ samples, which can be analyzed with EMD. Simply adding up the IMFs extracted from the different segments would induce boundary artifacts, however. This is illustrated in a simple example in Fig. 12.4.

Thus segmenting a time series into *non-overlapping* windows for further analysis leads to strong boundary artifacts. The latter can be avoided when the segments, respectively windows $s_m, m = 1 \ldots M$, each of length δ_m overlap by $\delta_m - \tau$ samples implying a step size τ. If, additionally, the window size δ_m is a multiple of the step size τ, i.e. if

$$q_m = \frac{\delta_m}{\tau} \in \mathbb{N} \qquad \forall m = 1, \ldots, M \tag{12.11}$$

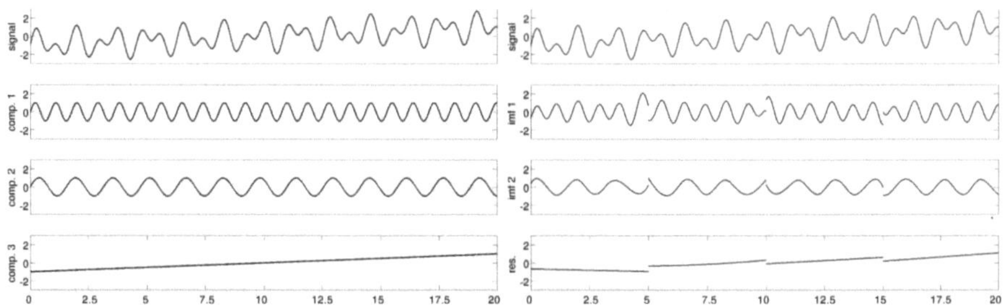

Fig. 12.4. *Left:* Toy signal (top trace) and underlying signal components: $\sin(7t)$ (second trace), $\sin(4t)$ (third trace) and the trend $0,1 \cdot t - 1$ (bottom trace). *Right:* EMD decomposition of the toy signal from Fig. 12.4. The time series has been segmented into 4 segments and decomposed with EMD. After joining the resulting IMFs together, boundary effects become clearly visible at $t = 5, 10$ and $t = 15$

holds, neighboring windows can be joined without having discontinuities or gaps at the boundaries. With this choice, every sample is represented q_m - times in overlapping windows corresponding to consecutive shifts of the time series by τ samples. Thus for each sample $x(t_n), n = 1, \ldots, N$ of the original time series, estimates are calculated within the different overlapping windows δ_m, at least after the first q_m iterations have been done. In a perfect world, all q_m estimates for a distinct sample $x(t_n)$ would be identical, but due to boundary artifacts of the EMD algorithm, they usually differ. Similar to EEMD, we take the mean value of all estimates of a distinct sample $x(t_n)$ as the resulting value of SEMD for this sample. If the conditions $\frac{\delta_m}{\tau} \in \mathbb{N}$ and $\frac{N}{\tau} \in \mathbb{N}$ hold with N the number of samples, the number M of overlapping windows s_m of lengths δ_m, that have to be calculated for the decomposition is given by:

$$M = \left(\frac{N - \delta_m}{\tau} \right) + 1 \tag{12.12}$$

A schematic illustration of the principle mode of operation of the SEMD algorithm is shown in Fig. 12.5 below. The time series in every segment s_m is decomposed by EMD into $J - 1$ IMFs $x_{m,j}(t)$ and a local residuum $r_m(t) \equiv x_{m,J}(t)$ according to

$$x_m(t) = \sum_{j=1}^{(J-1)} x_{m,j}(t) + r_m(t) \tag{12.13}$$

whereby the number of sifting steps is kept equal in all segments. Resulting IMFs are collected in a matrix with corresponding samples $x(t_n)$ forming the columns of the matrix with $q = \frac{\delta}{\tau}$ entries. Columns corresponding to the beginning or end of the time series are deficient, hence are omitted from further processing. This assures that all columns contain the same amount of information for estimating average IMF amplitudes at every time point t_n in each segment s_m. This finally yields for $n > \delta_m$ average IMFs according to (q - size of the ensemble)

$$\langle x_j(t_n) \rangle = \frac{1}{q} \sum_{m}^{m+q-1} x_{m,j}(t_n) \tag{12.14}$$

$$\langle r(t_n) \rangle = \frac{1}{q} \sum_{m}^{m+q-1} r_m(t_n) \tag{12.15}$$

$$m = \lfloor \frac{t_n - \delta}{\tau} \rfloor + 2 \tag{12.16}$$

Note that by the formulas 12.14, 12.15 and 12.16 the completeness of the decomposition process is warranted as for every sample $x(t_n)$ the mean value of q decompositions of conventional EMD, which in turn are complete, forms the result of the SEMD decomposition.

Problems arise if different numbers of IMFs results in different overlapping segments, or if the same intrinsic frequency appears in different IMFs in different overlapping segments. To avoid such complications, the number of sifting steps and IMFs is kept constant for all decompositions. Except for the stopping criterion, any EMD algorithm can be applied but EEMD has proven to yield best results mostly.

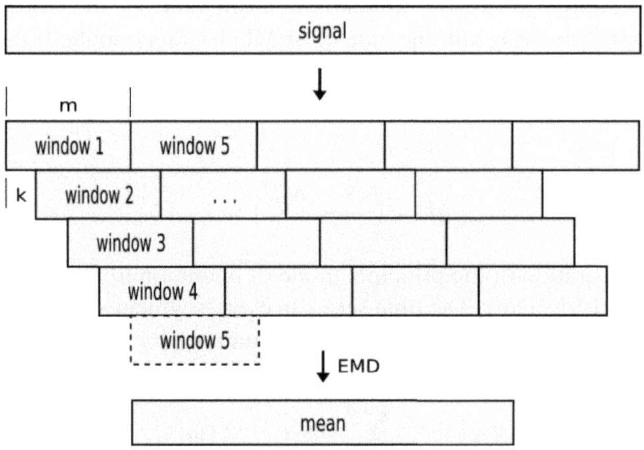

Fig. 12.5. Schema of the SEMD algorithm. The time series segments in the shifted windows are decomposed with EMD. IMFs and the residuum are determined finally by mode amplitudes which are averaged over corresponding samples in all windows. δ_m describes the window size and τ the step size.

12.3.2 *Properties of SEMD*

Contrary to global EMD, the local residua estimated with SEMD for every segment may turn into low frequency oscillations when joined together into an average global residuum. By choosing the segment size properly it may be controlled which local oscillations should appear as distinct IMFs and which should be absorbed as apparent local trends into the respective local residua. These *apparent* local trends, which combine to low frequency oscillations in the final average global residuum, may be down-sampled and subsequently analyzed with SEMD as well. This process can be repeated until finally a truly monotonous trend remains. Hence, this cascaded application of SEMD acts as a low frequency filter for long-term oscillations and trends as can be found, for example, in biomedical time series.

Similar to EEMD, also with SEMD an averaging over differently decomposed data sets is achieved. While, due to added noise, with EEMD a given sample is associated with different amplitudes, with SEMD the same amplitude is associated with different samples in different shifted segments. This latter behavior alleviates effects related with a non-unique data decomposition via EMD. Furthermore, artifacts resulting from end effects loose their impact via averaging. Finally, segmentation with proper window size and step size does not result in boundary artifacts after combination of the local IMFs. SEMD is furthermore similar to *local EMD* in that the stopping criterion needs to be valid only locally, i.e. within the window considered. This substantially reduces *over-sifting* and also reduces the number of necessary sifting steps until the stopping criterion is met locally. Finally note that while local IMFs fulfill all defining conditions, this is not necessarily true for the resulting average IMFs though the related deviations should always be small.

12.3.3 *Application of SEMD to Toy Data*

The following simulations use well-chosen toy examples for investigating the impact of the window size on the estimated residuum as well as on the reconstruction quality. The analysis was performed with toy data to be able to assess the results quantitatively. Note that the number of IMFs to be extracted is fixed to $\lfloor j \rfloor = \log_2(\delta_m)$ where δ_m designates the size of the m-th segment. This assures an identical number of extracted IMFs in every segment which is necessary for the construction of appropriate IMFs. Furthermore, the frequencies of the underlying component signals have been chosen sufficiently different to avoid problems with frequency resolution.

The toy signal $s(t) = \sum_i s^{(i)}(t)$ (see Fig. 12.6) consists of

- a sawtooth wave

$$s^{(1)}(t) = (1.5708)^{-1} \arcsin\left(\sin(699 \cdot t)\right),$$

- a sinusoid

$$s^{(2)}(t) = \sin(327 \cdot t),$$

- a cosine function with a time-dependent frequency

$$s^{(3)}(t) = cos\left(2 \cdot (t + 20)^2\right)$$

- a monotonous trend , which renders the superposition of the signals non-stationary

$$s^{(4)}(t) = 0.1 \cdot t - 1$$

The oscillatory components are chosen carefully regarding their frequency so that mode mixing does not occur. For the time series $N = 40,000$ data points were used with $\Delta t = t_{n+1} - t_n = 0.0005$. Note that the oscillating signal components have a period much smaller than the window sizes studied. Hence, these component signals should be reconstructed within the IMFs. The decomposition has been effected using a segment with $\delta = 2500$ samples each and varying step size τ and ensemble size q, respectively.

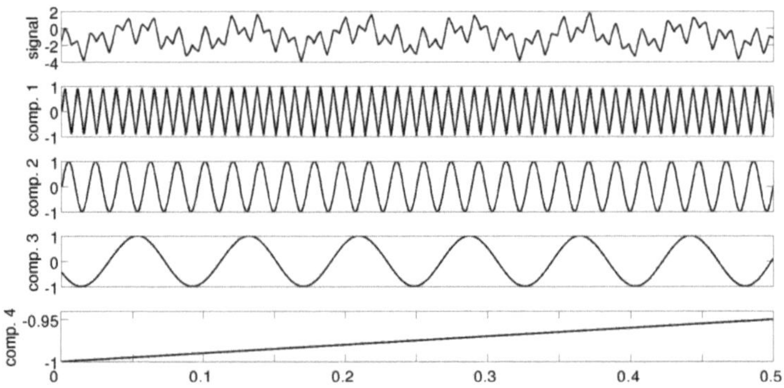

Fig. 12.6. A detail of the toy signal (top trace) consisting of a sawtooth wave, a sinusoid, a cosine function with time-dependent frequency and a monotonic function (lower traces).

12.3.3.1 Variable Step Size

The decomposition has been performed for all step sizes $\tau = 2, \ldots, 1250$ separately and the condition $q = \frac{\delta}{\tau} \in \mathbb{N}$ has been assured in each case. For the three oscillatory components and the residuum, the quality of the decomposition can be judged by estimating the Minkowski distance (MD_p) of order p between the original signal components $s^{(i)}(t)$ and their corresponding IMFs $x_j(t)$ or the residuum $r(t)$, respectively.

$$MD_p = \left(\sum_{n=1}^{N} \left| s_i(t_n) - x_i(t_n) \right|^p \right)^{\frac{1}{p}} \tag{12.17}$$

Using $p = 1$ results in what is called Manhattan distance or L_1 - norm which is related to an average reconstruction error $\langle RQE \rangle = N^{(-1)} \cdot MD_1$, while $p = 2$ provides an Euclidian distance or L_2 - norm, which is related to a mean squared error (MSE) via $MSE = N^{(-1)}(MD_2)^2$. The latter will be used throughout in the following to measure reconstruction quality RQ. Fig. 12.7 shows the MSE of components 1-3 and the monotonous trend as function of the ensemble size q.

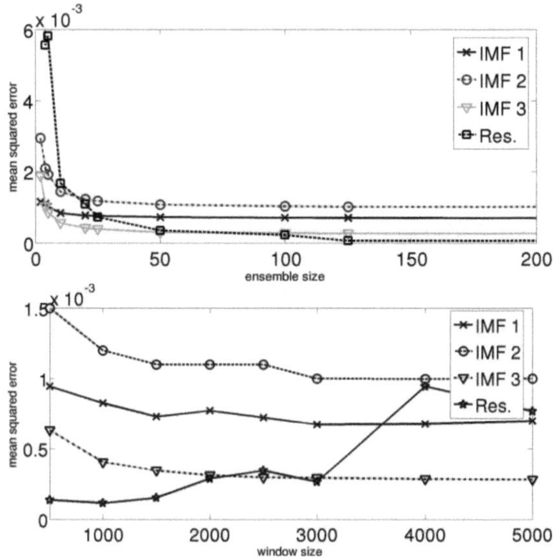

Fig. 12.7. *Top:* Mean square error (MSE) of components 1-3 and the monotonous trend estimated with SEMD for the toy signal given above as function of ensemble size q. The window size was $\delta_m = 2500$. *Bottom*: Mean square error of component 1-3 and the monotonic trend estimated with SEMD for the toy signal given above as function of the window size δ_m. The ensemble size was kept constant at $q = 50$.

With increasing ensemble size q or decreasing step size τ, the reconstruction quality improves. Starting with an ensemble size of about $q = 50$ estimates, the MSE stays roughly constant. Therefore the results show that for practical applications an ensemble size of about $q = 50$ is a good trade-off between reconstruction quality and computational load.

12.3.3.2 Variable Window Size

In the following we studied the influence of the window size on the reconstruction quality. The decomposition was applied for different segment sizes and a constant ensemble size $q = 50$. Fig. 12.7 shows the resulting reconstruction quality of component 1-3 and the monotonic trend as a function of the segment size.

Especially for $\delta = 500$ and $\delta = 1000$ the MSE is higher than for larger segment sizes, but with a value of $MSE \leq 0.0015$ the reconstruction still can be considered very satisfactory. With increasing window size, the reconstruction error decreases, which also holds true for all other tested numbers of estimates. Supposing that the artifacts caused by the boundary conditions have an almost constant penetration depth, for larger windows the percentage of corrupted estimates decreases. This effect may lead to a better reconstruction in case of a larger window size. Only the reconstruction error of the residuum increases for larger window sizes. The MSE of the monotonous trend obtained by SEMD fluctuates strongly, because the last component of the decomposition with SEMD is not necessarily a line after averaging over all suggestions for the data points. As the errors resulting from reconstructing the other components add up and reside in the residuum at the end, it is possible that the residuum is the most irregular component after the decomposition. In our example, the reconstruction of the residuum fluctuates with a larger 'amplitude' around the original monotonous function for larger window sizes, thus the error increases with window size (see fig. 12.8). The 'true' monotonous trend can be obtained by smoothing the data.

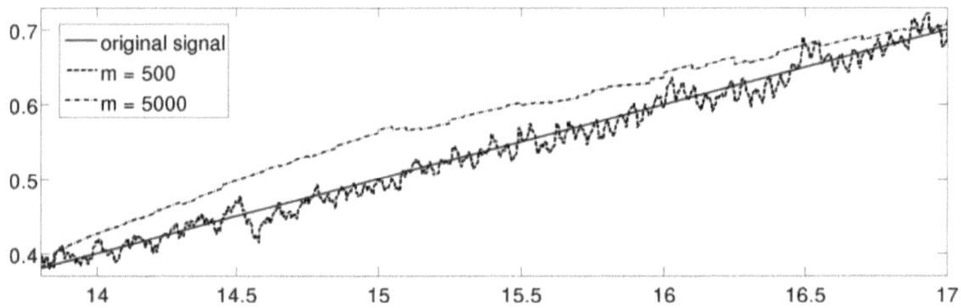

Fig. 12.8. The figure shows a detail of the original monotonic component of toy example two compared to the residuum of the decomposition via SEMD with window sizes $m = 5000$ and $m = 500$.

12.3.3.3 Segment Size and Periodic Components of the Residuum

The residuum of SEMD may contain periodic components if the periodicity is much larger than the window size used. Components with a periodicity much smaller than the window size are recognized as IMFs. To evaluate the behavior of SEMD in the intermediate range, a toy signal $x(t) = x_0 \cdot sin(2\pi \cdot t) + \epsilon(t)$ ($\Delta t = 0.0005$) consisting of a sinusoid with a period of $T = 2000$ data points and i.i.d. noise $\epsilon(t)$ with an amplitude $\epsilon_0 = 0.1x_0$ corresponding to 10% of the maximal amplitude x_0 of the sinusoid is decomposed with SEMD. Different segment sizes $\delta = 200, \ldots, 5000$ were used, whereas the ensemble size $q = 50$ was kept constant and the central

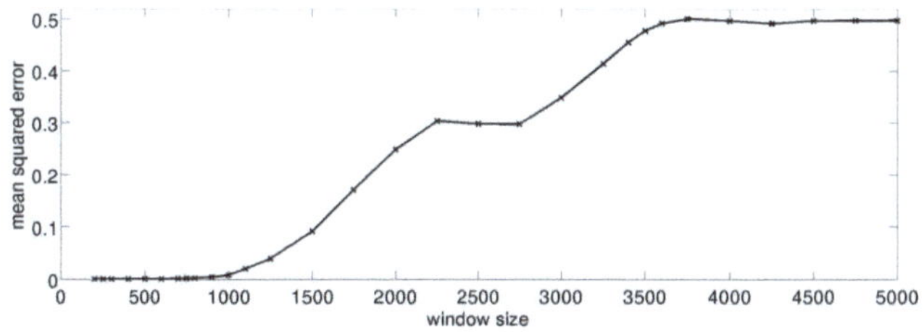

Fig. 12.9. The figure shows the mean squared error of the residuum compared to the sinusoid depending on the window size.

$26,000$ data points (of $40,000$ originally) were considered for the analysis. Eventually the MSE of the residuum compared to the sine function was computed, as can be seen in fig. 12.9.

For a larger window size, the MSE increases because the amplitude of the sine function decreases as it is split up into the residuum and an IMF. The sinusoid resides entirely in the residuum for a window size of about $\delta = 1000$ respectively $\delta = 0.5 \cdot T$ or smaller. The component is found completely in an IMF for ca. $\delta = 3500$ respectively $\delta = 1.75 \cdot T$ or larger. Depending on which oscillatory components of the original signal are supposed to appear in IMFs, an appropriate window size should be chosen.

12.3.4 Performance Evaluation of SEMD

In order to be able assessing the performance of SEMD, its results were compared to existing versions of EMD. Again the toy example consisting of a sawtooth wave, a sinusoid, a cosine function with a time-dependent frequency and a monotonic trend (see fig. 12.6) was used for demonstration purposes. The described time series was then decomposed with SEMD, whereas different window sizes ($\delta = 500$, $1000, 1500, 2000, 2500, 3000, 4000$ and 5000) and step sizes were used, so that the ensemble of suggestions $q = 50$ for each resulting date point remained constant. Figure 12.10 compares the dependence of the mean squared reconstruction error (MSE) of components $1 - 4$ of the toy example on window size for decompositions envoking SEMD and EMD, respectively.

For $\delta = 500$ and $\delta = 1000$ the reconstruction error is generally slightly higher and therefore the use of such small window sizes is not recommended. The MSE of IMF 1, 2 and 3 is smaller for the decomposition with SEMD and, concerning the residuum, mostly smaller for the decomposition with EMD. As has been said already, the reconstruction quality of the monotonous trend obtained by SEMD is worse than obtained by EMD for larger window sizes.

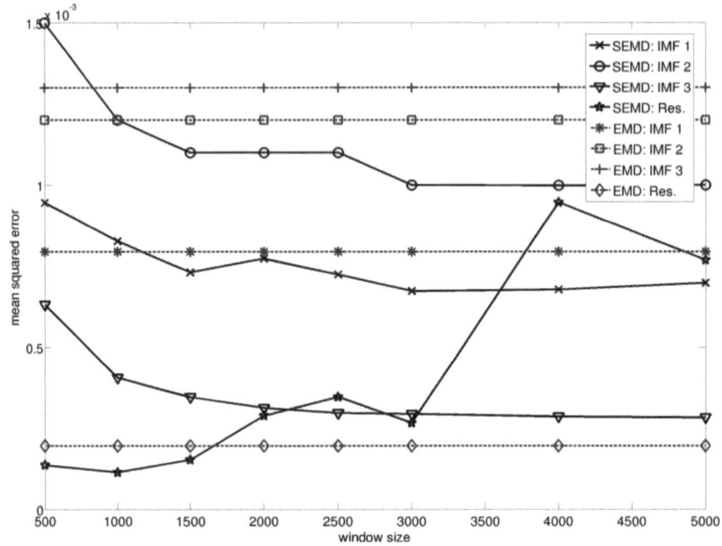

Fig. 12.10. The figure shows the reconstruction error of components $1-4$ of toy example two after the decomposition with SEMD and EMD.

Flandrin *et al.*, in a short note, already suggested, in a short note only, an online EMD [23]. Furthermore, because of the averaging operation, SEMD bears some resemblance to Ensemble EMD. Thus also these two methods have been evaluated and their respective MSEs are included for comparison as well.

Figure 12.11 shows the sum of the reconstruction errors of all four components of the toy example decomposed with EMD, SEMD, EMD online and Ensemble EMD for 50 iterations. That number was chosen for EEMD, as the ensemble of suggestions for one component decomposed with SEMD has been $q = 50$ as well in our tests.

As the reconstruction with EEMD exhibited strong mode mixing, a large reconstruction error resulted, and any comparison deemed inappropriate. Therefore fig. 12.11 also shows the reconstruction quality of the other three methods separately with higher resolution.

SEMD outperforms EMD and online EMD by far as is evident from its reconstruction error which is much smaller. Better results than with online EMD are still obtained with plain EMD. With an ensemble size of only $q = 50$, SEMD performs the most accurate decomposition of all tested methods and represents a high quality data analysis. Additionally, it is a true online method and constitutes a better and more sophisticated tool than the online EMD method proposed by [23].

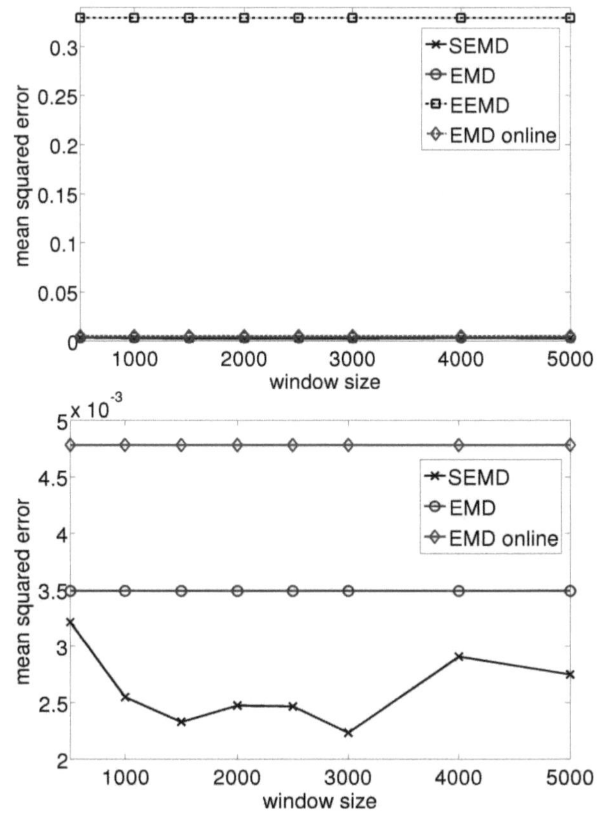

Fig. 12.11. *Top:* Reconstruction error of the four components obtained with SEMD, EMD, EEMD and online EMD summed up. *Bottom*: Reconstruction error of the four components obtained with SEMD, EMD and online EMD summed up.

12.4 Weighted Sliding EMD

Using EMD, and especially SEMD, it is important to estimate the error range caused by boundary effects. In order to be able to evaluate how large a part of the data set at the beginning and at the end is affected by boundary effects, this impact was studied. A number of data points residing near the beginning or end of a window are used by SEMD for the calculation of the resulting IMFs or the residuum, respectively. Thus the ensemble contains data points which are probably defective. In order to diminish reconstruction errors due to such boundary effects, we developed the weighted Sliding Empirical Mode Decomposition (wSEMD).

12.4.1 Error Range of EMD

In the following, the error range of EMD is studied depending on the frequency of the underlying components and the window size. In order to be able to quantify the reconstruction error, toy data examples were used. The results obtained with the toy example shown in fig. 12.6 are depicted in this section.

After the decomposition of a segment of the time series with EMD, the evolution of the error was calculated in the chosen window by subtracting the original component and the corresponding IMF, and recording the absolute value of the result. This time series was smoothed by averaging, which can be seen in fig. 12.12, where the crosses mark the error range. As a limiting value for the error spread, 2.5 times the average error of the middle part of the error time series was defined. In this case, the middle part of the time series means the time series without the first and last 200 data points, respectively.

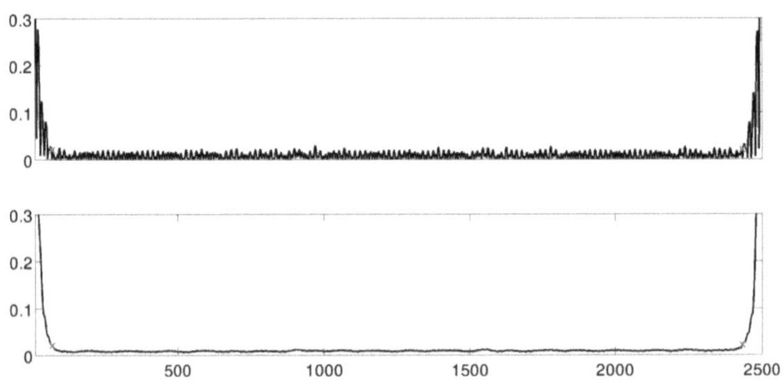

Fig. 12.12. The time series depicts the reconstruction error of the first component after EMD was applied to a window of size $\delta = 2500$ depending on the position within the window in the upper trace and the same time series after smoothing in the bottom trace. The red crosses mark the error range from the boundaries.

As the reconstruction error for the monotonous trend is lowest where the two lines intersect, and an error spread cannot be defined in that case, only components 1 - 3 were used for the analysis. The window sizes $\delta = 500, 1000, 1500, 2000, 2500, 3000, 4000$ and 5000 were considered. For every IMF and all window sizes a sample of $N = 1000$ different fragments of the data were tested regarding their error range. In the following diagram (fig. 12.13), average error ranges and their standard deviations at the beginning of the time intervals are shown as function of window size. They are only marginally different to the error ranges at the rear ends, which therefore are not shown.

For IMF 1 and 2 (not depicted) of the example, the error ranges remain constant, whereas for IMF 3 they are smaller for smaller window sizes, which is apparent in figure 12.13. That behavior can be caused by the evaluation method, since the reconstruction error is generally larger for smaller window sizes. Hence the limit value

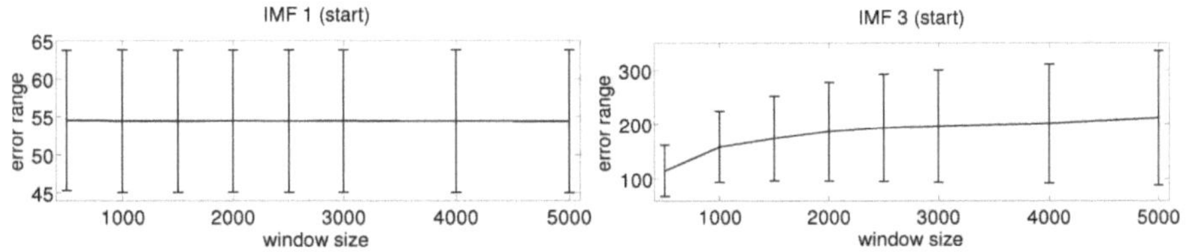

Fig. 12.13. The average error ranges and their standard deviations at the beginning of the windows for different window sizes and IMFs 1 and 3.

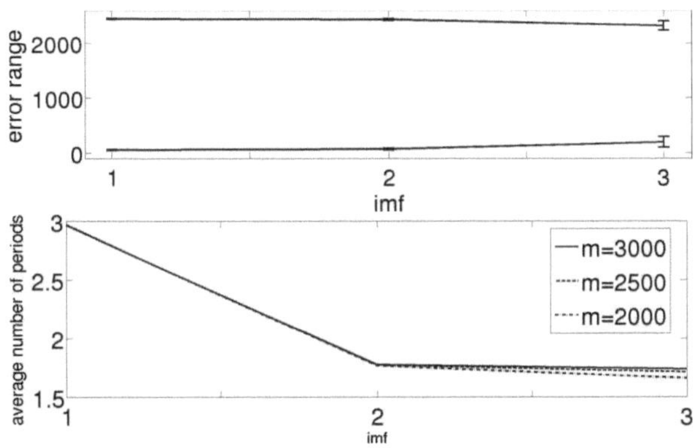

Fig. 12.14. *Top:* The average error ranges and their standard deviations at both ends of the windows for a constant window size $\delta = 2500$ and IMFs 1 - 3. *Bottom:* The average number of periods of the corresponding components affected by the error range for the window sizes $\delta = 2000, 2500$ and 3000. As the frequency of component 3 changes with time its average period was considered as reference.

for the error range is higher. Further it can be noted, that the lower the frequency, the wider the error spreads from the boundaries towards the center of the segment.

Figure 12.14 shows the error range at the beginning and end of the window for all studied IMFs respectively frequencies with a fixed window size $\delta = 2500$.

Again the frequency dependence of the error is apparent. Presumably the reason for this effect is that for higher frequencies fewer extrema and accordingly knots for the spline interpolation are available, i. e. the signal reconstruction is less accurate.

In figure 12.14 the error spread is shown as average number of periods of the corresponding frequency component which is affected for window sizes $\delta = 2000, 2500$ and 3000.

It is demonstrated that the error range is weaker than linearly dependent on the frequency and the number of periods affected by the error spread is not constant. The average number of periods affected by the error spread is only marginally dependent on the window size. The frequency of the third component changes over time. As a reference period of the original component, the average period of the time series was used, because the whole time series or all periods, respectively, were used for the calculation of the error spread.

Considering all results it can be concluded that the use of too small window sizes is not recommended for SEMD. Window sizes should amount to at least about $\delta =$ 1500 data points. For smaller window sizes, the error range does affect a too large part of the windows. Generally the error spread hardly depends on the frequency of the underlying component.

12.4.2 *The Principle of Weighted SEMD*

In order to suppress the influence of boundary effects on the resulting time series, weight functions are used in the following. After the decomposition with EMD, the IMFs and the residuum, estimated within every window, is multiplied by a weighting function. Calculating the average of the estimates q data points and q coefficients, respectively, are used, and, at the end, every data point is normalized by the reciprocal of the sum of all used weights. Therefore the amplitude of every IMF is preserved.

Considering our results concerning the error spread in the data windows, we used the following weighting functions

- a parabola $f(x) = x(1-x)$ with $x = \frac{1}{\delta}, \frac{2}{\delta}, \ldots, \frac{\delta}{\delta}$, where δ denotes the window size
- a Gaussian distribution generated by the *MATLAB* command *gausswin(δ)*, which generates a vector $w(n) = e^{-\frac{1}{2}\left(\alpha \frac{2n}{N}\right)^2}$ of rank δ where $-\frac{N}{2} \leq n \leq \frac{N}{2}$, $\alpha = 2.5$ and $N = \delta - 1$
- several step functions, which alternate between 0 near the boundaries and 1 in the central part of the time series

12.4.3 *Performance Evaluation of Weighted SEMD*

Again we used toy data to evaluate the performance of weighted SEMD. The results depicted were obtained with the example shown in 12.6.

In the following, IMF 1 generated by either EMD or weighted SEMD is illustrated. The algorithm wSEMD applied a step function with variable step length indicated by the parameter w (see fig. 12.15).

The parameter w indicates that w data points at both ends of each window were multiplied by 0 and the rest of the data points by 1. The parameter w was increased from $w = 25$ to $w = 225$. Because of the similarity of all diagrams which illustrate the reconstruction errors of the different IMFs, only the reconstruction error of IMF 1 is shown. The MSE is slightly increased for $\delta = 500$ and $\delta = 1000$, but also for $\delta = 2000$ in both diagrams. However, one has to take into account that the scale of the MSE shows a quite small margin. Therefore, also its dependence on the window size is only marginal. The reconstruction error of IMFs 1-3 generated with weighted

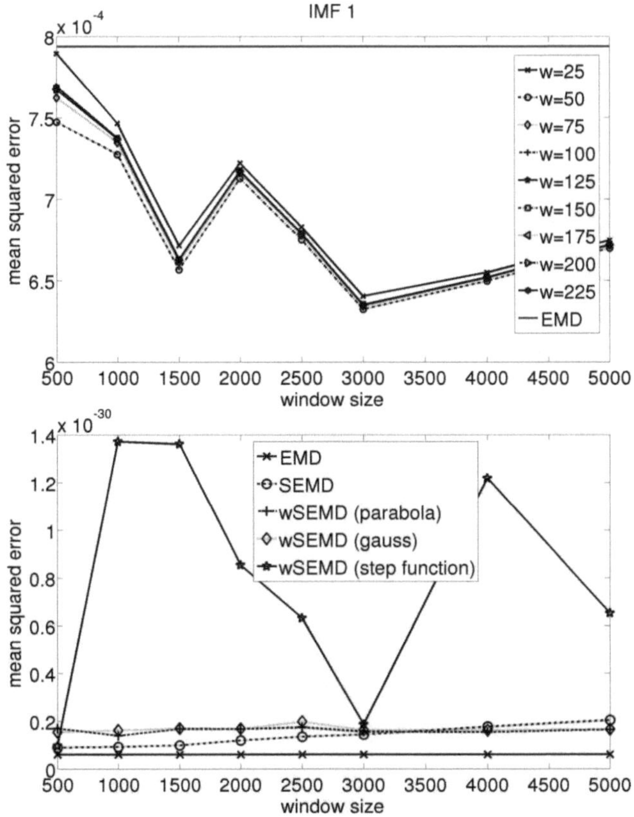

Fig. 12.15. *Top:* The mean squared error of IMF 1 obtained by SEMD weighted with different step functions respectively different parameters w and EMD. *Bottom:* The reconstruction error of the sum of all IMFs and the residuum generated with SEMD weighted with several functions (a step function with $w = 150$, a parabola and a Gaussian distribution), un-weighted SEMD and EMD compared to the original signal.

SEMD with the described weight function is generally smaller than with EMD and only slightly dependent on w.

The MSE of IMF 1 and the residuum is also shown depending on different weight functions, and compared to SEMD and EMD as can be seen in fig. 12.18.

For the step function, the parameter $w = 150$ was used, because it yielded good results while still using a sufficient amount of estimates for the algorithm. The MSE functions for IMF 1, 2 and 3 (not depicted) generated by SEMD with different weight functions are all very similar. The evolution of the error resembles the one of SEMD, but the error itself is smaller. The reconstruction with EMD yields the largest error for most window sizes. In order to obtain a good result in terms of reconstruction error, any of the tested weight functions can be used.

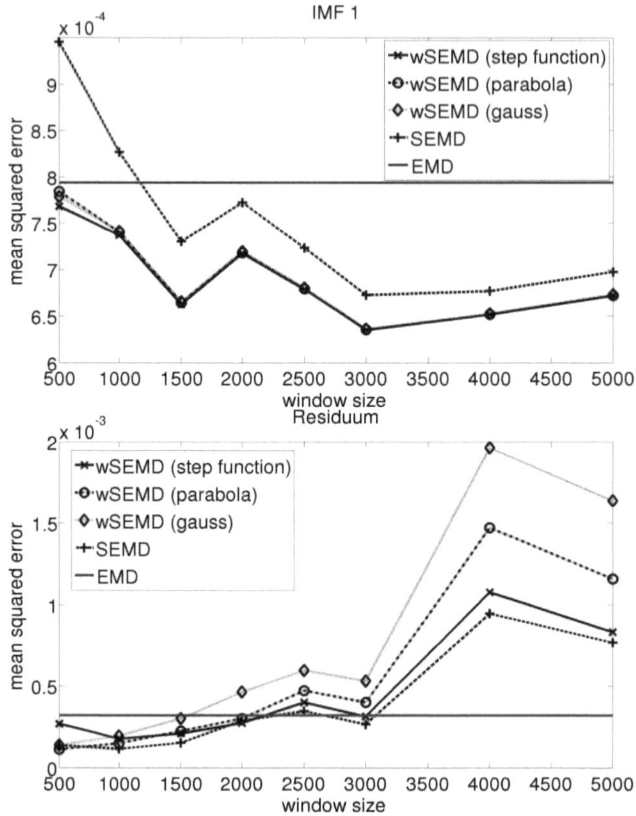

Fig. 12.16. *Top:* The reconstruction error of IMF 1, which was generated by EMD, SEMD and SEMD with different weight functions (namely a step function with $w = 150$, a parabola and a Gaussian distribution). *Bottom:*The reconstruction error of the residuum of toy example two, which was generated by EMD, SEMD and SEMD with different weight functions (namely a step function with $w = 150$, a parabola and a Gaussian distribution).

Applying a weighting function does not guarantee an improvement of the reconstruction of the residuum, though. This is because the error is minimal, when two lines intersect. However the intersection does not necessarily occur in the middle of the window. Furthermore, the residuum gained with (weighted) SEMD is not necessarily a monotonous function after averaging. Therefore the reconstruction of the trend is mostly best with EMD. For the IMFs, the use of weighted SEMD can be recommended strongly as the best results are achieved that way. Anyway, a true online analysis of a given time series can only be achieved with SEMD and the less satisfactory reconstruction of the residuum has to be accepted as a little grain of salt in the otherwise very satisfactory and robust performance of the SEMD algorithm.

12.4.4 Completeness

After applying weighted SEMD, we studied the completeness of the decomposition into IMFs and the residuum as obtained with weighted SEMD compared to the original signal. Again the MSE of the sum of all IMFs and the residuum of applying SEMD weighted by a step function with $w = 150$, a parabolic weight function and a Gaussian weight function, un-weighted SEMD and a plain EMD algorithm is analyzed (see fig. 12.15).

Only the decomposition with SEMD where a step function was used for weighting shows a complex behavior of the error function with larger errors which, however, in absolute value are still vanishingly small. The completeness of the other decompositions with weighted SEMD and SEMD is in accord with EMD. Contrary to SEMD, the reconstruction error of SEMD weighted with the parabola and the Gaussian distribution does not increase but stays constant. However, the reconstruction error of SEMD is smaller, hence the reconstruction is more complete, for small window sizes. Generally, EMD reaches a higher completeness than all other tested methods. The averaging, which is done using (weighted) SEMD, is possibly the reason for the slightly decreased completeness of the decompositions. But quantitatively the numerical value of the error is vanishingly small, hence the decomposition still can be considered complete in every respect.

12.4.5 Examination of the IMF Criteria

Further it has to be verified if the components which were obtained by a decomposition with SEMD, weighted SEMD or Ensemble EMD truly represent IMFs in the sense that they obey the conditions to be fulfilled by an IMF by definition. First the mean value of an IMF has to be zero everywhere, second the number of local extremes and zero crossings can differ by one, at most. Once more toy data was used in order to examine if the two conditions are met.

Evaluating the results one has to keep in mind, that the number of sifting steps is kept constant in the algorithm so that SEMD, weighted SEMD and also Ensemble EMD work properly (see 12.13). This method was first introduced in [24], which means that also for plain EMD the stopping criterion respectively the IMF criteria are not necessarily fulfilled.

12.4.5.1 Mean Value

Figure 12.17 shows the absolute mean value of IMFs 1 and 3 depending on the ensemble size of the decomposition.

In most cases SEMD and especially weighted SEMD produces an IMF that meets the first criterion even better than EMD as soon as an ensemble size of ca. 50 is utilized. The average value fluctuates only slightly beginning at $E = 50$ which indicates again that a decomposition with that parameter value is a good trade-off

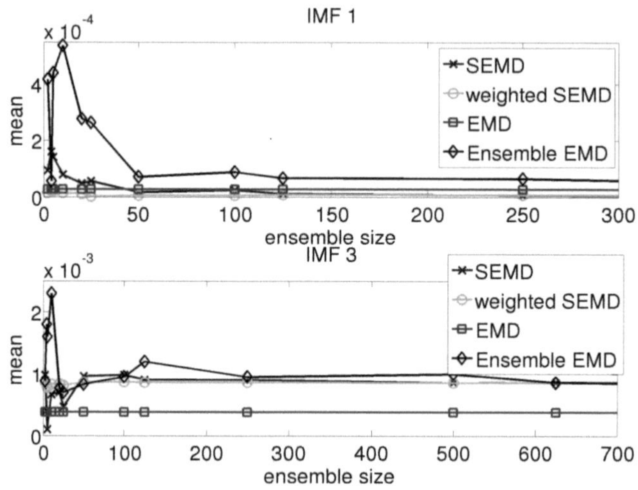

Fig. 12.17. *Top:* The mean value of IMF 1 obtained with SEMD, weighted SEMD, EMD and Ensemble EMD of the toy example depending on the ensemble size is depicted. *Bottom:* The mean value of IMF 3 obtained with SEMD, weighted SEMD, EMD and Ensemble EMD of the toy example depending on the ensemble size is depicted.

between computational load and reconstruction quality. The right subfigure 12.17 shows one of the examples in which EMD reaches the best result. However, the average value of the (weighted) SEMD decomposition is still quite small and barely higher than the EMD result.

12.4.5.2 Number of Extrema and Zero Crossings

The difference between the number of extrema and zero crossings depending on the ensemble size is depicted in fig. 12.18 for IMF 1 and 3 of the toy example.

Whereas the EMD decomposition generally fulfills the second criterion for all IMFs of the toy data examples, Ensemble EMD fluctuates strongly in some cases even for large ensemble sizes (not depicted). These large values of the difference originate from the potential ruggedness of the signals after the calculation of the average time series. SEMD and weighted SEMD, however, show only for small ensemble sizes a difference larger than 1 between the extrema and zero crossings. Especially weighted SEMD meets the IMF requirement already for very small ensemble sizes E, therefore it is to be preferred. In summary, the IMF constraints are very well satisfied by the extracted components, hence the results of the (weighted) SEMD decompositions really can be considered true IMFs.

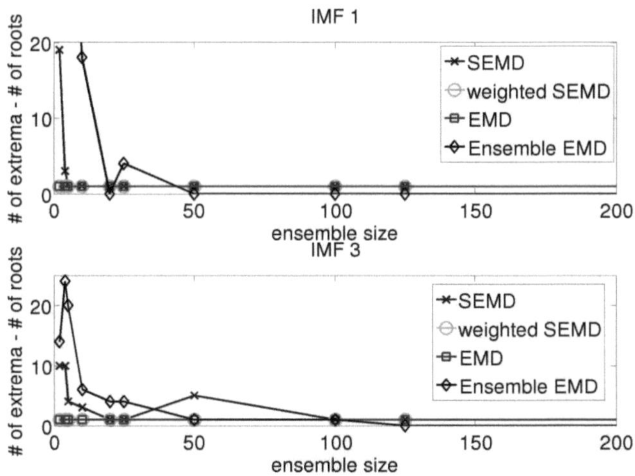

Fig. 12.18. *Top:* The difference between the number of extrema and zero crossings of IMF 1 obtained with SEMD, weighted SEMD, EMD and Ensemble EMD of the toy example depending on the ensemble size is depicted. *Bottom:* The difference between the number of extrema and zero crossings of IMF 3 obtained with SEMD, weighted SEMD, EMD and Ensemble EMD of the toy example depending on the ensemble size is depicted.

12.5 Analysis of Brain Status Data

After a thorough evaluation of all properties of the newly proposed algorithms, the following section discusses applications of SEMD and wSEMD to brain status data and compares them with respective results from EEMD.

12.5.1 EEMD Applied to Brain Status Data

Non-stationary time series are difficult to deal with using classical analysis methods. However, EMD decomposes any time series into stationary IMFs and a non-stationary residuum. EEMD is a noise-assisted method to improve sifting [19]. In practice, EEMD works by repeatedly adding white noise to the data set in each iteration and decomposing the noisy data into IMFs. Finally, an ensemble average of the respective IMFs is calculated yielding the final result. In practice, EEMD helps to avoid over-sifting and mode-mixing.

12.5.1.1 EEMD Decomposition of Brain Signals

Brain status data, especially ABP and ICP, have been recorded synchronously from a patient. Fig. 12.19 shows the signal recorded for 14 h with a sampling rate of 1 Hz. It illustrates the non-stationary component, while the high-frequency oscillatory components become visible from an expanded segment of 300 s duration.

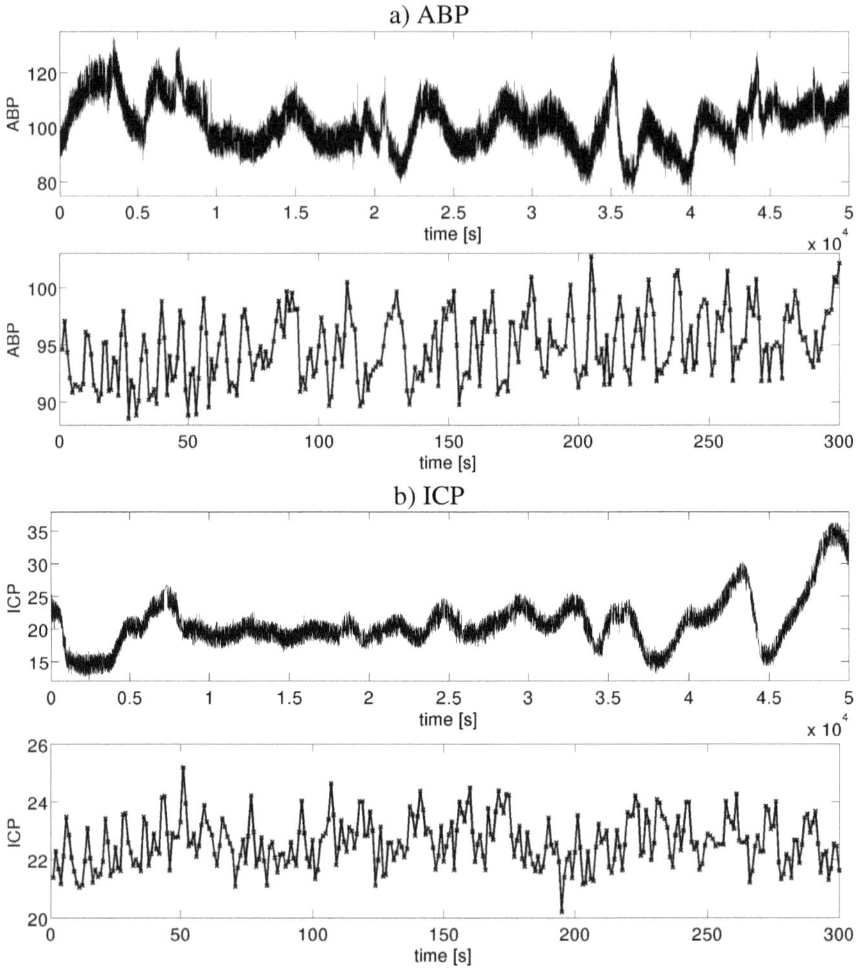

Fig. 12.19. Arterial blood pressure (a, *top*) and intracranial pressure (b, *top*) recorded for 14 h with a sampling rate of 1 Hz. Expanded view of the first 300 s segment of the ABP (a, *bottom*) and ICP (b, *bottom*) recording.

Plain EMD or EEMD was applied to these signals to extract their IMFs and residues. Only the first and last IMF are shown in Fig. 12.20 and 12.21 together with the original signal and the residuum. For better comparison, IMFs obtained with either algorithm are presented in corresponding figures. Both algorithms use the following parameters if not stated otherwise:

- The envelops are estimated using cubic splines.
- Boundary artifacts are avoided by adding data points as proposed in [24].
- The number of sifting steps is fixed to $N_{sifting} = 10$.

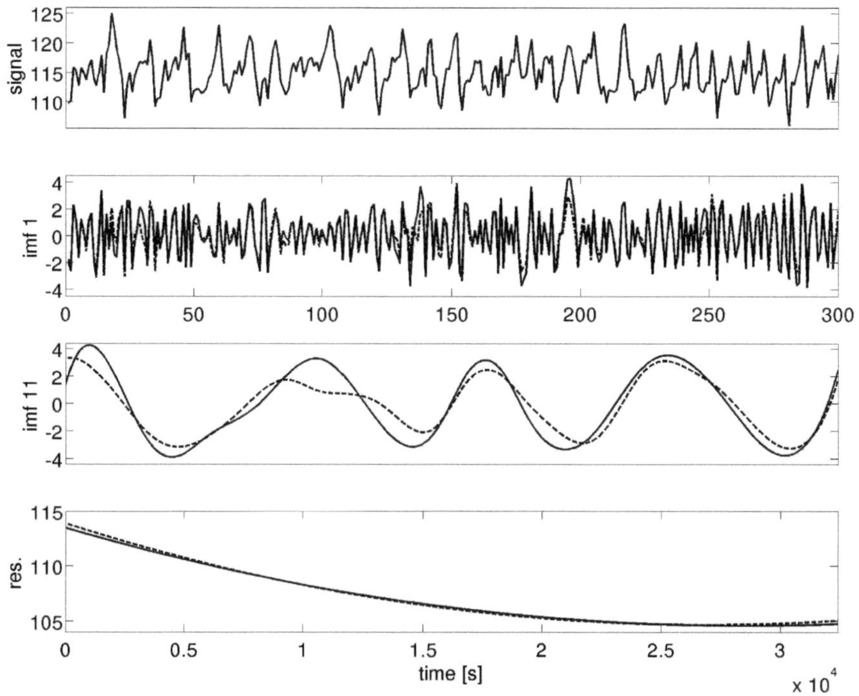

Fig. 12.20. A 300 s recording of the ABP signal and the first IMF extracted with EMD (full line) and EEMD (broken line), respectively. The bottom traces illustrate a low frequency IMF and the residuum on a time scale of 9 hours.

- The number of signal components is chosen to $j = \lfloor \log_2(N) \rfloor$ with N the number of data points sampled.
- EEMD used normally distributed noise and an IMFs were estimated as an ensemble average over 100 realizations.

The first extracted IMF in every case reflects the breathing mode with an average period of $\langle T_B \rangle = 4$ [s] corresponding to an average frequency of $\langle \nu_B \rangle = 0.25[Hz]$. The corresponding Hilbert-Huang transforms (see Fig. 12.22) exhibit fluctuations of their instantaneous frequencies around the average frequency $\langle \nu_B \rangle$ due to the low sampling rate employed. Average frequencies and signal periods for the $300[s]$ segment of both pressure time series are collected in Tab. 12.1. Generally, an EEMD decomposition results in smoother IMFs than plain EMD as can be seen from a comparison (see Tab. 12.2) of average frequencies $\langle \nu \rangle$ and related standard deviations σ_ν, deduced by applying a Hilbert-Huang transform (HHT) to the data, for some IMFs extracted from both ABP and ICP time series. The example demonstrates that physiologically meaningful signal components can be extracted as single IMFs with either EMD or EEMD. Note that because of the low sampling rate faster modes like, for example, the heart beat cannot be resolved.

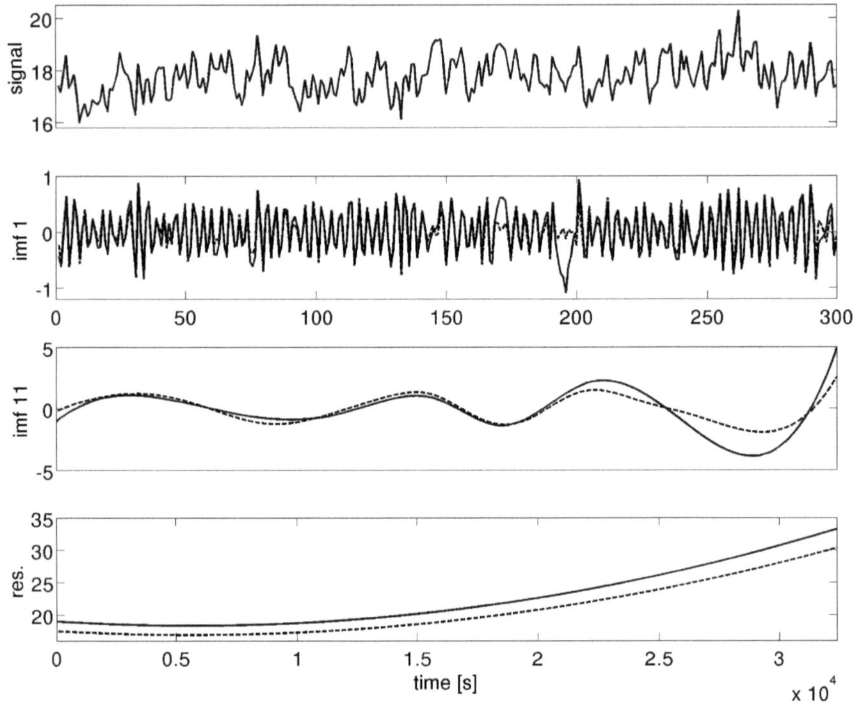

Fig. 12.21. A 300 s recording of the ICP signal and the first IMF extracted with EMD (full line) and EEMD (broken line), respectively. The bottom traces illustrate a low frequency IMF and the residuum on a time scale of 9 hours.

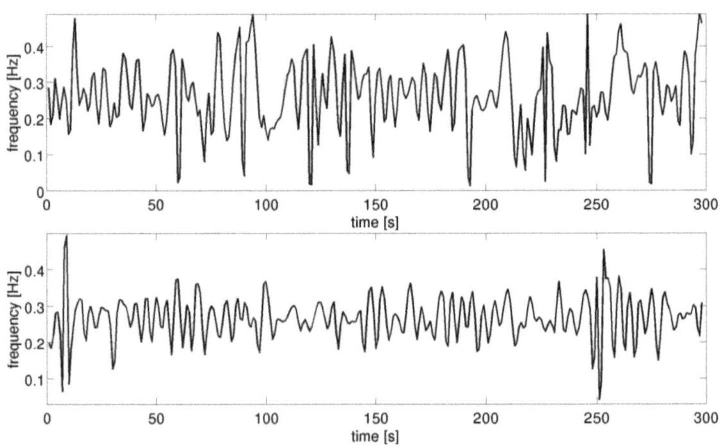

Fig. 12.22. The time-frequency spectrum resulting from a Hilbert-Transformation of the first IMF extracted with EEMD from the recorded ABP signal (*top*) and the recorded ICP signal (*bottom*).

Table 12.1. Average frequency $\langle \nu \rangle$ and related average period $\langle \tau \rangle$ for a $\Delta t = 300$ s long segment of both pressure time series recorded

	$\langle \nu \rangle$[Hz]	$\langle \tau \rangle$[s]
ABP	0,2620	3,8168
ICP	0,2664	3,7538

Table 12.2. Average frequencies $\langle \nu \rangle$ and related standard deviations σ_ν of some IMFs extracted with plain EMD and EEMD

	$\langle \nu \rangle$ [Hz]		σ_ν [Hz]	
	EMD	EEMD	EMD	EEMD
ABP: IMF 1	0,2687	0,2620	0,1047	0,0995
ABP: IMF 3	0,0680	0,0715	0,0558	0,0549
ABP: IMF 10	0,0066	0,0046	0,0560	0,0464
ICP: IMF 1	0,2596	0,2664	0,0728	0,0587
ICP: IMF 3	0,0617	0,0634	0,0424	0,0169
ICP: IMF 8	0,0288	0,0204	0,1146	0,0950
ICP: IMF 10	0,0086	0,0066	0,0645	0,0560

12.5.1.2 EEMD Detrending of Brain Signals

EEMD can be used to remove *non-stationary* contributions, called trends, from any given time series. When analyzing brain status data, the non-stationary contributions correspond to long-term trends of the patient's brain status which are of interest during neuromonitoring. This detrending of the recorded time series can be achieved by subtracting from the recorded time series $x^{(c)}(t)$ the residuum $r(t)$ and, possibly, some slowly varying IMFs $x_j(t), \;\; j = J, J-1, \ldots, J-k$. The resulting signal

$$x^{(c)}_{stat}(t) := x^{(c)}(t) - r(t) = \sum_{j=1}^{(J-1-k)} x_j(t) \tag{12.18}$$

is then checked for stationarity, and the number of IMFs being summed up finally is adjusted accordingly. As an example, recorded ABP and ICP time series $x^{(c)}(t), \;\; c = ABP, ICP$ are considered encompassing $N = 50000$ samples collected with a sampling rate of $(\Delta t)^{-1} = 0.2$ Hz. The resulting sum of 14 IMFs is oscillating around zero but some slow trends with long periods which are present, suggest that not all non-stationary contributions have been removed. This can be checked by monitoring the constancy of the mean of the $\delta_S = \frac{N}{S} \in \mathbb{N}$ samples of $x^{(c)}(t)$ in S non-overlapping time windows. These mean values are estimated as

$$\langle x^{(c)}_s \rangle(s) = \frac{1}{M} \sum_{m=1}^{M} x^{(c)}_m(t + m \cdot (s-1)), \;\; s = \{1, 2, \ldots, S\} \tag{12.19}$$

Finally the $(J - k|k = \{1, 2, \ldots\})$ lowest frequency IMFs, from all J IMFs extracted, are additionally subtracted from the recorded data $x^{(c)}(t)$, leaving only the high-frequency IMFs to form the de-trended, stationary time series :

$$x^{(c)}_{J-k}(t) = \sum_{j=1}^{J-k} x_j^{(c)}(t) = x^{(c)}(t) - \left[r(t) + \sum_{j=J-k}^{J} x_j^{(c)}(t) \right]$$

Again the mean of the resulting signal, i.e. the sum of the stationary high-frequency IMFs, is estimated in the S time windows and the whole procedure is repeated for varying the number S of segments chosen. Note that IMFs 12 and 13 show very low amplitudes only, hence have been neglected from further consideration. Fig. 12.23 shows the standard deviations of the resulting series of mean values of the recorded ICP time series as function of the number S of segments.

The diagrams corroborate an apparent time dependence of the mean values estimated within the time windows. Depending on the chosen segment length, low frequency stationary oscillations are transformed into seemingly non-stationary trends which determine the mean value of the time series in the considered segment. Only with very few segments, encompassing a large number of samples, a convergence of the resulting mean values towards weak stationarity can be observed. In summary, stationarity cannot be defined uniquely within the realm of SEMD, rather it strongly depends on the segment length chosen and the lowest frequency oscillations present in the recorded signals. This is obvious from IMF 7 and IMF 8 which contain oscillations with a period of $T_7 = 240$ s and $T_8 = 600$ s, respectively, which correspond to relevant physiological signals, namely the so-called Mayer waves [25].

12.5.2 SEMD *Applied to Brain Status Data*

In this section the potential of *SEMD* to analyze and decompose real brain status data will be explored. The time series concern arterial blood pressure (ABP), intracranial pressure (ICP) and partial oxygen pressure (TiPO2) recordings sampled continuously over many days with sampling rates of $(\Delta t)^{-1} = 0.2[Hz]$ and $(\Delta t)^{-1} = 1[Hz]$, respectively. First, the dependence of the decomposition onto segment size and step size will be investigated. Next it will be studied how low frequency oscillations can be separated via *SEMD* into the residuum. The latter often correspond to non-stationary signal components which are hard to deal with using classical signal processing paradigms. This *detrending*, applying EMD, has been discussed in the literature already by [26] and [23]. However, if not only a monotonous residuum but also certain low frequency oscillations need to be separated out, a subsequent sophisticated analysis of the estimated IMFs is necessary as discussed above. To the contrary, *SEMD* provides a very simple and efficient way to achieve this goal by simply varying the segment size accordingly.

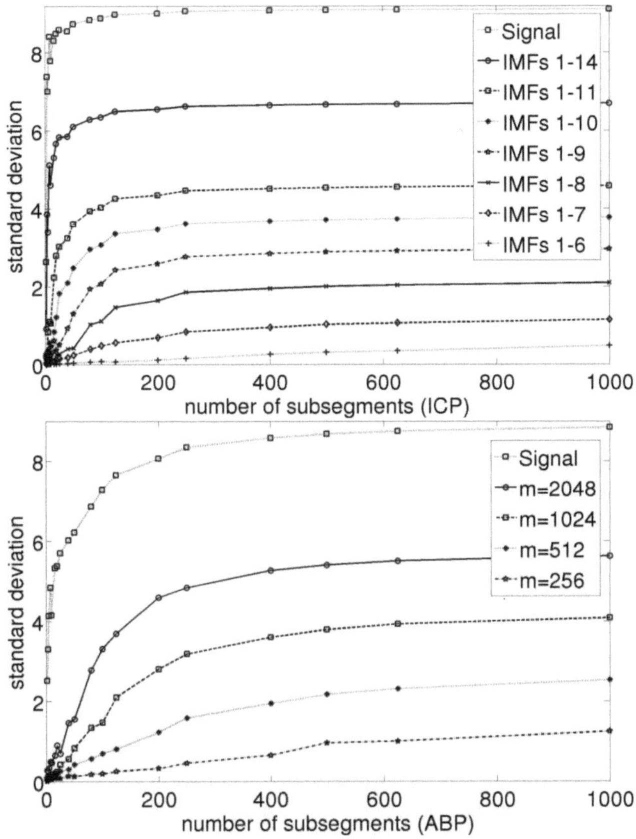

Fig. 12.23. *Top:* Standard deviation of mean values of $x^{ICP}(t)$ as function of S after various subtracted "residual" IMFs. *Bottom:* Dependence of the standard deviations of the mean values on the number S of subsegments for the original ABP time series and its difference to the residues estimated with *SEMD*

12.5.2.1 Step Size and Window Size

Analyzing biomedical signals, their component signals are usually not know with sufficient precision. Hence reconstruction quality needs to be assessed despite the fact that no proper reference signal is available as is the case with toy data examples discussed above. Following, apparent residues will be estimated with *SEMD* varying the step size τ but keeping the window size δ_m fixed. Next, these apparent residues corresponding to subsequent step sizes will be subtracted sample by sample. The squared differences summed and divided by the number of samples then yield the mean squared reconstruction error (MSRE), i. e.

$$MSRE = \frac{1}{N} \sum_{n=1}^{N} \Big(r_{\mathcal{T}}(t_n) - r_{\mathcal{T}+1}(t_n) \Big)^2 \tag{12.20}$$

Here N designates the number of samples, $r_{\mathcal{T}}(t)$ represents the "apparent" residuum for step size $\mathcal{T}$ and $r_{\mathcal{T}+1}(t)$ the corresponding "apparent" residuum for the subsequent step size $\mathcal{T} + 1$. The window size $\delta = 256, 512, 1024, 2048$ used is always a multiple of the step size considered. In the following examples (see Fig. 12.24), a window size $\delta = 2048$ and a step size $\mathcal{T} = 8$ was used for illustrative purposes.

The MSRE for two corresponding residues $\mathcal{T}, \mathcal{T} + 1$ has been estimated for the ABP, ICP and TiPO2 time series using segment sizes $\delta = 256$, 512, 1024 and $\delta = 2048$. Fig. 12.25 illustrates the MSRE for segment size $\delta = 1024$. Similar graphs result for all other segment sizes and are omitted here.

Under all conditions investigated, the MSRE as function of the step size quickly converges for small step sizes towards values close to zero. Hence, the estimated residues are very similar under such conditions. Clearly, the size q of the ensemble is the key quantity determining the MSRE. An ensemble size $16 \leq q \leq 64$ provides a good trade-off between reconstruction error (MSRE) and computational load.

12.5.2.2 *SEMD* Detrending of Brain Status Data

The algorithm *SEMD* can also be used to remove non-stationary components from the recorded signals. Which underlying components contribute to the residuum, hence should be separated, can be controlled simply and efficiently by the segment size. The following study applies very small step sizes merely to achieve optimal results. In *on-line* applications such small step sizes are usually impractical. The investigations use ABP and ICP time series with $N = 50000$ samples, a variable segment size and step sizes $\mathcal{T} = 2$. After decomposing the signal with *SEMD*, the residuum is subtracted from the original time series. Fig. 12.26 presents results of such a detrending of an ABP time series using *SEMD* with a segment size $\delta = 2048$ and a step size $\mathcal{T} = 2$. The related Fig. 12.25 presents the mean and standard deviation of the difference between the residuum and the original ABP time series for different step sizes and a segment size $\delta = 2048$. Both parameters fluctuate only slightly at small step sizes, indicating that a high reconstruction quality could be achieved robustly.

To test for stationarity, mean values of the original time series and the difference of the original time series and the estimated residuum are followed over time by estimating them in subsequent subsegments s of variable segment size $\delta_s = 256$, 512, 1024 and 2048. Fig. 12.23 exhibits the dependence of the standard deviations of the fluctuations around their mean values on the number of subsegments S.

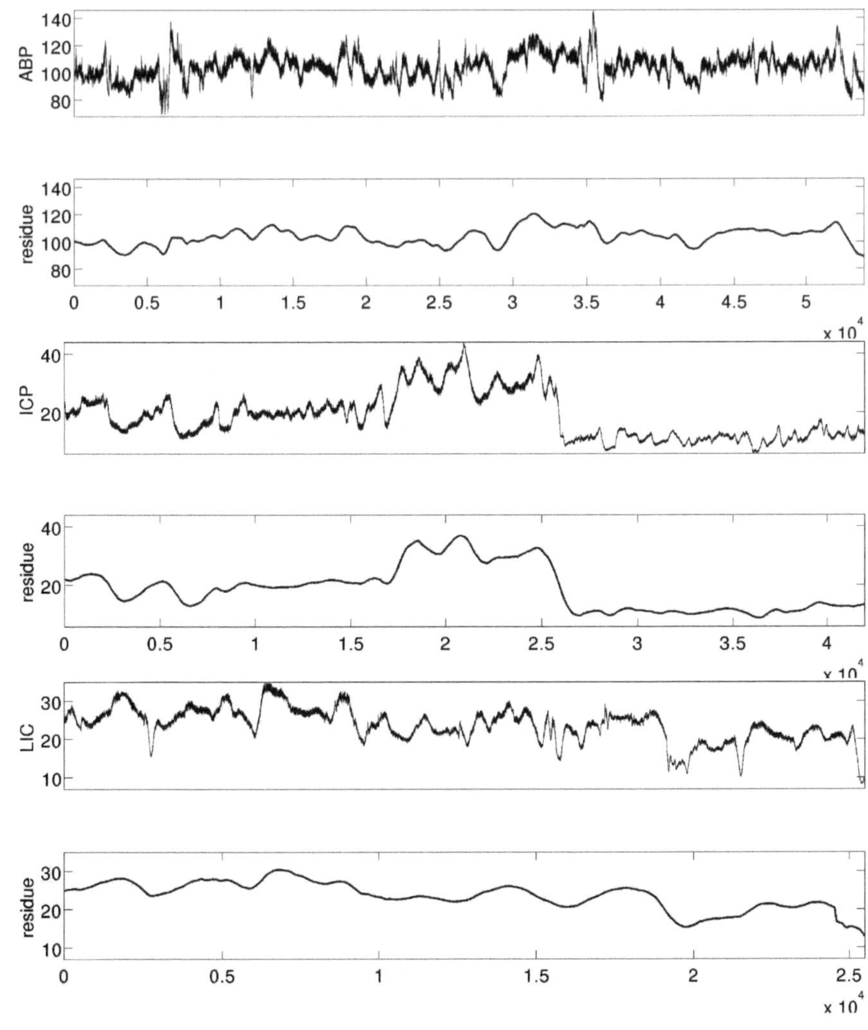

Fig. 12.24. Arterial blood pressure (ABP) (*top*), intracranial pressure (ICP) (*middle*) and Partial oxygen pressure (TiPO2) (*bottom*) time series and corresponding residua estimated with *SEMD*, $\delta = 2048, \tau = 8$.

The results prove a substantial reduction of the non-stationary components in the difference signal. But weak stationarity is only reached if the number S of sub-segments stays small in accord with results obtained earlier already. However, the amount of non-stationary signal parts removed with *SEMD* is much larger than when the residues are estimated with EMD and then subtracted from the original time series. Again a proper choice of the segment size is critical not to remove informative oscillations from the recordings.

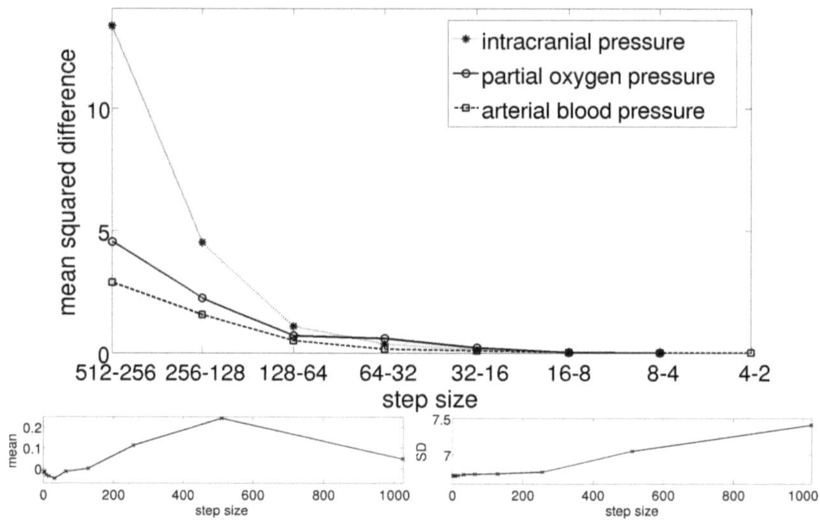

Fig. 12.25. *Top:* Mean square distance MSD of two neighboring residues i, j of the ABP, ICP and TiPO2 time series estimated with *SEMD* using a segment size $\delta = 1024$. *Bottom :* Mean (*left*) and standard deviation (*right*) of the ABP time series after detrending as function of the step size and a segment size $\delta = 2048$

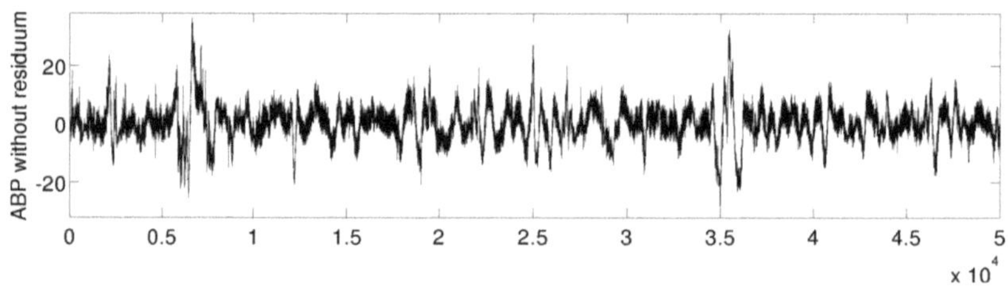

Fig. 12.26. Sum of all IMFs estimated from the ABP time series using *SEMD* with $\delta = 2048$ and $\tau = 2$.

12.5.2.3 *SEMD* versus EEMD Detrending

The following study compares the residues estimated with *SEMD* with the residues plus the sum of low frequency IMFs estimated with EEMD. First, the mean square difference (MSD) between the residues estimated with either *SEMD* or EEMD is calculated according to

$$MSD_1 =\cdot \frac{1}{N} \sum_{n=1}^{N} (r_s(t_n) - r_e(t_n))^2 \qquad (12.21)$$

Here N denotes the total number of samples, $r_s(t_n)$ represents the *apparent* residuum estimated with *SEMD* and $r_e(t_n)$ the corresponding *apparent* residuum estimated with EEMD. Next, the *apparent* residuum $r_e(t_n)$ and the lowest frequency IMF are added and the sum is subtracted from the *apparent* residuum $r_s(t_n)$. Next, the *apparent* residuum $r_e(t_n)$ and the two lowest frequency IMFs are added and the sum is subtracted from the *apparent* residuum $r_s(t_n)$. This process is iterated until the resulting MSD is negligible.

$$MSD_{k+2} = \frac{1}{N}\sum_{n=1}^{N}\left(r_s(t_n) - \left(r_e(t_n) + \sum_{k=0}^{K} x_{14-K}(t_n)\right)\right)^2 \qquad (12.22)$$

The result of this iteration ($K = \{0, 1, \ldots, 13\}$) is summarized in Fig. 12.27 in case of the ABP time series using segment sizes $\delta = 512$, 1024, 2048 and a step size $\tau = 2$.

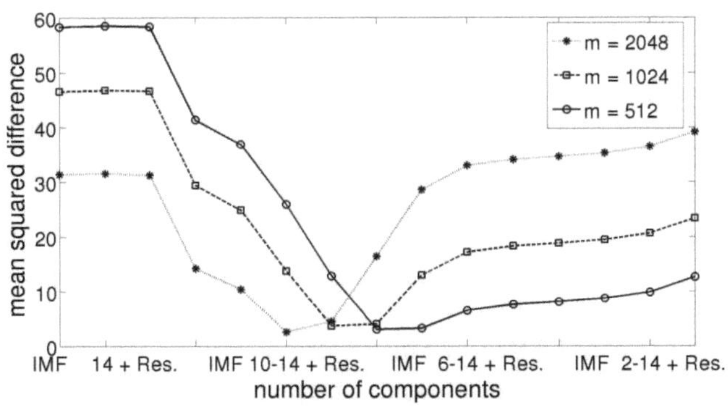

Fig. 12.27. Mean square difference MSD_{n+2} of the decomposition of an ABP time series with either *SEMD* or EMD for segment sizes $\delta = 512, 1024, 2048$ and a step size $\tau = 2$ used within *SEMD*. Note that to the residuum estimated with EMD a variable amount of IMFs is added successively.

It becomes obvious that the *apparent* residuum estimated with *SEMD* using a segment size $\delta = 2048$ and the *apparent* residuum plus IMFs 10 to 14, all estimated from applying EEMD, are very similar. Much the same holds true in the following constellations: segment size $\delta = 1024$ and $r_e(t_n) + \mathrm{IMF}_9 + \ldots + \mathrm{IMF}_{14}$ and segment size $\delta = 512$ and $r_e(t_n) + \mathrm{IMF}_8 + \ldots + \mathrm{IMF}_{14}$. Fig. 12.28 illustrates the result for the latter constellation.

The results corroborate that *SEMD* yields a decomposition well in accord with standard EMD. However, *SEMD* offers the additional advantage of being a true *on-line* algorithm which, furthermore, is based on an ensemble of estimates. It thus provides an equally robust estimate of underlying intrinsic mode functions as EEMD. It

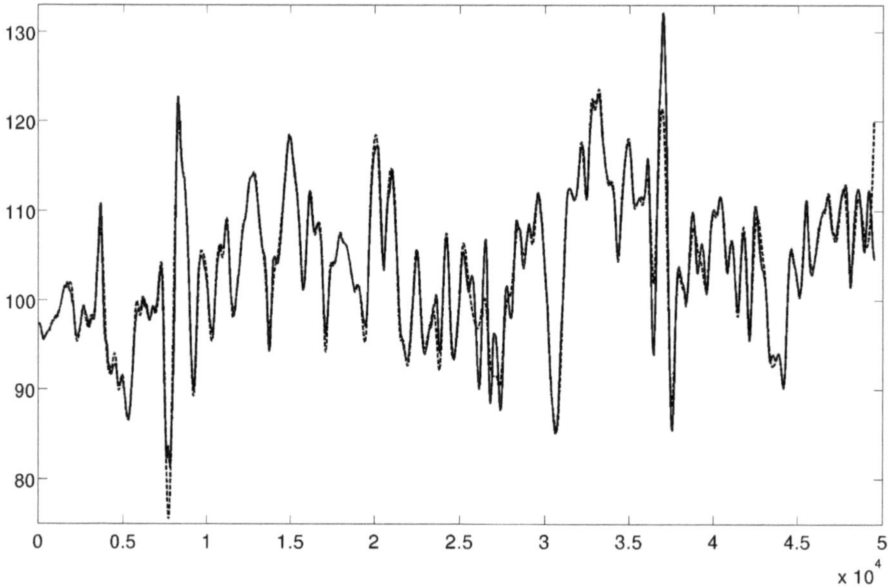

Fig. 12.28. Residuum $r_s(t)$ (full line) estimated with *SEMD* using a window size $\delta = 512$ and the sum of residuum plus low frequency IMFs, i.e. $r_e(t) + IMF_8 + \ldots + IMF_{14}$, estimated with EMD (broken line).

is also as flexible as standard EMD in detrending applications, and allows to extract the stationary part of originally non-stationary biomedical time series data. Hence, for practical applications *SEMD* is to be preferred.

Acknowledgment. Support by the German Academic Exchange Service (DAAD), the Fundação para a Ciência e a Tecnologia (FCT), the Spanish Ministerio de Ciencia y Inovación and the Granada Excellence Network of Innovation Laboratories (GENIL) is gratefully acknowledged.

References

1. Huang, N.E., Shen, Z., Long, S.R., Wu, M.L., Shih, H.H., Zheng, Q., Yen, N.C., Tung, C.C., Liu, H.H.: Proc. Roy. Soc. London A 454, 903 (1998)
2. Georgia, M.A.D., Deogaonkar, A.: The Neurologist 11(1), 45 (2005)
3. Sahuquillo, J.: European Journal of Anaesthesiology 42(suppl. 42), 83 (2008)
4. Brawanski, A., Faltermeier, R., Rothoerl, R.D., Woertgen, C.: Journal of Cerebral Blood Flow & Metabolism 22, 605 (2002)
5. Ursino, M., Lodi, C.: Journal of Applied Physiology 82(4), 1256 (1997)
6. Ursino, M., Minassian, A.T., Lodi, C.A., Beydon, L.: American journal of physiology. Heart and Circulatory Physiology 279(5), H2439 (2000)
7. Jung, A., Faltermeier, R., Rothoer, R., Brawanski, A., Math, J.: J. Math. Biol. 51(5), 491 (2005)

8. Böhm, M.: Workshop Report VI Graduate College 638, Nonlinearity and Nonequilibrium in Condensed Matter (2006)
9. Lo, M.T., Hu, K., Liu, Y., Peng, C.K., Novak, V.: EURASIP J. Adv. Sig. Proc. (2008); doi:10.1155/2008/785243
10. Ursino, M.: Annals of Biomedical Engineering 16, 379 (1988)
11. Purves, M.J.: The Physiology of the Cerebral Circulation. Cambirdge University Press (1972)
12. Sklar, F.H., Elashvili, I.: Journal of Neurosurgery 47, 670 (1977)
13. Fridén, H.G., Ekstedt, J.: Neurosurgery 13(4), 351 (1983)
14. Böhm, M., Stadlthanner, K., Gruber, P., Theis, F.J., Tomé, A.M., Teixeira, A.R., Gronwald, W., Kalbitzer, H.R., Lang, E.W.: IEEE Trans. Biomed. Engineering 53, 810 (2006)
15. Quiroga, R.Q., Arnhold, J., Grassberger, P.: Phys. Rev. E 61, 5142 (2000)
16. Quiroga, R.Q., Kraskov, A., Kreuz, T., Grassberger, P.: Phys. Rev. E 65, 041903 (2002)
17. Hunag, N.E., Shen, S.S.P.: Hilbert Huang Transform and its Applications. World Scientific (2005)
18. Hunag, N.E., Attoh-Okine, N.: Hilbert Huang Transform in Engineering. Taylor & Francis (2005)
19. Wu, Z., Huang, N.E.: Adv. Adaptive Data Analysis 1(1), 1 (2009)
20. Niazy, R.K., Beckmann, C.F., Brady, J.M., Smith, S.M.: Advances Adaptive Data Analysis 1(2), 231 (2009)
21. Faltermeier, R., Zeiler, A., Keck, I.R., Tomé, A.M., Brawanski, A., Lang, E.W.: In: Proc. IJCNN 2010, pp. 978–971. IEEE (2010) ISBN:978-1-4244-6917-8
22. Rilling, G., Flandrin, P., Goncalès, P.: In: Proc. 6th IEEE-EURASIP Workshop on Nonlinear Signal and Image Processing (2003)
23. Flandrin, P., Gonçalvès, P., Rilling, G.: In: EUSIPCO 2004, pp. 1581–1584 (2004)
24. Wu, Z.: `http://rcada.ncu.edu.tw/eemd.m`
25. Julien, C.: Cardiovascular Research 70, 12 (2006)
26. Wu, M.C., Struzik, Z.R., Watanabe, E., Yamamoto, Y., Hu, C.K.: In: AIP Conf. Proc., vol. 922, pp. 573–576 (2007)

Author Index

MIX
Papier aus verantwortungsvollen Quellen
Paper from responsible sources
FSC® C105338

If you have any concerns about our products,
you can contact us on
ProductSafety@springernature.com

In case Publisher is established outside the EU,
the EU authorized representative is:
Springer Nature Customer Service Center GmbH
Europaplatz 3, 69115 Heidelberg, Germany

Printed by Libri Plureos GmbH
in Hamburg, Germany